An Introduction to Digital Video

For Chrissie

An Introduction to Digital Video

Second Edition

John Watkinson

Focal Press
OXFORD AUCKLAND BOSTON JOHANNESBURG MELBOURNE NEW DELHI

8DP
801–2041
essional Publishing Ltd

R A member of the Reed Elsevier plc group

First published 1994
Reprinted 1995, 1998, 1999
Second edition 2001

British Library Cataloguing in Publication Data
Watkinson, John
    An introduction to digital video
    1. Digital video
    I. Title
    621.3'8833

Library of Congress Cataloguing in Publication Data
A catalogue record for this book is available from the Library of Congress

ISBN 0 240 51637 0

For information on all Focal Press publications visit our website at
www.focalpress.com

Composition by Genesis Typesetting, Rochester, Kent
Printed and bound in Great Britain

Contents

Chapter 3 Conversion   114

Chapter 4 Digital video processing   156

Preface

Digital video was conceptually made possible by the invention of pulse code modulation, but the technology required to implement it did not then exist. Experiments were made with digital video from the late 1950s onwards, but the cost remained prohibitive for commercial applications until sufficient advances in microelectronics and high-density recording had been made. Early digital broadcast and production equipment was large and expensive and was handled by specialists, but now the technology has advanced to the point where traditional analog equipment has been replaced by digital equipment on economic grounds. Digital video equipment has become the norm in broadcast and production applications. This book is designed to introduce in an understandable manner the theory behind digital video and to illustrate it with current practice. Whilst some aspects of the subject are complex, and are generally treated mathematically, such treatments are not appropriate for a mainstream book and have been replaced wherever possible by plain English supported by diagrams. Many of the explanations here have evolved during the many training courses and lectures I have given on the subject to students from all technical backgrounds. No great existing knowledge of digital techniques is assumed, as every topic here evolves from first principles. As an affordable introductory book, this volume concentrates on essentials and basics. Readers requiring an in-depth treatment of the subject are recommended to the companion work *The Art of Digital Video*.

John Watkinson
Burghfield Common, England

Chapter 1

Introducing digital video

1.1 Video as data

The most exciting aspects of digital video are the tremendous possibilities which were not available with analog technology. Error correction, compression, motion estimation and interpolation are difficult or impossible in the analog domain, but are straightforward in the digital domain. Once video is in the digital domain, it becomes data, and only differs from generic data in that it needs to be reproduced with a certain timebase.

The worlds of digital video, digital audio, communication and computation are closely related, and that is where the real potential lies. The time when television was a specialist subject which could evolve in isolation from other disciplines has gone. Video has now become a branch of information technology (IT); a fact which is reflected in the approach of this book.

Systems and techniques developed in other industries for other purposes can be used to store, process and transmit video. IT equipment is available at low cost because the volume of production is far greater than that of professional video equipment. Disk drives and memories developed for computers can be put to use in video products. Communications networks developed to handle data can happily carry digital video and accompanying audio over indefinite distances without quality loss. Techniques such as ADSL allow compressed digital video to travel over a conventional telephone line to the consumer.

As the power of processors increases, it becomes possible to perform under software control processes which previously required dedicated hardware. This causes a dramatic reduction in hardware cost. Inevitably the very nature of video equipment and the ways in which it is used is changing along with the manufacturers who supply it. The computer industry is competing with traditional manufacturers, using the economics of mass production.

Tape is a linear medium and it is necessary to wait for the tape to wind to a desired part of the recording. In contrast, the head of a hard disk drive can access any stored data in milliseconds. This is known in computers as direct access and in television as non-linear access. As a result the non-linear editing workstation based on hard drives has eclipsed the use of videotape for editing.

Digital TV Broadcasting uses coding techniques to eliminate the interference, fading and multipath reception problems of analog broadcasting. At the same time, more efficient use is made of available bandwidth.

One of the fundamental requirements of computer communication is that it is bidirectional. When this technology becomes available to the consumer, services such as video-on-demand and interactive video become available. Television programs may contain metadata which allows the viewer rapidly to access web sites relating to items mentioned in the program. When the TV set is a computer there is no difficulty in displaying both on the same screen.

The hard drive-based consumer video recorder gives the consumer more power. A consumer with random access may never watch another TV commercial again. The consequences of this technology are far-reaching.

Figure 1.1 shows what the television set of the future may look like. MPEG-compressed signals may arrive in real time by terrestrial or satellite broadcast, via a cable, or on media such as DVD. The TV set is simply a display, and the heart of the system is a hard drive-based server. This can be used to time shift broadcast programs, to skip commercial breaks or to assemble requested movies transmitted in non-real time at low bit rates. If equipped with a web browser, the server may explore the web looking for material which is of the same kind the viewer normally watches. As the cost of storage falls, the server may download this material speculatively.

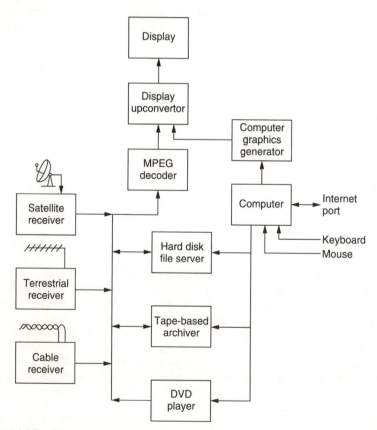

Figure 1.1 The TV set of the future may look something like this.

Note that when the hard drive is used to time shift or record, it simply stores the MPEG bitstream. On playback the bitstream is decoded and the picture quality will be unimpaired. The generation loss due to using an analog VCR is eliminated.

Ultimately digital technology will change the nature of television broadcasting out of recognition. Once the viewer has non-linear storage technology and electronic program guides, the traditional broadcaster's transmitted schedule is irrelevant. Increasingly viewers will be able to choose what is watched and when, rather than the broadcaster deciding for them. The broadcasting of conventional commercials will cease to be effective when viewers have the technology to skip them. Anyone with a web site which can stream video can become a broadcaster.

1.2 What is a video signal?

When a two-dimensional image changes with time the basic information is three-dimensional. An analog electrical waveform is two-dimensional in that it carries a voltage changing with respect to time. In order to convey three-dimensional picture information down a two-dimensional channel it is necessary to resort to scanning. Instead of attempting to convey the brightness of all parts of a picture at once, scanning conveys the brightness of a single point which moves with time.

The scanning process converts spatial resolution on the image into the temporal frequency domain. The higher the resolution of the image, the more lines are necessary to resolve the vertical detail. The line rate is increased along with the number of cycles of modulation which need to be carried in each line. If the frame rate remains constant, the bandwidth goes up as the square of the resolution.

In an analog system, the video waveform is conveyed by some infinite variation of a continuous parameter such as the voltage on a wire or the strength or frequency of flux on a tape. In a recorder, distance along the medium is a further, continuous, analog of time. It does not matter at what point a recording is examined along its length, a value will be found for the recorded signal. That value can itself change with infinite resolution within the physical limits of the system.

Those characteristics are the main weakness of analog signals. Within the allowable bandwidth, *any* waveform is valid. If the speed of the medium is not constant, one valid waveform is changed into another valid waveform; a timebase error cannot be detected in an analog system. In addition, a voltage error simply changes one valid voltage into another; noise cannot be detected in an analog system. Noise might be suspected, but how is one to know what proportion of the received signal is noise and what is the original? If the transfer function of a system is not linear, distortion results, but the distorted waveforms are still valid; an analog system cannot detect distortion. Again distortion might be suspected, but it is impossible to tell how much of the energy at a given frequency is due to the distortion and how much was actually present in the original signal.

It is a characteristic of analog systems that degradations cannot be separated from the original signal, so nothing can be done about them. At the end of a system a signal carries the sum of all degradations introduced at each stage through which it passed. This sets a limit to the number of stages through which a signal can be passed before it is useless. Alternatively, if many stages are envisaged, each piece of equipment must be far better than necessary so that the signal is still acceptable at the end. The equipment will naturally be more expensive.

Digital video is simply an alternative means of carrying a moving image. Although there are a number of ways in which this can be done, there is one system, known as pulse code modulation (PCM) which is in virtually universal use.[1] Figure 1.2 shows how PCM works. Instead of being continuous, the time axis is represented in a discrete, or stepwise manner. The video waveform is not carried by continuous representation, but by measurement at regular intervals. This process is called sampling and the frequency with which samples are taken is called the sampling rate or sampling frequency F_s.

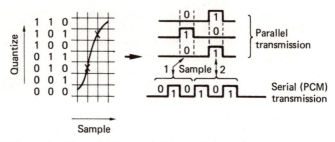

Figure 1.2 When a signal is carried in numerical form, either parallel or serial, the mechanisms described in the text ensure that the only degradation is in the conversion process.

In analog video systems, the time axis is sampled into frames, and the vertical axis is sampled into lines. Digital video simply adds a third sampling process along the lines. Each sample still varies infinitely as the original waveform did. To complete the conversion to PCM, each sample is then represented to finite accuracy by a discrete number in a process known as quantizing.

It is common to make the sampling rate a whole multiple of the line rate. Samples are then taken in the same place on every line. If this is done, a monochrome digital image becomes a rectangular array of points at which the brightness is stored as a number. The points are known as picture cells, pixels or pels. As shown in Figure 1.3, the array will generally be arranged with an even spacing between pixels, which are in rows and columns. Obviously the finer the pixel spacing, the greater the resolution of the picture will be, but the amount of data needed to store one picture will increase as the square of the resolution, and with it the costs.

At the ADC (analog-to-digital convertor), every effort is made to rid the sampling clock of jitter, or time instability, so every sample is taken at an exactly even time step. Clearly, if there is any subsequent timebase error, the instants at which samples arrive will be changed and the effect can be detected. If samples arrive at some destination with an irregular timebase, the effect can be eliminated by storing the samples temporarily in a memory and reading them out using a stable, locally generated clock. This process is called timebase correction and all properly engineered digital video systems must use it.

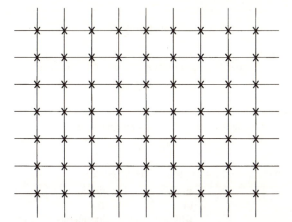

Figure 1.3 A picture can be stored digitally by representing the brightness at each of the above points by a binary number. For a colour picture each point becomes a vector and has to describe the brightness, hue and saturation of that part of the picture. Samples are usually but not always formed into regular arrays of rows and columns, and it is most efficient if the horizontal and vertical spacing are the same.

Those who are not familiar with digital principles often worry that sampling takes away something from a signal because it is not taking notice of what happened between the samples. This would be true in a system having infinite bandwidth, but no analog signal can have infinite bandwidth. All analog signal sources from cameras and so on have a resolution or frequency response limit, as indeed do devices such as CRTs and human vision. When a signal has finite bandwidth, the rate at which it can change is limited, and the way in which it changes becomes predictable. When a waveform can only change between samples in one way, it is then only necessary to convey the samples and the original waveform can be unambiguously reconstructed from them. A more detailed treatment of the principle will be given in Chapter 3.

As stated, each sample is also discrete, or represented in a stepwise manner. The magnitude of the sample, which will be proportional to the voltage of the video signal, is represented by a whole number. This process is known as quantizing and results in an approximation, but the size of the error can be controlled until it is negligible. The link between video quality and sample resolution is explored in Chapter 3. The advantage of using whole numbers is that they are not prone to drift.

If a whole number can be carried from one place to another without numerical error, it has not changed at all. By describing video waveforms numerically, the original information has been expressed in a way which is more robust.

Essentially, digital video carries the images numerically. Each sample is (in the case of luminance) an analog of the brightness at the appropriate point in the image.

1.3 Why binary?

Arithmetically, the binary system is the simplest numbering scheme possible. Figure 1.4(a) shows that there are only two symbols: 1 and 0. Each symbol is a

---

What is binary?

(a) Mathematically:
     The simplest numbering scheme possible, there are only two symbols:

1 and 0

Logically:
A system of thought in which there are only two states:

True and False

(b) Binary information is **not** subject to misinterpretation

Black	**White**
In	**Out**
Guilty	**Innocent**

(c) Variables or non-binary terms:

Somewhat	**Undecided**
Probably	**Not proven**
Grey	**Under par**

Figure 1.4 Binary digits (a) can only have two values. At (b) are shown some everyday binary terms, whereas (c) shows some terms which cannot be expressed by a binary digit.

binary digit, abbreviated to *bit*. One bit is a datum and many bits are data. Logically, binary allows a system of thought in which statements can only be true or false.

The great advantage of binary systems is that they are the most resistant to misinterpretation. In information terms they are *robust*. Figure 1.4(b) shows some binary terms and (c) some non-binary terms for comparison. In all real processes, the wanted information is disturbed by noise and distortion, but with only two possibilities to distinguish, binary systems have the greatest resistance to such effects.

Figure 1.5(a) shows an ideal binary electrical signal is simply two different voltages: a high voltage representing a true logic state or a binary 1 and a low voltage representing a false logic state or a binary 0. The ideal waveform is also shown at (b) after it has passed through a real system. The waveform has been considerably altered, but the binary information can be recovered by comparing the voltage with a threshold which is set half-way between the ideal levels. In this way any received voltage which is above the threshold is considered a 1 and any voltage below is considered a 0. This process is called slicing, and can reject significant amounts of unwanted noise added to the signal. The signal will be carried in a channel with finite bandwidth, and this limits the slew rate of the signal; an ideally upright edge is made to slope.

Noise added to a sloping signal (c) can change the time at which the slicer judges that the level passed through the threshold. This effect is also eliminated when the output of the slicer is reclocked. Figure 1.5(d) shows that however many stages the binary signal passes through, the information is unchanged except for a delay.

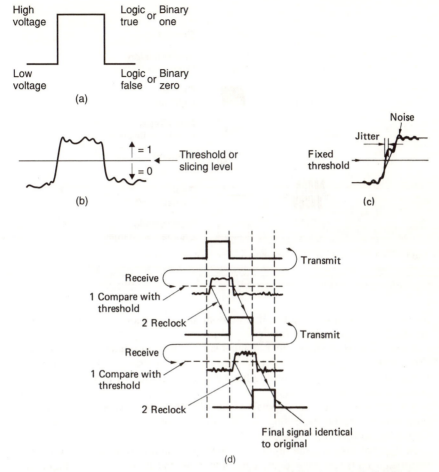

Figure 1.5 An ideal binary signal (a) has two levels. After transmission it may look like (b), but after slicing the two levels can be recovered. Noise on a sliced signal can result in jitter (c), but reclocking combined with slicing makes the final signal identical to the original as shown in (d).

Of course, an excessive noise could cause a problem. If it had sufficient level and the correct sense or polarity, noise could cause the signal to cross the threshold and the output of the slicer would then be incorrect. However, as binary has only two symbols, if it is known that the symbol is incorrect, it need only be set to the other state and a perfect correction has been achieved. Error correction really is as trivial as that, although determining which bit needs to be changed is somewhat harder.

Figure 1.6 shows that binary information can be represented by a wide range of real phenomena. All that is needed is the ability to exist in two states. A switch can be open or closed and so represent a single bit. This switch may control the voltage in a wire which allows the bit to be transmitted. In an optical system, light may be transmitted or obstructed. In a mechanical system, the presence or absence of some feature can denote the state of a bit. The presence or absence of

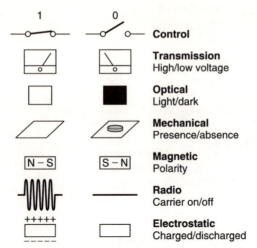

Figure 1.6 A large number of real phenomena can be used to represent binary data.

a radio carrier can signal a bit. In a random access memory (RAM), the state of an electric charge stores a bit.

Figure 1.6 also shows that magnetism is naturally binary as two stable directions of magnetization are easily arranged and rearranged as required. This is why digital magnetic recording has been so successful: it is a natural way of storing binary signals.

The robustness of binary signals means that bits can be packed more densely onto storage media, increasing the performance or reducing the cost. In radio signalling, lower power can be used.

In decimal systems, the digits in a number (counting from the right, or least significant end) represent ones, tens, hundreds and thousands, etc. Figure 1.7 shows that in binary, the bits represent one, two, four, eight, sixteen, etc. A multi-digit binary number is commonly called a word, and the number of bits in the word is called the wordlength. The right-hand bit is called the least significant bit (LSB) whereas the bit on the left-hand end of the word is called the most significant bit (MSB). Clearly more digits are required in binary than in decimal, but they are more easily handled. A word of eight bits is called a byte, which is a contraction of 'by eight'.

Figure 1.7 also shows some binary numbers and their equivalent in decimal. The radix point has the same significance in binary: symbols to the right of it represent one half, one quarter and so on.

Binary words can have a remarkable range of meanings. They may describe the magnitude of a number such as an audio sample or an image pixel or they may specify the address of a single location in a memory. In all cases the possible range of a word is limited by the wordlength. The range is found by raising two to the power of the wordlength. Thus a four-bit word has sixteen combinations, and could address a memory having sixteen locations. A sixteen-bit word has 65 536 combinations. Figure 1.8(a) shows some examples of wordlength and resolution.

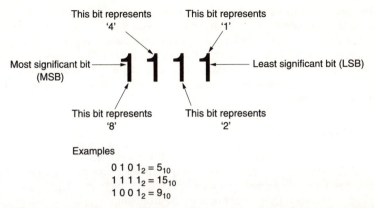

Examples

$$0\ 1\ 0\ 1_2 = 5_{10}$$
$$1\ 1\ 1\ 1_2 = 15_{10}$$
$$1\ 0\ 0\ 1_2 = 9_{10}$$

Figure 1.7 In a binary number, the digits represent increasing powers of two from the LSB. Also defined here are MSB and wordlength. When the wordlength is eight bits, the word is a byte. Binary numbers are used as memory addresses, and the range is defined by the address wordlength. Some examples are shown here.

Wordlength and resolution

The wordlength determines the possible range of values:

Wordlength	Range
1	$2\ (2^1)$
2	$4\ (2^2)$
3	$8\ (2^3)$
•	
•	
•	
8	$256\ (2^8)$
•	
10	$1024\ (2^{10})$
•	
•	
•	
•	
16	$65\,536\ (2^{16})$

(a)

Round numbers in binary

10000000000_2	= 1024	= 1 k (kilo in computers)
1 k × 1 k	= 1 M	(Mega)
1 M × 1 k	= 1 G	(Giga)
1 M × 1 M	= 1 T	(Tera)

(b)

Figure 1.8 The wordlength of a sample controls the resolution as shown in (a). In the same way the ability to address memory locations is also determined as in (b).

The capacity of memories and storage media is measured in bytes, but to avoid large numbers, kilobytes, megabytes and gigabytes are often used. A ten-bit word has 1024 combinations, which is close to one thousand. In digital terminology, 1 K is defined as 1024, so a kilobyte of memory contains 1024 bytes. A megabyte (1 MB) contains 1024 kilobytes and would need a twenty-bit address. A gigabyte contains 1024 megabytes and would need a thirty-bit address. Figure 1.8(b) shows some examples.

1.4 Colour

Colorimetry will be treated in Chapter 2 and it is intended to introduce only the basics here. Colour is created in television by the additive mixing in the display of three primary colours, red, green and blue. Effectively the display needs to be supplied with three video signals, each representing a primary colour. Since practical colour cameras generally also have three separate sensors, one for each primary colour, a camera and a display can be directly connected. *RGB* consists of three parallel signals each having the same spectrum, and is used where the highest accuracy is needed. *RGB* is seldom used for broadcast applications because of the high cost.

If *RGB* is used in the digital domain, it will be seen from Figure 1.9 that each image consists of three superimposed layers of samples, one for each primary colour. The pixel is no longer a single number representing a scalar brightness value, but a vector which describes in some way the brightness, hue and saturation of that point in the picture. In *RGB*, the pixels contain three unipolar numbers representing the proportion of each of the three primary colours at that point in the picture.

Some saving of bit rate can be obtained by using colour difference working. The human eye relies on brightness to convey detail, and much less resolution is needed in the colour information. Accordingly *R*, *G* and *B* are matrixed together to form a luminance (and monochrome-compatible) signal *Y* which needs full

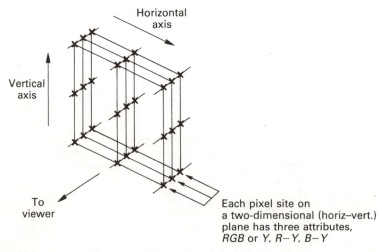

Figure 1.9 In the case of component video, each pixel site is described by three values and so the pixel becomes a vector quantity.

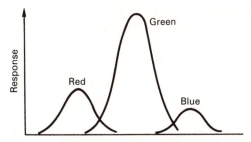

Figure 1.10 The response of the human eye to colour is not uniform.

bandwidth. The eye is not equally sensitive to the three primary colours, as can be seen in Figure 1.10 and so the luminance signal is a weighted sum.

The matrix also produces two colour difference signals, $R-Y$ and $B-Y$. Colour difference signals do not need the same bandwidth as Y, because the eye's acuity does not extend to colour vision. One half or one quarter of the bandwidth will do depending on the application.

In the digital domain, each pixel again contains three numbers, but one of these is a unipolar number representing the luminance and the other two are bipolar numbers representing the colour difference values. As the colour difference signals need less bandwidth, in the digital domain this translates to the use of a lower data rate, typically between one half and one sixteenth the bit rate of the luminance.

1.5 Why digital?

There are two main answers to this question, and it is not possible to say which is the most important, as it will depend on one's standpoint.

(a) The quality of reproduction of a well-engineered digital video system is independent of the medium and depends only on the quality of the conversion processes and of any compression scheme.

(b) The conversion of video to the digital domain allows tremendous opportunities which were denied to analog signals.

Someone who is only interested in picture quality will judge the former the most relevant. If good-quality convertors can be obtained, all the shortcomings of analog recording and transmission can be eliminated to great advantage. One's greatest effort is expended in the design of convertors, whereas those parts of the system which handle data need only be workmanlike. When a digital recording is copied, the same numbers appear on the copy: it is not a dub, it is a clone. If the copy is undistinguishable from the original, there has been no generation loss. Digital recordings can be copied indefinitely without loss of quality. This is, of course, wonderful for the production process, but when the technology becomes available to the consumer the issue of copyright becomes of great importance.

In the real world everything has a cost, and one of the greatest strengths of digital technology is low cost. When the information to be recorded is discrete numbers, they can be packed densely on the medium without quality loss. Should some bits be in error because of noise or dropout, error correction can restore the original value. Digital recordings take up less space than analog recordings for the same or better quality. Digital circuitry costs less to manufacture because more functionality can be put in the same chip.

Digital equipment can have self-diagnosis programs built-in. The machine points out its own failures so the the cost of maintenance falls. A small operation may not need maintenance staff at all; a service contract is sufficient. A larger organization will still need maintenance staff, but they will be fewer in number and their skills will be oriented more to systems than devices.

1.6 Some digital video processes outlined

Whilst digital video is a large subject, it is not necessarily a difficult one. Every process can be broken down into smaller steps, each of which is relatively easy to follow. The main difficulty with study is to appreciate where the small steps fit into the overall picture. Subsequent chapters of this book will describe the key processes found in digital technology in some detail, whereas this chapter

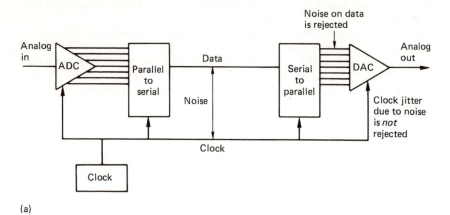

(a)

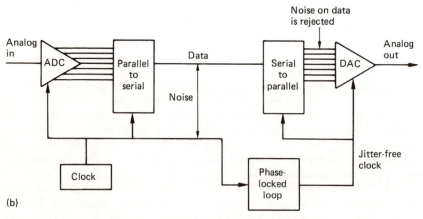

(b)

Figure 1.11 In (a) two convertors are joined by a serial link. Although simple, this system is deficient because it has no means to prevent noise on the clock lines causing jitter at the receiver. In (b) a phase-locked loop is incorporated, which filters jitter from the clock.

illustrates why these processes are necessary and shows how they are combined in various ways in real equipment. Once the general structure of digital devices is appreciated, the following chapters can be put in perspective.

Figure 1.11(a) shows a minimal digital video system. This is no more than a point-to-point link which conveys analog video from one place to another. It consists of a pair of convertors and hardware to serialize and de-serialize the samples. There is a need for standardization in serial transmission so that various devices can be connected together. These standards for digital interfaces are described in Chapter 9.

Analog video entering the system is converted in the analog-to-digital convertor (ADC) to samples which are expressed as binary numbers. A typical sample would have a wordlength of eight bits. The sample is connected in parallel into an output register which controls the cable drivers. The cable also carries the sampling rate clock. The data are sent to the other end of the line where a slicer reflects noise picked up on each signal. Sliced data are then loaded into a receiving register by the clock, and sent to the digital-to-analog convertor (DAC), which converts the sample back to an analog voltage.

As Figure 1.5 showed, noise can change the timing of a sliced signal. Whilst this system rejects noise which threatens to change the numerical value of the samples, it is powerless to prevent noise from causing jitter in the receipt of the sample clock. Noise on the clock means that samples are not converted with a regular timebase and the impairment caused can be noticeable.

The jitter problem is overcome in Figure 1.11(b) by the inclusion of a phase-locked loop which is an oscillator that synchronizes itself to the *average* frequency of the clock but which filters out the instantaneous jitter.

The system of Figure 1.11 is extended in Figure 1.12 by the addition of some random access memory (RAM). The operation of RAM is described in Chapter 2. What the device does is determined by the way in which the RAM address is controlled. If the RAM address increases by one every time a sample from the ADC is stored in the RAM, a recording can be made for a short period until the RAM is full. The recording can be played back by repeating the address sequence at the same clock rate but reading the memory into the DAC. The result is generally called a framestore.[2] If the memory capacity is increased, the device can be used for recording. At a rate of 200 million bits per second, each frame

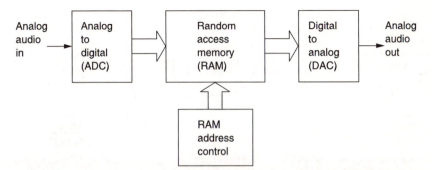

Figure 1.12 In the framestore, the recording medium is a random access memory (RAM). Recording time available is short compared with other media, but access to the recording is immediate and flexible as it is controlled by addressing the RAM.

needs a megabyte of memory and so the RAM recorder will be restricted to a fairly short playing time.

Using compression, the playing time of a RAM-based recorder can be extended. For predetermined images such as test patterns and station IDs, read only memory (ROM) can be used instead as it is non-volatile.

1.7 Time compression and expansion

Data files such as computer programs are simply lists of instructions and have no natural time axis. In contrast, audio and video data are sampled at a fixed rate and need to be presented to the viewer at the same rate. In audiovisual systems the audio also needs to be synchronized to the video. Continuous bitstreams at a fixed bit rate are difficult for generic data recording and transmission systems to handle. Most digital recording and transmission systems work on blocks of data which can be individually addressed and/or routed. The bit rate may be fixed at the design stage at a value which may be too low or too high for the audio or video data to be handled.

The solution is to use time compression or expansion. Figure 1.13 shows a RAM which is addressed by binary counters which periodically overflow to zero and start counting again, giving the RAM a ring structure. If write and read addresses increment at the same speed, the RAM becomes a fixed data delay as the addresses retain a fixed relationship. However, if the read address clock runs at a higher frequency but in bursts, the output data are assembled into blocks with spaces in between. The data are now time compressed. Instead of being an unbroken stream which is difficult to handle, the data are in blocks with convenient pauses in between them. In these pauses numerous processes can take place. A hard disk might move its heads to another track. In all types of recording and transmission, the time compression of the samples allows time for synchronizing patterns, subcode and error-correction words to be inserted.

Subsequently, any time compression can be reversed by time expansion. This requires a second RAM identical to the one shown. Data are written into the RAM in bursts, but read out at the standard sampling rate to restore a continuous bitstream. In a recorder, the time-expansion stage can be combined with the timebase correction stage so that speed variations in the medium can be

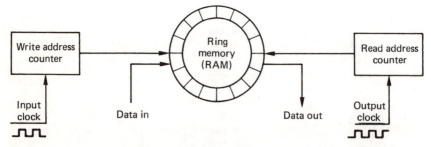

Figure 1.13 If the memory address is arranged to come from a counter which overflows, the memory can be made to appear circular. The write address then rotates endlessly, overwriting previous data once per revolution. The read address can follow the write address by a variable distance (not exceeding one revolution) and so a variable delay takes place between reading and writing.

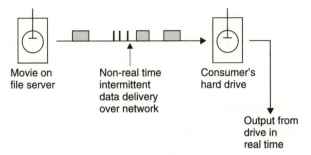

Figure 1.14 In non-real-time transmission, the data are transferred slowly to a storage medium which then outputs real-time data. Movies can be downloaded to the home in this way.

eliminated at the same time. The use of time compression is universal in digital recording and widely used in transmission. In general the *instantaneous* data rate in the channel is not the same as the original rate although clearly the *average* rate must be the same.

Where the bit rate of the communication path is inadequate, transmission is still possible, but not in real time. Figure 1.14 shows that the data to be transmitted will have to be written in real time on a storage device such as a disk drive, and the drive will then transfer the data at whatever rate is possible to another drive at the receiver. When the transmission is complete, the second drive can then provide the data at the corrrect bit rate.

In the case where the available bit rate is higher than the correct data rate, the same configuration can be used to copy an audio or video data file faster than real

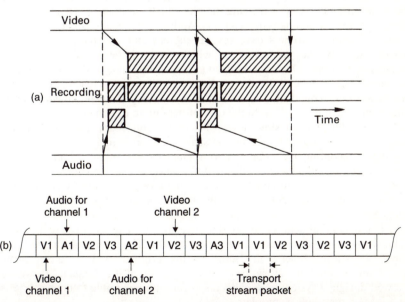

Figure 1.15 (a) Time compression is used to shorten the length of track needed by the video. Heavily time-compressed audio samples can then be recorded on the same track using common circuitry. In MPEG, multiplexing allows data from several TV channels to share one bitstream (b).

time. Another application of time compression is to allow several streams of data to be carried along the same channel in a technique known as *multiplexing*. Figure 1.15 shows some examples. At (a) multiplexing allows audio and video data to be recorded on the same heads in a digital video recorder such as DVC. At (b), several TV channels are multiplexed into one MPEG transport stream.

1.8 Error correction and concealment

All practical recording and transmission media are imperfect. Magnetic media, for example, suffer from noise and dropouts. In a digital recording of binary data, a bit is either correct or wrong, with no intermediate stage. Small amounts of noise are rejected, but inevitably, infrequent noise impulses cause some individual bits to be in error. Dropouts cause a larger number of bits in one place to be in error. An error of this kind is called a burst error. Whatever the medium and whatever the nature of the mechanism responsible, data are either recovered correctly or suffer some combination of bit errors and burst errors. In optical disks, random errors can be caused by imperfections in the moulding process, whereas burst errors are due to contamination or scratching of the disk surface.

The visibility of a bit error depends upon which bit of the sample is involved. If the LSB of one sample was in error in a detailed, contrast picture, the effect would be totally masked and no-one could detect it. Conversely, if the MSB of one sample was in error in a flat field, no-one could fail to notice the resulting spot. Clearly a means is needed to render errors from the medium inaudible. This is the purpose of error correction.

In binary, a bit has only two states. If it is wrong, it is only necessary to reverse the state and it must be right. Thus the correction process is trivial and perfect. The main difficulty is in identifying the bits which are in error. This is done by coding the data by adding redundant bits. Adding redundancy is not confined to digital technology, airliners have several engines and cars have twin braking systems. Clearly the more failures which have to be handled, the more redundancy is needed. If a four-engined airliner is designed to fly normally with one engine failed, three of the engines have enough power to reach cruise speed, and the fourth is redundant. The amount of redundancy is equal to the amount of failure which can be handled. In the case of the failure of two engines, the plane can still fly, but it must slow down; this is graceful degradation. Clearly the chances of a two-engine failure on the same flight are remote.

In digital recording, the amount of error which can be corrected is proportional to the amount of redundancy, and it will be shown in Chapter 6 that within this limit, the samples are returned to exactly their original value. Consequently *corrected* samples are undetectable. If the amount of error exceeds the amount of redundancy, correction is not possible, and, in order to allow graceful degradation, concealment will be used. Concealment is a process where the value of a missing sample is estimated from those nearby. The estimated sample value is not necessarily exactly the same as the original, and so under some circumstances concealment can be audible, especially if it is frequent. However, in a well-designed system, concealments occur with negligible frequency unless there is an actual fault or problem.

Concealment is made possible by rearranging the sample sequence prior to recording. This is shown in Figure 1.16 where odd-numbered samples are

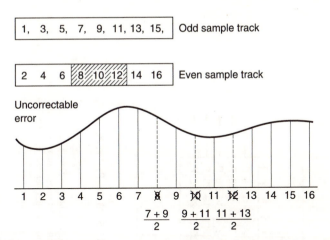

Figure 1.16 In cases where the error correction is inadequate, concealment can be used provided that the samples have been ordered appropriately in the recording. Odd and even samples are recorded in different places as shown here. As a result an uncorrectable error causes incorrect samples to occur singly, between correct samples. In the example shown, sample 8 is incorrect, but samples 7 and 9 are unaffected and an approximation to the value of sample 8 can be had by taking the average value of the two. This interpolated value is substituted for the incorrect value.

separated from even-numbered samples prior to recording. The odd and even sets of samples may be recorded in different places on the medium, so that an uncorrectable burst error affects only one set. On replay, the samples are recombined into their natural sequence, and the error is now split up so that it results in every other sample being lost in a two-dimensional structure. The picture is now described half as often, but can still be reproduced with some loss of accuracy. This is better than not being reproduced at all even if it is not perfect. Many digital video recorders use such an odd/even distribution for concealment. Clearly if any errors are fully correctable, the distribution is a waste of time; it is only needed if correction is not possible.

The presence of an error-correction system means that the video (and audio) quality is independent of the medium/head quality within limits. There is no point in trying to assess the health of a machine by watching a monitor or listening to the audio, as this will not reveal whether the error rate is normal or within a whisker of failure. The only useful procedure is to monitor the frequency with which errors are being corrected, and to compare it with normal figures.

1.9 Product codes

Digital systems such as broadcasting, optical disks and magnetic recorders are prone to burst errors. Adding redundancy equal to the size of expected bursts to every code is inefficient. Figure 1.17(a) shows that the efficiency of the system can be raised using interleaving. Sequential samples from the ADC are assembled into codes, but these are not recorded/transmitted in their natural sequence. A number of sequential codes are assembled along rows in a

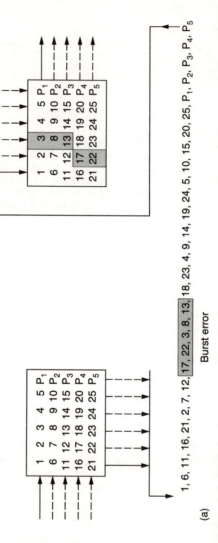

(a)

Burst error

1, 6, 11, 16, 21, 2, 7, 12, 17, 22, 3, 8, 13, 18, 23, 4, 9, 14, 19, 24, 5, 10, 15, 20, 25, P_1, P_2, P_3, P_4, P_5

Figure 1.17(a) Interleaving is essential to make error-correction schemes more efficient. Samples written sequentially in rows into a memory have redundancy P added to each row. The memory is then read in columns and the data sent to the recording medium. On replay the non-sequential samples from the medium are de-interleaved to return them to their normal sequence. This breaks up the burst error (shaded) into one error symbol per row in the memory, which can be corrected by the redundancy P.

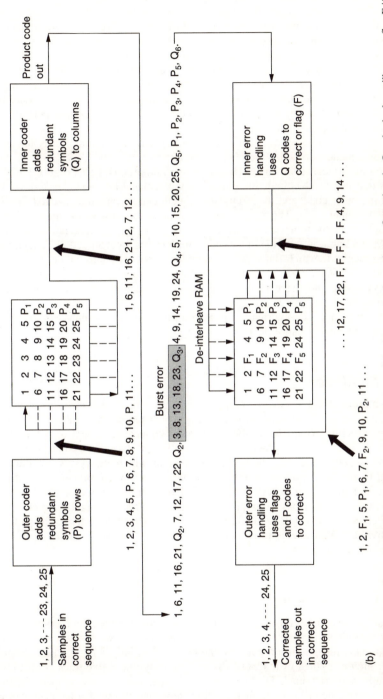

Figure 1.17(b) In addition to the redundancy P on rows, inner redundancy Q is also generated on columns. On replay, the Q code checker will pass on flag F if it finds an error too large to handle itself. The flags pass through the de-interleave process and are used by the outer error correction to identify which symbol in the row needs correcting with P redundancy. The concept of crossing two codes in this way is called a product code.

(b)

memory. When the memory is full, it is copied to the medium by reading down columns. Subsequently, the samples need to be de-interleaved to return them to their natural sequence. This is done by writing samples from tape into a memory in columns, and when it is full, the memory is read in rows. Samples read from the memory are now in their original sequence so there is no effect on the information. However, if a burst error occurs as is shown shaded on the diagram, it will damage sequential samples in a vertical direction in the de-interleave memory. When the memory is read, a single large error is broken down into a number of small errors whose size is exactly equal to the correcting power of the codes and the correction is performed with maximum efficiency.

An extension of the process of interleave is where the memory array has not only rows made into codewords but also columns made into codewords by the addition of vertical redundancy. This is known as a product code. Figure 1.17(b) shows that in a product code the redundancy calculated first and checked last is called the outer code, and the redundancy calculated second and checked first is called the inner code. The inner code is formed along tracks on the medium. Random errors due to noise are corrected by the inner code and do not impair the burst-correcting power of the outer code. Burst errors are declared uncorrectable by the inner code which flags the bad samples on the way into the de-interleave memory. The outer code reads the error flags in order to locate the erroneous data. As it does not have to compute the error locations, the outer code can correct more errors.

The interleave, de-interleave, time-compression and timebase-correction processes inevitably cause delay.

1.10 Shuffling

When a product code-based recording suffers an uncorrectable error the result is a rectangular block of failed sample values which require concealment. Such a regular structure would be visible even after concealment, and an additional process is necessary to reduce the visibility. Figure 1.18 shows that a shuffle process is performed prior to product coding in which the pixels are moved around the picture in a pseudo-random fashion. The reverse process is used on replay, and the overall effect is nullified. However, if an uncorrectable error occurs, this will only pass through the de-shuffle and so the regular structure of the failed data blocks will be randomized. The errors are spread across the picture as individual failed pixels in an irregular structure. Chapter 11 treats shuffling techniques in more detail.

1.11 Channel coding

In most recorders used for storing digital information, the medium carries a track which reproduces a single waveform. Clearly data words representing video contain many bits and so they have to be recorded serially, a bit at a time. Some media, such as optical or magnetic disks, have only one active track, so it must be totally self-contained. DVTRs may have one, two or four tracks read or written simultaneously. At high recording densities, physical tolerances cause

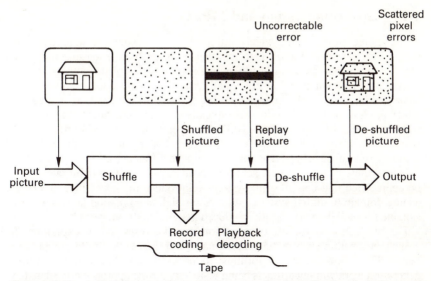

Figure 1.18 The shuffle before recording and the corresponding de-shuffle after playback cancel out as far as the picture is concerned. However, a block of errors due to dropout only experiences the de-shuffle, which spreads the error randomly over the screen. The pixel errors are then easier to correct.

phase shifts, or timing errors, between tracks and so it is not possible to read them in parallel. Each track must still be self-contained until the replayed signal has been timebase corrected.

Recording data serially is not as simple as connecting the serial output of a shift register to the head. In digital video, samples may contain strings of identical bits. If a shift register is loaded with such a sample and shifted out serially, the output stays at a constant level for the period of the identical bits, and nothing is recorded on the track. On replay there is nothing to indicate how many bits were present, or even how fast to move the medium. Clearly, serialized raw data cannot be recorded directly, it has to be modulated into a waveform which contains an embedded clock irrespective of the values of the bits in the samples. On replay a circuit called a data separator can lock to the embedded clock and use it to separate strings of identical bits.

The process of modulating serial data to make it self-clocking is called channel coding. Channel coding also shapes the spectrum of the serialized waveform to make it more efficient. With a good channel code, more data can be stored on a given medium. Spectrum shaping is used in optical disks to prevent the data from interfering with the focus and tracking servos, and in hard disks and in certain DVTRs to allow rerecording without erase heads.

Channel coding is also needed to broadcast digital television signals where shaping of the spectrum is an obvious requirement to avoid interference with other services.

The techniques of channel coding for recording are covered in detail in Chapter 6, whereas the modulation schemes for digital television are described in Chapter 9.

1.12 Video compression and MPEG

In its native form, digital video suffers from an extremely high data rate, particularly in high definition. One approach to the problem is to use compression which reduces that rate significantly with a moderate loss of subjective quality of the picture. The human eye is not equally sensitive to all spatial frequencies, so some coding gain can be obtained by using fewer bits to describe the frequencies which are less visible. Video images typically contain a great deal of redundancy where flat areas contain the same pixel value repeated many times. Furthermore, in many cases there is little difference between one picture and the next, and compression can be achieved by sending only the differences.

Whilst these techniques may achieve considerable reduction in bit rate, it must be appreciated that compression systems reintroduce the generation loss of the analog domain to digital systems. As a result high compression factors are only suitable for final delivery of fully produced material to the viewer.

For production purposes, compression may be restricted to exploiting the redundancy within each picture individually and then with a mild compression factor. This allows simple algorithms to be used and also permits multiple-generation work without artifacts being visible. A similar approach is used in disk-based workstations. Where off-line editing is used (see Chapter 13) higher compression factors can be employed as the impaired pictures are not seen by the viewer.

Clearly, a consumer DVTR needs only single-generation operation and has simple editing requirements. A much greater degree of compression can then be used, which may take advantage of redundancy between fields. The same is true for broadcasting, where bandwidth is at a premium. A similar approach may be used in disk-based camcorders which are intended for ENG purposes.

The future of television broadcasting (and of any high-definition television) lies completely in compression technology. Compression requires an encoder prior to the medium and a compatible decoder after it. Extensive consumer use of compression could not occur without suitable standards. The ISO-MPEG coding standards were specifically designed to allow wide interchange of compressed video data. Digital television broadcasting and the digital video disk both use MPEG standard bitstreams.

Figure 1.19 shows that the output of a single compressor is called an *elementary stream*. In practice audio and video streams of this type can be combined using multiplexing. The *program stream* is optimized for recording and is based on blocks of arbitrary size. The *transport stream* is optimized for transmission and is based on blocks of constant size.

In production equipment such as workstations and VTRs which are designed for editing, the MPEG standard is less useful and many successful products use non-MPEG compression.

Compression and the corresponding decoding are complex processes and take time, adding to existing delays in signal paths. Concealment of uncorrectable errors is also more difficult on compressed data.

1.13 Disk-based recording

The magnetic disk drive was perfected by the computer industry to allow rapid random access to data, and so it makes an ideal medium for editing. As will be

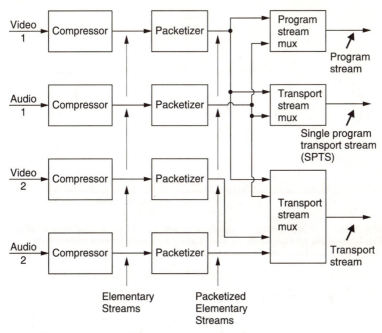

Figure 1.19 The bitstream types of MPEG-2. See text for details.

seen in Chapter 7, the heads do not touch the disk, but are supported on a thin air film which gives them a long life but which restricts the recording density. Thus disks cannot compete with tape for lengthy recordings, but for short-duration work such as commercials or animation they have no equal. The data rate of digital video is too high for a single disk head, and so a number of solutions have been explored. One obvious solution is to use compression, which cuts the data rate and extends the playing time. Another approach is to operate a large array of conventional drives in parallel. The highest-capacity magnetic disks are not removable from the drive.

The disk drive provides intermittent data transfer owing to the need to reposition the heads. Figure 1.20 shows that disk-based devices rely on a quantity of RAM acting as a buffer between the real-time video environment and the intermittent data environment.

Figure 1.21 shows the block diagram of a camcorder based on hard disks and compression. The recording time and picture quality will not compete with full bandwidth tape-based devices, but following acquisition the disks can be used directly in an edit system, allowing a useful time saving in ENG applications.

Development of the optical disk was stimulated by the availability of low-cost lasers. Optical disks are available in many different types, some which can only be recorded once, some which are erasable. These will be contrasted in Chapter 7. Optical disks have in common the fact that access is generally slower than with magnetic drives and it is difficult to obtain high data rates, but most of them are removable and can act as interchange media.

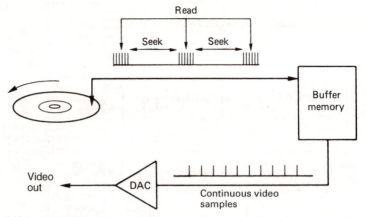

Figure 1.20 In a hard disk recorder, a large-capacity memory is used as a buffer or timebase corrector between the convertors and the disk. The memory allows the convertors to run constantly despite the interruptions in disk transfer caused by the head moving between tracks.

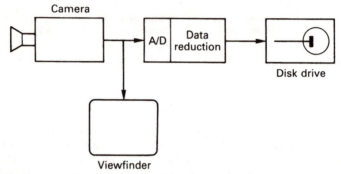

Figure 1.21 In a disk-based camcorder, the PCM data rate from the camera is too high for direct recording on disk. Data reduction is used to cut the bit rate and extend playing time. If a standard file structure is used, disks may physically be transferred to an edit system after recording.

1.14 Rotary-head digital recorders

The rotary-head recorder has the advantage that the spinning heads create a high head-to-tape speed offering a high bit rate recording without high tape speed. Whilst mechanically complex, the rotary-head transport has been raised to a high degree of refinement and offers the highest recording density and thus lowest cost per bit of all digital recorders.[3]

Digital VTRs segment incoming fields into several tape tracks and invisibly reassemble them in memory on replay in order to keep the tracks reasonably short.

Figure 1.22 shows a representative block diagram of a DVTR. Following the convertors, a compression process may be found. In an uncompressed recorder, there will be distribution of odd and even samples and a shuffle process for concealment purposes. An interleaved product code will be formed prior to the channel coding stage which produces the recorded waveform. On replay the data

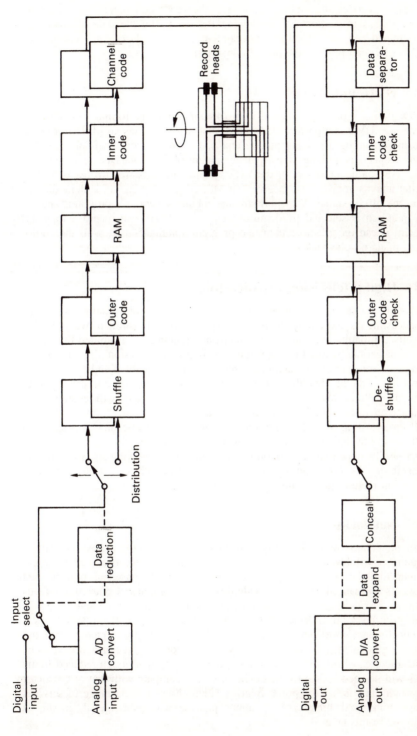

Figure 1.22 Block diagram of a DVTR. Note optional data reduction unit which may be used to allow a common transport to record a variety of formats.

separator decodes the channel code and the inner and outer codes perform correction as in section 1.9. Following the de-shuffle the data channels are recombined and any necessary concealment will take place. Any compression will be decoded prior to the output convertors.

1.15 DVD and DVHS

DVD (digital video disk) and DVHS (digital VHS) are formats intended for home use. DVD is a prerecorded optical disk which carries an MPEG program stream containing a moving picture and one or more audio channels. DVD in many respects is a higher density development of Compact Disc technology.

DVHS is a development of the VHS system which records MPEG data. In a digital television broadcast receiver, an MPEG transport stream is demodulated from the off-air signal. Transport streams contain several TV programs and the one required is selected by demultiplexing. As DVHS can record a transport stream, it can record more than one program simultaneously, with the choice being made on playback.

1.16 Digital television broadcasting

Although it has given good service for many years, analog television broadcasting is extremely inefficient because the transmitted waveform is directly compatible with the CRT display, and nothing is transmitted during the blanking periods whilst the beam retraces. Using compression, digital modulation and error-correction techniques, the same picture quality can be obtained in a fraction of the bandwidth of analog. Pressure on spectrum use from other uses such as cellular telephones will only increase and this will ensure rapid changeover to digital television and radio broadcasts.

If broadcasting of high-definition television is ever to become widespread it will do so via digital compressed transmission as the bandwidth required will otherwise be hopelessly uneconomic. In addition to conserving spectrum, digital transmission is (or should be) resistant to multipath reception and gives consistent picture quality throughout the service area.

1.17 Networks

Communications networks allow transmission of data files whose content or meaning is irrelevant to the transmission medium. These files can therefore contain digital video. Video production systems can be based on high bit rate networks instead of traditional video-routing techniques. Contribution feeds between broadcasters and station output to transmitters no longer requires special-purpose links. Video delivery is also possible on the Internet. As a practical matter, most Internet users suffer from a relatively limited bit rate and if moving pictures are to be sent, a very high compression factor will have to be used. Pictures with a relatively small number of lines and much reduced frame rates will be needed. Whilst the quality does not compare with that of broadcast television, this is not the point. Internet video allows a wide range of services which traditional broadcasting cannot provide and phenomenal growth is expected in this area.

References

1. Devereux, V.G., Pulse code modulation of video signals: 8 bit coder and decoder. *BBC Res. Dept. Rept.*, **EL−42**, No.25 (1970)
2. Pursell, S. and Newby, H., Digital frame store for television video. *SMPTE Journal*, **82**, 402−403 (1973)
3. Baldwin, J.L.E., Digital television recording – history and background. *SMPTE Journal*, **95**, 1206−1214 (1986)

Chapter 2

Video principles

2.1 The eye

All television signals ultimately excite some response in the eye and the viewer can only describe the result subjectively. Familiarity with the operation and limitations of the eye is essential to an understanding of television principles.

The simple representation of Figure 2.1 shows that the eyeball is nearly spherical and is swivelled by muscles so that it can track movement. This has a large bearing on the way moving pictures are reproduced. The space between the cornea and the lens is filled with transparent fluid known as *aqueous humour*. The remainder of the eyeball is filled with a transparent jelly known as *vitreous*

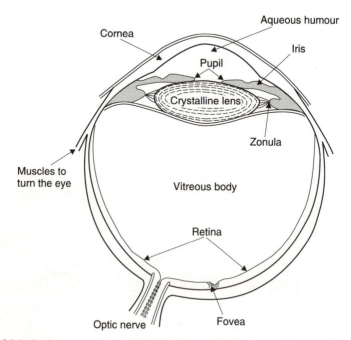

Figure 2.1 A simple representation of an eyeball; see text for details.

humour. Light enters the cornea, and the the amount of light admitted is controlled by the pupil in the iris. Light entering is involuntarily focused on the retina by the lens in a process called *visual accommodation.*

The retina is responsible for light sensing and contains a number of layers. The surface of the retina is covered with arteries, veins and nerve fibres and light has to penetrate these in order to reach the sensitive layer. This contains two types of discrete receptors known as *rods* and *cones* from their shape. The distribution and characteristics of these two receptors are quite different. Rods dominate the periphery of the retina whereas cones dominate a central area known as the *fovea* outside which their density drops off. Vision using the rods is monochromatic and has poor resolution but remains effective at very low light levels, whereas the cones provide high resolution and colour vision but require more light.

The cones in the fovea are densely packed and directly connected to the nervous system allowing the highest resolution. Resolution then falls off away from the fovea. As a result the eye must move to scan large areas of detail. The image perceived is not just a function of the retinal response, but is also affected by processing of the nerve signals. The overall acuity of the eye can be displayed as a graph of the response plotted against the degree of detail being viewed. Image detail is generally measured in lines per millimetre or cycles per picture height, but this takes no account of the distance from the image to the eye. A better unit for eye resolution is one based upon the subtended angle of detail as this will be independent of distance. Units of cycles per degree are then appropriate. Figure 2.2 shows the response of the eye to static detail. Note that the response to very low frequencies is also attenuated. An extension of this characteristic allows the vision system to ignore the fixed pattern of shadow on the retina due to the nerves and arteries.

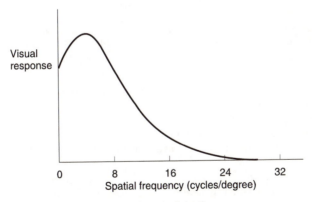

Figure 2.2 Response of the eye to different degrees of detail.

The retina does not respond instantly to light, but requires between 0.15 and 0.3 second before the brain perceives an image. The resolution of the eye is primarily a spatio-temporal compromise. The eye is a spatial sampling device; the spacing of the rods and cones on the retina represents a spatial sampling frequency. The measured acuity of the eye exceeds the value calculated from the sample site spacing because a form of oversampling is used.

The eye is in a continuous state of unconscious vibration called saccadic motion. This causes the sampling sites to exist in more than one location, effectively increasing the spatial sampling rate provided there is a temporal filter which is able to integrate the information from the various different positions of the retina.

This temporal filtering is responsible for 'persistence of vision'. Flashing lights are perceived to flicker until the critical flicker frequency (CFF) is reached; the light appears continuous for higher frequencies. The CFF is not constant but varies with brightness. Note that the field rate of European television at 50 fields per second is marginal with bright images. Film projected at 48 Hz works because cinemas are darkened and the screen brightnes is actually quite low. Figure 2.3 shows the two-dimensional or spatio-temporal response of the eye.

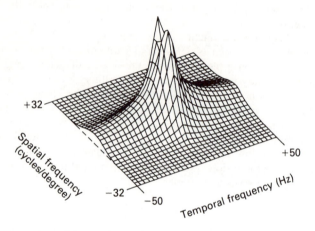

Figure 2.3 The response of the eye shown with respect to temporal and spatial frequencies. Note that even slow relative movement causes a serious loss of resolution. The eye tracks moving objects to prevent this loss.

If the eye were static, a detailed object moving past it would give rise to temporal frequencies, as Figure 2.4(a) shows. The temporal frequency is given by the detail in the object, in lines per millimetre, multiplied by the speed. Clearly a highly detailed object can reach high temporal frequencies even at slow speeds, yet Figure 2.3 shows that the eye cannot respond to high temporal frequencies.

However, the human viewer has an interactive visual system which causes the eyes to track the movement of any object of interest. Figure 2.4(b) shows that when eye tracking is considered, a moving object is rendered stationary with respect to the retina so that temporal frequencies fall to zero and much the same acuity to detail is available despite motion. This is known as dynamic resolution and it's how humans judge the detail in real moving pictures. Dynamic resolution will be considered in section 2.10.

The contrast sensitivity of the eye is defined as the smallest brightness difference which is visible. In fact the contrast sensitivity is not constant, but increases proportionally to brightness. Thus whatever the brightness of an object, if that brightness changes by about 1 per cent it will be equally detectable.

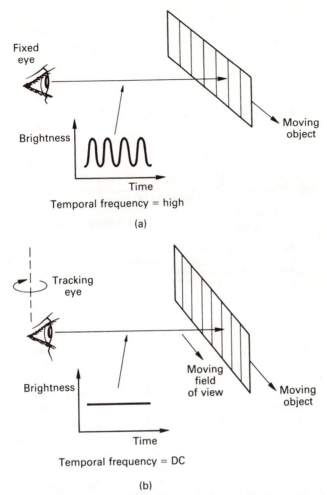

Figure 2.4 In (a) a detailed object moves past a fixed eye, causing temporal frequencies beyond the response of the eye. This is the cause of motion blur. In (b) the eye tracks the motion and the temporal frequency becomes zero. Motion blur cannot then occur.

2.2 Gamma

The true brightness of a television picture can be affected by electrical noise on the video signal. As contrast sensitivity is proportional to brightness, noise is more visible in dark picture areas than in bright areas. For economic reasons, video signals have to be made non-linear to render noise less visible. An inverse gamma function takes place at the camera so that the video signal is non-linear for most of its journey. Figure 2.5 shows a reverse gamma function. As a true power function requires infinite gain near black, a linear segment is substituted. It will be seen that contrast variations near black result in larger signal amplitude than variations near white. The result is that noise picked up by the video signal has less effect on dark areas than on bright areas. After a gamma function at the display, noise at near-black levels is compressed with respect to noise at near-

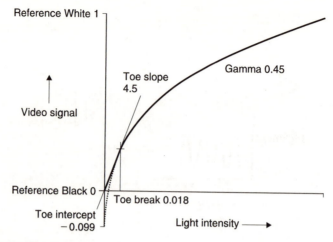

Figure 2.5 CCIR Rec.709 reverse gamma function used at camera has a straight-line approximation at the lower part of the curve to avoid boosting camera noise. Note that the output amplitude is greater for modulation near black.

white levels. Thus a video transmission system using gamma has a lower perceived noise level than one without. Without gamma, vision signals would need around 30 dB better signal-to-noise ratio for the same perceived quality and digital video samples would need five or six extra bits.

In practice the system is not rendered perfectly linear by gamma correction and a slight overall exponential effect is usually retained in order further to reduce the effect of noise in darker parts of the picture. A gamma correction factor of 0.45 may be used to achieve this effect. Clearly, image data which are intended to be displayed on a video system must have the correct gamma characteristic or the grey scale will not be correctly reproduced.

As all television signals, analog and digital, are subject to gamma correction, it is technically incorrect to refer to the Y signal as luminance, because this parameter is defined as linear in colorimetry. It has been suggested that the term *luma* should be used to describe luminance which has been gamma corrected.

In a CRT (cathode ray tube) the relationship between the tube drive voltage and the phosphor brightness is not linear, but an exponential function where the power is known as *gamma*. The power is the same for all CRTs as it is a function of the physics of the electron gun and it has a value of around 2.8. It is a happy but pure coincidence that the gamma function of a CRT follows roughly the same curve as human contrast sensitivity.

Consequently if video signals are predistorted at source by an inverse gamma, the gamma characteristic of the CRT will linearize the signal. Figure 2.6 shows the principle. CRT gamma is not a nuisance, but is actually used to enhance the noise performance of a system. If the CRT had no gamma characteristic, a gamma circuit would have been necessary ahead of it. As all standard video signals are inverse gamma processed, it follows that if a non-CRT display such as a plasma or LCD device is to be used, some gamma conversion will be required at the display.

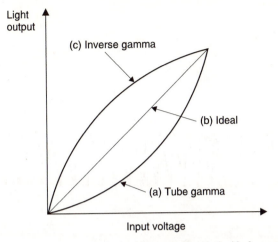

Figure 2.6 The non-linear characteristic of tube (a) contrasted with the ideal response (b). Non-linearity may be opposed by gamma correction with a response (c).

2.3 Scanning

Figure 2.7(a) shows that the monochrome camera produces a *video signal* whose voltage is a function of the image brightness at a single point on the sensor. This voltage is converted back to the brightness of the same point on the display. The points on the sensor and display must be scanned synchronously if the picture is to be re-created properly. If this is done rapidly enough it is largely invisible to the eye. Figure 2.7(b) shows that the scanning is controlled by a triangular or *sawtooth* waveform in each dimension which causes a constant speed forward scan followed by a rapid return or *flyback*. As the horizontal scan is much more rapid than the vertical scan the image is broken up into lines which are not quite horizontal.

In the example of Figure 2.7(b), the horizontal scanning frequency or *line rate*, F_h, is an integer multiple of the vertical scanning frequency or *frame rate* and a *progressive scan* system results in which every frame is identical. Figure 2.7(c)

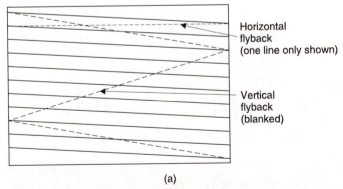

(a)

Figure 2.7 Scanning converts two-dimensional images into a signal which can be sent electrically. In (a) the scanning of camera and display must be identical.

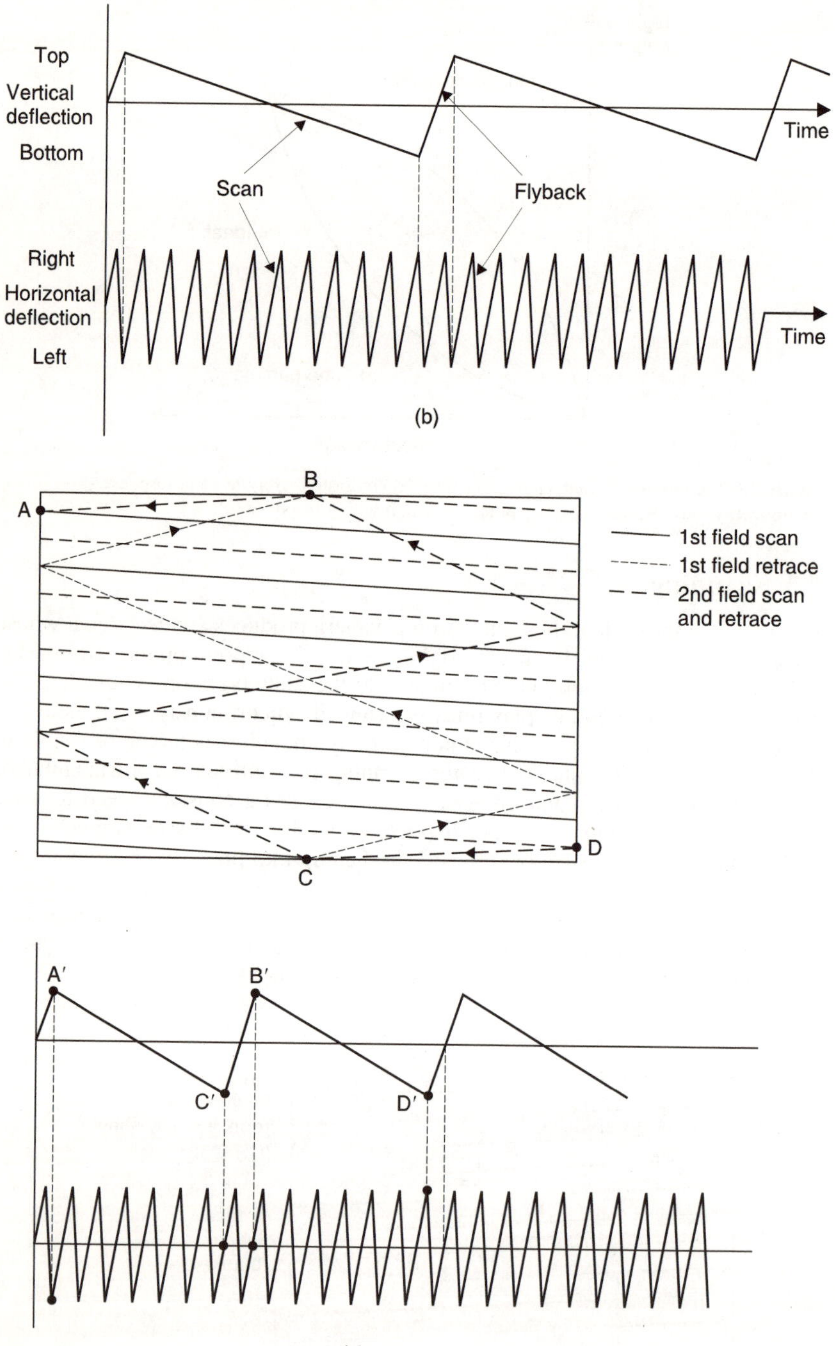

(b)

1st field scan
1st field retrace
2nd field scan
and retrace

(c)

Figure 2.7 (*Continued*) The scanning is controlled by horizontal and vertical sawtooth waveforms (b). Where two vertical scans are needed to complete a whole number of lines, the scan is interlaced as shown in (c). The frame is now split into two fields.

shows an *interlaced scan* system in which there is an integer number of lines in *two* vertical scans or *fields*. The first field begins with a full line and ends on a half line and the second field begins with a half line and ends with a full line. The lines from the two fields interlace or mesh on the screen. Current analog broadcast systems such as PAL and NTSC use interlace, although in MPEG systems it is not necessary. The additional complication of interlace has both merits and drawbacks which will be discussed in section 2.11.

2.4 Synchronizing

The *synchronizing* or sync system must send timing information to the display alongside the video signal. In very early television equipment this was achieved using two quite separate or *non-composite* signals. Figure 2.8(a) shows one of the first (US) television signal standards in which the video waveform had an amplitude of 1 V peak-to-peak and the sync signal had an amplitude of 4 V peak-to-peak. In practice, it was more convenient to combine both into a single electrical waveform then called *composite video* which carries the synchronizing information as well as the scanned brightness signal. The single signal is effectively shared by using some of the flyback period for synchronizing.

The 4 V sync signal was attenuated by a factor of ten and added to the video to produce a 1.4 V peak-to-peak signal. This was the origin of the 10:4 video:sync relationship of US analog television practice. Later the amplitude was reduced to 1 V peak-to-peak so that the signal had the same range as the original non-composite video. The 10:4 ratio was retained. As Figure 2.8(b) shows, this ratio results in some rather odd voltages and to simplify matters, a new unit called the *IRE* unit (after the Institute of Radio Engineers) was devised. Originally this was defined as 1 per cent of the video voltage swing, independent of the actual amplitude in use, but it came in practice to mean 1 per cent of 0.714 V. In European analog systems shown in Figure 2.8(c) the messy numbers were avoided by using a 7:3 ratio and the waveforms are always measured in milliVolts. Whilst such a signal was originally called composite video, today it would be referred to as monochrome video or *Ys*, meaning luma carrying syncs although in practice the *s* is often omitted.

Figure 2.8(d) shows how the two signals are separated. The voltage swing needed to go from black to peak white is less than the total swing available. In a standard analog video signal the maximum amplitude is 1 V peak-to-peak. The upper part of the voltage range represents the variations in brightness of the image from black to white. Signals below that range are 'blacker than black' and cannot be seen on the display. These signals are used for synchronizing.

Figure 2.9(a) shows the line synchronizing system part-way through a field or frame. The part of the waveform which corresponds to the forward scan is called the *active line* and during the active line the voltage represents the brightness of the image. In between the active line periods are *horizontal blanking intervals* in which the signal voltage will be at or below black. Figure 2.9(b) shows that in some systems the active line voltage is superimposed on a *pedestal* or *black level set-up* voltage of 7.5 *IRE*. The purpose of this set-up is to ensure that the blanking interval signal is below black on simple displays so that it is guaranteed to be invisible on the screen. When set-up is used, black level and blanking level differ by the pedestal height. When set-up is not used, black level and blanking level are one and the same.

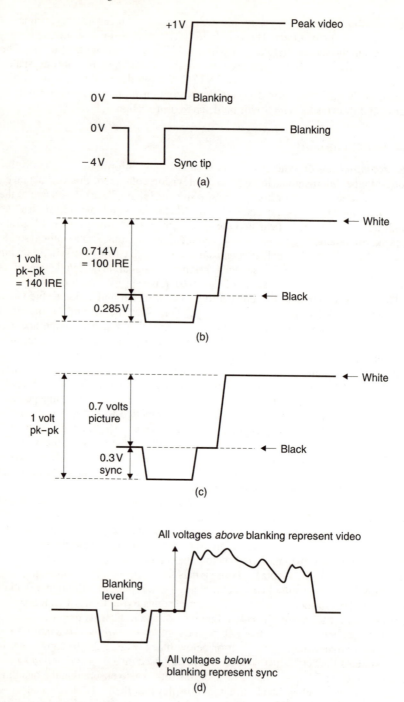

Figure 2.8 Early video used separate vision and sync signals shown in (a). The US one volt video waveform in (b) has 10:4 video/sync ratio. (c) European systems use 7:3 ratio to avoid odd voltages. (d) Sync separation relies on two voltage ranges in the signal.

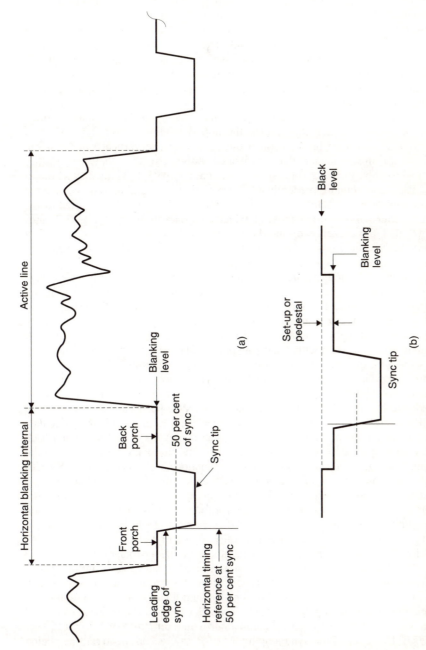

Figure 2.9 (a) Part of a video waveform with important features named. (b) Use of pedestal or set-up.

The blanking period immediately after the active line is known as the *front porch*, which is followed by the *leading edge of sync*. When the leading edge of sync passes through 50 per cent of its own amplitude, the horizontal retrace pulse is considered to have occurred. The flat part at the bottom of the horizontal sync pulse is known as *sync tip* and this is followed by the trailing edge of sync which returns the waveform to blanking level. The signal remains at blanking level during the *back porch* during which the display completes the horizontal flyback. The sync pulses have sloping edges because if they were square they would contain high frequencies which would go outside the allowable channel bandwidth on being broadcast.

The vertical synchronizing system is more complex because the vertical flyback period is much longer than the horizontal line period and horizontal synchronization must be maintained throughout it. The vertical synchronizing pulses are much longer than horizontal pulses so that they are readily distinguishable. Figure 2.10(a) shows a simple approach to vertical synchronizing. The signal remains predominantly at sync tip for several lines to indicate the vertical retrace, but returns to blanking level briefly immediately prior to the leading edges of the horizontal sync, which continues throughout. Figure 2.10(b) shows that the presence of interlace complicates matters, as in one vertical interval the vertical sync pulse coincides with a horizontal sync pulse whereas in the next the vertical sync pulse occurs half-way down a line.

In practice the long vertical sync pulses were found to disturb the average signal voltage too much and to reduce the effect extra *equalizing pulses* were put in, half-way between the horizontal sync pulses. The horizontal timebase system can ignore the equalizing pulses because it contains a flywheel circuit which only expects pulses roughly one line period apart. Figure 2.10(c) shows the final result of an interlaced system with equalizing pulses. The vertical blanking interval can be seen, with the vertical pulse itself towards the beginning.

In digital video signals it is possible to synchronize simply by digitizing the analog sync pulses. However, this is inefficient because many samples are needed to describe them. In practice the analog sync pulses are used to generate timing reference signals (TRS) which are special codes inserted in the video data which indicate the picture timing. In a manner analogous to the analog approach of dividing the video voltage range into two, one for syncs, the solution in the digital domain is the same: certain bit combinations are reserved for TRS codes and these cannot occur in legal video. TRS codes are detailed in Chapter 9.

It is essential accurately to extract the timing or synchronizing information from a sync or Ys signal in order to control some process such as the generation of a digital sampling clock. Figure 2.11(a) shows a block diagram of a simple sync separator. The first stage will generally consist of a black level clamp which stabilizes the DC conditions in the separator. Figure 2.11(b) shows that if this is not done the presence of a DC shift on a sync edge can cause a timing error.

The sync time is defined as the instant when the leading edge passes through the 50 per cent level. The incoming signal should ideally have a sync amplitude of either 0.3 V peak-to-peak or 40 *IRE*, in which case it can be *sliced* or converted to a binary waveform by using a comparator with a reference of either 0.15 V or 20 *IRE*. However, if the sync amplitude is for any reason incorrect, the slicing level will be wrong. Figure 2.11(a) shows that the solution is to measure both blanking and sync tip voltages and to derive the slicing level from them with a potential divider. In this way the slicing level will always be 50 per cent of the

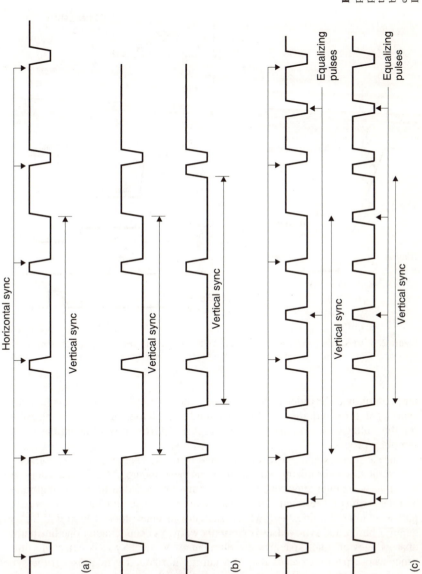

Figure 2.10 (a) A simple vertical pulse is longer than a horizontal pulse. (b) In an interlaced system there are two relationships between H and V. (c) The use of equalizing pulses to balance the DC component of the signal.

(a)

Horizontal sync

Vertical sync

Vertical sync

(b)

Vertical sync

Vertical sync

(c)

Equalizing
pulses

Vertical sync

Equalizing
pulses

Vertical sync

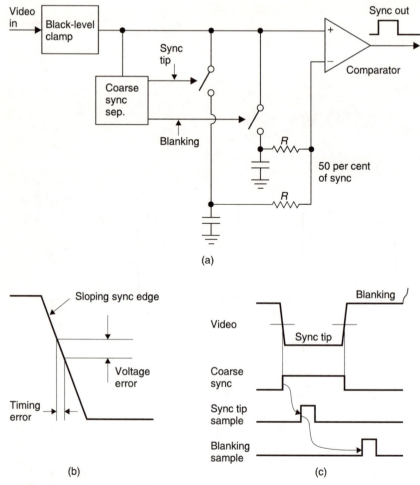

Figure 2.11 (a) Sync separator block diagram; see text for details. (b) Slicing at the wrong level introduces a timing error. (c) The timing of the sync separation process.

input amplitude. In order to measure the sync tip and blanking levels, a coarse sync separator is required, which is accurate enough to generate sampling pulses for the voltage measurement system. Figure 2.11(c) shows the timing of the sampling process.

Once a binary signal has been extracted from the analog input, the horizontal and vertical synchronizing information can be separated. All falling edges are potential horizontal sync leading edges, but some are due to equalizing pulses and these must be rejected. This is easily done because equalizing pulses occur part-way down the line. A flywheel oscillator or phase-locked loop will lock to genuine horizontal sync pulses because they always occur exactly one line period apart. Edges at other spacings are eliminated. Vertical sync is detected with a timer whose period exceeds that of a normal horizontal sync pulse. If the sync waveform is still low when the timer expires, there must be a vertical pulse present. Once again a phase-locked loop may be used which will continue to run if the input is noisy or disturbed. This may take the form of a counter which counts the number of lines in a frame before resetting.

The sync separator can determine which type of field is beginning because in one the vertical and horizontal pulses coincide whereas in the other the vertical pulse begins in the middle of a line.

2.5 Bandwidth and definition

As the conventional analog television picture is made up of lines, the line structure determines the *definition* or the fineness of detail which can be portrayed in the vertical axis. The limit *is* reached in theory when alternate lines show black and white. In a 625-line picture there are roughly 600 unblanked lines. If 300 of these are white and 300 are black then there will be 300 complete cycles of detail in one picture height. One unit of resolution, which is a unit of spatial frequency, is c/ph or cycles per picture height. In practical displays the contrast will have fallen to virtually nothing at this ideal limit and the resolution actually achieved is around 70 per cent of the ideal, or about 210 c/ph. The degree to which the ideal is met is known as the *Kell factor* of the display.

Definition in one axis is wasted unless it is matched in the other and so the horizontal axis should be able to offer the same performance. As the aspect ratio of conventional television is 4:3 then it should be possible to display 400 cycles in one picture width, reduced to about 300 cycles by the Kell factor. As part of the line period is lost due to flyback, 300 cycles per picture width becomes about 360 cycles per line period.

In 625-line television, the frame rate is 25 Hz and so the line rate F_h will be:

$$F_h = 625 \times 25 = 15\,625\,\text{Hz}$$

If 360 cycles of video waveform must be carried in each line period, then the bandwidth required will be given by:

$$15\,625 \times 360 = 5.625\,\text{MHz}$$

In the 525-line system, there are roughly 500 unblanked lines allowing 250 c/ph theoretical definition, or 175 lines allowing for the Kell factor. Allowing for the aspect ratio, equal horizontal definition requires about 230 cycles per picture width. Allowing for horizontal blanking this requires about 280 cycles per line period.

In 525-line video, $F_h = 525 \times 30 = 15\,750\,\text{Hz}$ Thus the bandwidth required is:

$$15\,750 \times 280 = 4.4\,\text{MHz}$$

If it is proposed to build a high-definition television system, one might start by doubling the number of lines and hence double the definition. Thus in a 1250-line format about 420 c/ph might be obtained. To achieve equal horizontal definition, bearing in mind the aspect ratio is now 16:9, then nearly 750 cycles per picture width will be needed. Allowing for horizontal blanking, then around 890 cycles per line period will be needed. The line frequency is now given by:

$$F_h = 1250 \times 25 = 31250\,\text{Hz}$$

and the bandwidth required is given by

$$31\,250 \times 890 = 28\,\text{MHz}$$

Note the dramatic increase in bandwidth. In general the bandwidth rises as the square of the resolution because there are more lines and more cycles needed in each line. It should be clear that, except for research purposes, high-definition television will never be broadcast as a conventional analog signal because the bandwidth required is simply uneconomic. If and when high-definition broadcasting becomes common, it will be compelled to use digital compression techniques to make it economic.

2.6 Aperture effect

The aperture effect will show up in many aspects of television in both the sampled and continuous domains. The image sensor has a finite aperture function. In tube cameras and in CRTs, the beam will have a finite radius with a Gaussian distribution of energy across its diameter. This results in a Gaussian spatial frequency response. Tube cameras often contain an *aperture corrector* which is a filter designed to boost the higher spatial frequencies that are attenuated by the Gaussian response. The horizontal filter is simple enough, but the vertical filter will require line delays in order to produce points above and below the line to be corrected. Aperture correctors also amplify aliasing products and an overcorrected signal may contain more vertical aliasing than resolution.

Some digital-to-analog convertors keep the signal constant for a substantial part of or even the whole sample period. In CCD cameras, the sensor is split into elements which may almost touch in some cases. The element integrates light falling on its surface. In both cases the aperture will be rectangular. The case where the pulses have been extended in width to become equal to the sample period is known as a zero-order hold system and has a 100 per cent aperture ratio.

Rectangular apertures have a $\sin x/x$ spectrum which is shown in Figure 2.12. With a 100 per cent aperture ratio, the frequency response falls to a null at the sampling rate, and as a result is about 4 dB down at the edge of the baseband.

The temporal aperture effect varies according to the equipment used. Tube cameras have a long integration time and thus a wide temporal aperture. Whilst this reduces temporal aliasing, it causes smear on moving objects. CCD cameras do not suffer from lag and as a result their temporal response is better. Some CCD cameras deliberately have a short temporal aperture as the time axis is resampled by a shutter. The intention is to reduce smear, hence the popularity of such devices for sporting events, but there will be more aliasing on certain subjects.

The eye has a temporal aperture effect which is known as persistence of vision, and the phosphors of CRTs continue to emit light after the electron beam has passed. These produce further temporal aperture effects in series with those in the camera.

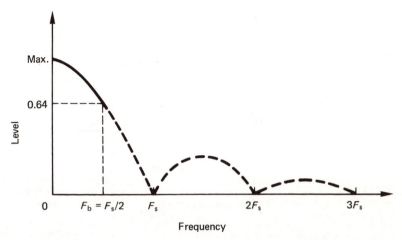

Figure 2.12 Frequency response with 100 per cent aperture nulls at multiples of sampling rate. Area of interest is up to half sampling rate.

2.7 Colour

Colour vision is made possible by the cones on the retina which occur in three different types, responding to different colours. Figure 2.13 shows that human vision is restricted to range of light wavelengths from 400 nanometres to 700 nanometres. Shorter wavelengths are called ultra-violet and longer wavelengths are called infra-red. Note that the response is not uniform, but peaks in the area of green. The response to blue is very poor and makes a nonsense of the traditional use of blue lights on emergency vehicles.

Figure 2.14 shows an approximate response for each of the three types of cone. If light of a single wavelength is observed, the relative responses of the three sensors allows us to discern what we call the colour of the light. Note that at both ends of the visible spectrum there are areas in which only one receptor responds; all colours in those areas look the same. There is a great deal of variation in receptor response from one individual to the next and the curves used in television are the average of a great many tests. In a surprising number of people

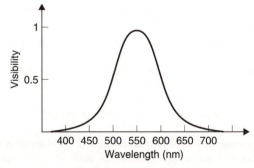

Figure 2.13 The luminous efficiency function shows the response of the HVS to light of different wavelengths.

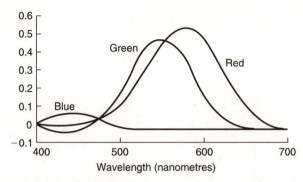

Figure 2.14 All human vision takes place over this range of wavelengths. The response is not uniform, but has a central peak. The three types of cone approximate to the three responses shown to give colour vision.

the single receptor zones are extended and discrimination between, for example, red and orange is difficult.

The full resolution of human vision is restricted to brightness variations. Our ability to resolve colour details is only about a quarter of that.

The triple receptor characteristic of the eye is extremely fortunate as it means that we can generate a range of colours by adding together light sources having just three different wavelengths in various proportions. This process is known as *additive colour matching* which should be clearly distinguished from the subtractive colour matching that occurs with paints and inks. Subtractive matching begins with white light and selectively removes parts of the spectrum by filtering. Additive matching uses coloured light sources which are combined.

An effective colour television system can be made in which only three pure or single-wavelength colours or *primaries* can be generated. The primaries need to be similar in wavelength to the peaks of the three receptor responses, but need not be identical. Figure 2.15 shows a rudimentary colour television system. Note that the colour camera is in fact three cameras in one, where each is fitted with a different coloured filter. Three signals, *R*, *G* and *B* must be transmitted to the display which produces three images that must be superimposed to obtain a colour picture.

In practice the primaries must be selected from available phosphor compounds. Once the primaries have been selected, the proportions needed to reproduce a given colour can be found using a colorimeter. Figure 2.16 shows a colorimeter which consists of two adjacent white screens. One screen is illuminated by three light sources, one of each of the selected primary colours. Initially, the second screen is illuminated with white light and the three sources are adjusted until the first screen displays the same white. The sources are then calibrated. Light of a single wavelength is then projected on the second screen. The primaries are once more adjusted until both screens appear to have the same colour. The proportions of the primaries are noted. This process is repeated for the whole visible spectrum, resulting in *colour mixture curves* shown in Figure 2.17. In some cases it will not be possible to find a match because an impossible negative contribution is needed. In this case we can simulate a negative

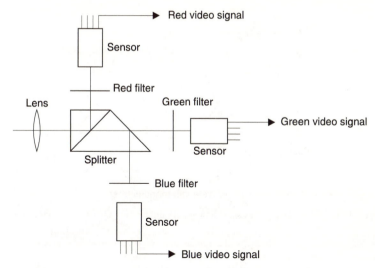

Figure 2.15 Simple colour television system. Camera image is split by three filters. Red, green and blue video signals are sent to three primary coloured displays whose images are combined.

contribution by shining some primary colour on the test screen until a match is obtained. If the primaries were ideal, monochromatic or single-wavelength sources, it would be possible to find three wavelengths at which two of the primaries were completely absent. However, practical phosphors are not monochromatic, but produce a distribution of wavelengths around the nominal value, and in order to make them spectrally pure other wavelengths have to be subtracted.

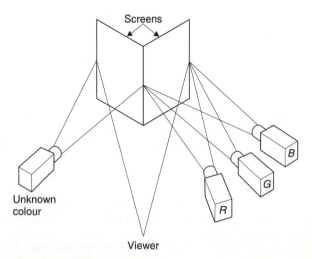

Figure 2.16 Simple colorimeter. Intensities of primaries on the right screen are adjusted to match the test colour on the left screen.

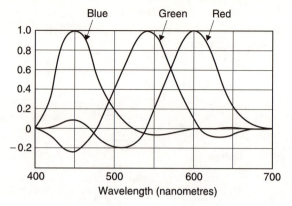

Figure 2.17 Colour mixture curves show how to mix primaries to obtain any spectral colour.

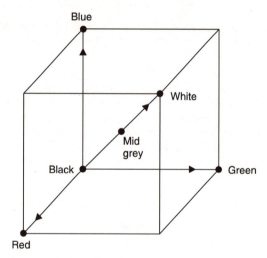

Figure 2.18 *RGB* colour space is three-dimensional and not easy to draw.

The colour-mixing curves dictate what the response of the three sensors in the colour camera must be. The primaries are determined in this way because it is easier to make camera filters to suit available CRT phosphors rather than the other way round.

As there are three signals in a colour television system, they can only be simultaneously depicted in three dimensions. Figure 2.18 shows the *RGB* colour space which is basically a cube with black at the origin and white at the diagonally opposite corner. Figure 2.19 shows the colour mixture curves plotted in *RGB* space. For each visible wavelength a vector exists whose direction is determined by the proportions of the three primaries. If the brightness is allowed to vary this will affect all three primaries and thus the length of the vector in the same proportion.

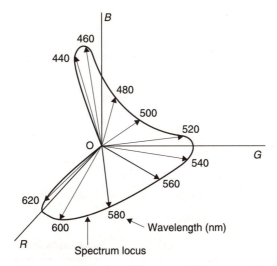

Figure 2.19 Colour mixture curves plotted in *RGB* space result in a vector whose locus moves with wavelength in three dimensions.

Depicting and visualizing the *RGB* colour space is not easy and it is also difficult to take objective measurements from it. The solution is to modify the diagram to allow it to be rendered in two dimensions on flat paper. This is done by eliminating luminance (brightness) changes and depicting only the colour at constant brightness. Figure 2.20(a) shows how a constant luminance unit plane intersects the *RGB* space at unity on each axis. At any point on the plane the three components add up to one. A two-dimensional plot results when vectors representing all colours intersect the plane. Vectors may be extended if necessary to allow intersection. Figure 2.20(b) shows that the 500 nm vector has to be produced (extended) to meet the unit plane, whereas the 580 nm vector naturally intersects. Any colour can now uniquely be specified in two dimensions.

The points where the unit plane intersects the axes of *RGB* space form a triangle on the plot. The horseshoe-shaped locus of pure spectral colours goes outside this triangle because, as was seen above, the colour mixture curves require negative contributions for certain colours.

Having the spectral locus outside the triangle is a nuisance, and a larger triangle can be created by postulating new coordinates called *X*, *Y* and *Z* representing hypothetical primaries that cannot exist. This representation is shown in Figure 2.20(c).

The Commission Internationale d'Eclairage (CIE) standard *chromaticity diagram* shown in Figure 2.20(d) is obtained in this way by projecting the unity luminance plane onto the *X*, *Y* plane. This projection has the effect of bringing the red and blue primaries closer together. Note that the curved part of the locus is due to spectral or single-wavelength colours. The straight base is due to non-spectral colours obtained by additively mixing red and blue.

As negative light is impossible, only colours within the triangle joining the primaries can be reproduced and so practical television systems cannot reproduce all possible colours. Clearly, efforts should be made to obtain primaries which embrace as large an area as possible. Figure 2.21 shows how the colour range or

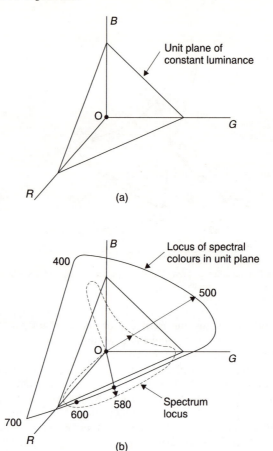

Figure 2.20 (a) A constant luminance plane intersects *RGB* space, allowing colours to be studied in two dimensions only. (b) The intersection of the unit plane by vectors joining the origin and the spectrum locus produces the locus of spectral colours which requires negative values of *R*, *G* and *B* to describe it.

gamut of television compares with paint and printing inks and illustrates that the comparison is favourable. Most everyday scenes fall within the colour gamut of television. Exceptions include saturated turquoise, spectrally pure iridescent colours formed by interference in a duck's feathers or reflections in Compact Discs. For special purposes displays have been made having four primaries to give a wider colour range, but these are uncommon.

Figure 2.22 shows the primaries initially selected for NTSC. However, manufacturers looking for brighter displays substituted more efficient phosphors having a smaller colour range. This was later standardized as the SMPTE C phosphors which were also adopted for PAL.

Whites appear in the centre of the chromaticity diagram corresponding to roughly equal amounts of primary colour. Two terms are used to describe colours: *hue* and *saturation*. Colours having the same hue lie on a straight line between the white point and the perimeter of the primary triangle. The saturation

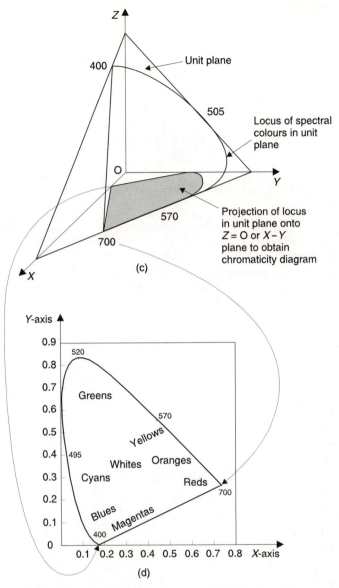

Figure 2.20 (*Continued*) In (c) a new coordinate system, *X*, *Y*, *Z*, is used so that only positive values are required. The spectrum locus now fits entirely in the triangular space where the unit plane intersects these axes. To obtain the CIE chromaticity diagram (d), the locus is projected onto the *X–Y* plane.

of the colour increases with distance from the white point. As an example, pink is a desaturated red.

The apparent colour of an object is also a function of the illumination. The 'true colour' will only be revealed under ideal white light which in practice is uncommon. An ideal white object reflects all wavelengths equally and simply

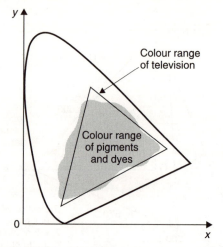

Figure 2.21 Comparison of the colour range of television and printing.

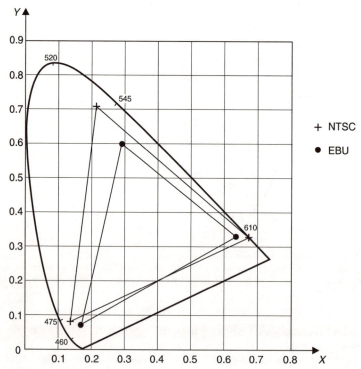

Figure 2.22 The primary colours for NTSC were initially as shown. These were later changed to more efficient phosphors which were also adopted for PAL. See text.

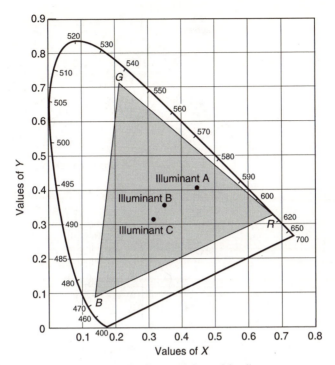

Figure 2.23 Position of three common illuminants on chromaticity diagram.

takes on the colour of the ambient illumination. Figure 2.23 shows the location of three 'white' sources or *illuminants* on the chromaticity diagram. Illuminant A corresponds to a tungsten filament lamp, illuminant B to midday sunlight and illuminant C to typical daylight which is bluer because it consists of a mixture of sunlight and light scattered by the atmosphere. In everyday life we accommodate automatically to the change in apparent colour of objects as the sun's position or the amount of cloud changes and as we enter artificially lit buildings, but colour cameras accurately reproduce these colour changes. Attempting to edit a television program from recordings made at different times of day or indoors and outdoors would result in obvious and irritating colour changes unless some steps are taken to keep the white balance reasonably constant.

2.8 Colour displays

In order to display colour pictures, three simultaneous images must be generated, one for each primary colour. The colour CRT does this geometrically. Figure 2.24(a) shows that three electron beams pass down the tube from guns mounted in a triangular or delta array. Immediately before the tube face is mounted a perforated metal plate known as a shadow mask. The three beams approach holes in the shadow mask at a slightly different angle and so fall upon three different areas of phosphor which each produce a different primary colour. The sets of three phosphors are known as triads. Figure 2.24(b) shows an alternative arrangement in which the three electron guns are mounted in a straight line and

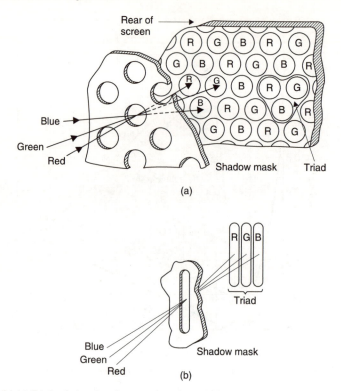

Figure 2.24 (a) Triads of phosphor dots are triangular and electron guns are arranged in a triangle. (b) Inline tube has strips of phosphor side by side.

the shadow mask is slotted and the triads are rectangular. This is known as a PIL (precision-in-line) tube. The triads can easily be seen upon close inspection of an operating CRT.

In the plasma display the source of light is an electrical discharge in a gas at low pressure. This generates ultra-violet light which excites phosphors in the same way that a fluorescent light operates. Each pixel consists of three such elements, one for each primary colour. Figure 2.25 shows that the pixels are controlled by arranging the discharge to take place between electrodes which are arranged in rows and columns.

The advantage of the plasma display is that it can be made perfectly flat and it is very thin, even in large screen sizes. There is a size limit in CRTs beyond which they become very heavy. Plasma displays allow this limit to be exceeded.

The great difficulty with the plasma display is that the relationship between light output and drive voltage is highly non-linear. Below a certain voltage there is no discharge at all. Consequently the only way that the brightness can be varied is to modulate the time for which the discharge takes place. The electrode signals are pulse width modulated.

Eight-bit digital video has 256 different brightnesses and it is difficult to obtain such a scale by pulse width modulation as the increments of pulse length would

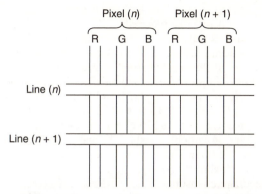

Figure 2.25 When a voltage is applied between a line or row electrode and a pixel electrode, a plasma discharge occurs. This excites a phosphor to produce visible light.

need to be generated by a clock of fantastic frequency. It is common practice to break up the picture period into many pulses, each of which is modulated in width. Despite this, plasma displays often display contouring or posterizing, indicating a lack of sufficient brightness levels. Multiple pulse drive also has some temporal effects which may be visible on moving material unless motion compensation is used.

2.9 Colour difference signals

There are many different ways in which television signals can be carried and these will be considered here. A monochrome camera produces a single luma signal Y or Ys whereas a colour camera produces three signals, or *components*, R, G and B which are essentially monochrome video signals representing an image in each primary colour. In some systems sync is present on a separate signal (*RGBS*), rarely is it present on all three components, whereas most commonly it is only present on the green component leading to the term *RGsB*. The use of the green component for sync has led to suggestions that the components should be called *GBR*. As the original and long-standing term *RGB* or *RGsB* correctly reflects the sequence of the colours in the spectrum it remains to be seen whether *GBR* will achieve common usage. Like luma, *RGsB* signals may use 0.7 or 0.714 V signals, with or without set-up.

RGB and Y signals are incompatible, yet when colour television was introduced it was a practical necessity that it should be possible to display colour signals on a monochrome display and vice versa.

Creating or *transcoding* a luma signal from R, Gs and B is relatively easy. Figure 2.13 showed the spectral response of the eye which has a peak in the green region. Green objects will produce a larger stimulus than red objects of the same brightness, with blue objects producing the least stimulus. A luma signal can be obtained by adding R, G and B together, not in equal amounts, but in a sum which is *weighted* by the relative response of the eye. Thus:

$$Y = 0.299\,R + 0.587\,G + 0.114\,B$$

Syncs may be regenerated, but will be identical to those on the *Gs* input and when added to *Y* result in *Ys* as required.

If *Ys* is derived in this way, a monochrome display will show nearly the same result as if a monochrome camera had been used in the first place. The results are not identical because of the non-linearities introduced by gamma correction.

As colour pictures require three signals, it should be possible to send *Ys* and two other signals which a colour display could arithmetically convert back to *R*, *G* and *B*. There are two important factors which restrict the form which the other two signals may take. One is to achieve reverse compatibility. If the source is a monochrome camera, it can only produce *Ys* and the other two signals will be completely absent. A colour display should be able to operate on the *Ys* signal only and show a monochrome picture. The other is the requirement to conserve bandwidth for economic reasons.

These requirements are met by sending two *colour difference signals* along with *Ys*. There are three possible colour difference signals, $R - Y$, $B - Y$ and $G - Y$. As the green signal makes the greatest contribution to *Y*, then the amplitude of $G - Y$ would be the smallest and would be most susceptible to noise. Thus $R - Y$ and $B - Y$ are used in practice as Figure 2.26 shows.

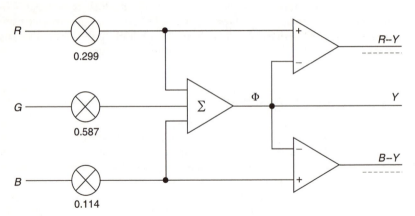

Figure 2.26 Colour components are converted to colour difference signals by the transcoding shown here.

R and *B* are readily obtained by adding *Y* to the two colour difference signals. *G* is obtained by rearranging the expression for *Y* above such that:

$$G = \frac{Y - 0.3R - 0.11B}{0.59}$$

If a colour CRT is being driven, it is possible to apply inverted luma to the cathodes and the $R - Y$ and $B - Y$ signals directly to two of the grids so that the tube performs some of the matrixing. It is then only necessary to obtain $G - Y$ for the third grid, using the expression:

$$G - Y = -0.51 (R - Y) - 0.186 (B - Y)$$

If a monochrome source having only a *Ys* output is supplied to a colour display, *R* – *Y* and *B* – *Y* will be zero. It is reasonably obvious that if there are no colour difference signals the colour signals cannot be different from one another and *R* = *G* = *B*. As a result the colour display can produce only a neutral picture.

The use of colour difference signals is essential for compatibility in both directions between colour and monochrome, but it has a further advantage that follows from the way in which the eye works. In order to produce the highest resolution in the fovea, the eye will use signals from all types of cone, regardless of colour. In order to determine colour the stimuli from three cones must be compared. There is evidence that the nervous system uses some form of colour difference processing to make this possible. As a result, the acuity of the human eye is available only in monochrome. Differences in colour cannot be resolved so well. A further factor is that the lens in the human eye is not achromatic and this means that the ends of the spectrum are not well focused. This is particularly noticeable on blue.

If the eye cannot resolve colour very well there is no point is expending valuable bandwidth sending high-resolution colour signals. Colour difference working allows the luma to be sent separately at full bandwidth. This determines the subjective sharpness of the picture. The colour difference signals can be sent with considerably reduced bandwidth, as little as one quarter that of luma, and the human eye is unable to tell.

In practice, analog component signals are never received perfectly, but suffer from slight differences in relative gain. In the case of *RGB* a gain error in one signal will cause a colour cast on the received picture. A gain error in *Y* causes no colour cast and gain errors in *R*–*Y* or *B*–*Y* cause much smaller perceived colour casts. Thus colour difference working is also more robust than *RGB* working.

The overwhelming advantages obtained by using colour difference signals mean that in broadcast and production facilities *RGB* is seldom used. The outputs from the *RGB* sensors in the camera are converted directly to *Y*, *R* – *Y* and *B* – *Y* in the camera control unit and output in that form. Standards exist for both analog and digital colour difference signals to ensure compatibility between equipment from various manufacturers. The M-II and Betacam formats record analog colour difference signals, and there are a number of colour difference digital formats.

Whilst signals such as *Y*, *R*, *G* and *B* are *unipolar* or positive only, it should be stressed that colour difference signals are *bipolar* and may meaningfully take on levels below zero volts.

The wide use of colour difference signals has led to the development of test signals and equipment to display them. The most important of the test signals are the ubiquitous *colour bars*. Colour bars are used to set the gains and timing of signal components and to check that matrix operations are performed using the correct weighting factors. The origin of the colour bar test signal is shown in Figure 2.27. In *100 per cent amplitude bars*, peak amplitude binary *RGB* signals are produced, having one, two and four cycles per screen width. When these are added together in a weighted sum, an eight-level luma staircase results because of the unequal weighting. The matrix also produces two colour difference signals, *R* – *Y* and *B* – *Y* as shown. Sometimes *75 per cent amplitude bars* are generated by suitably reducing the *RGB* signal amplitude. Note that in both cases the colours are fully saturated; it is only the brightness which is reduced to 75 per

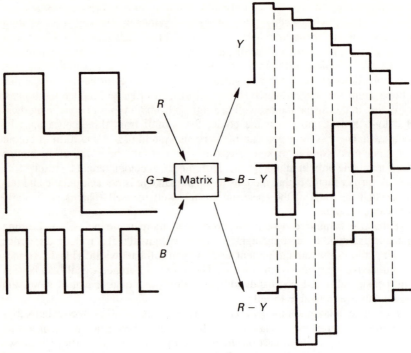

Figure 2.27 Origin of colour difference signals representing colour bars. Adding *R*, *G* and *B* according to the weighting factors produces an irregular luminance staircase.

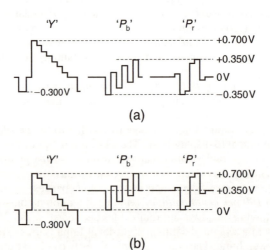

Figure 2.28 (a) 100 per cent colour bars represented by SMPTE/EBU standard colour difference signals. (b) Level comparison is easier in waveform monitors if the *B–Y* and *R–Y* signals are offset upwards.

cent. Sometimes the white bar of a 75 per cent bar signal is raised to 100 per cent to make calibration easier. Such a signal is sometimes erroneously called a 100 per cent bar signal.

Figure 2.28(a) shows an SMPTE/EBU standard colour difference signal set in which the signals are called Ys, P_b and P_r. 0.3 V syncs are on luma only and all three video signals have a 0.7 V peak-to-peak swing with 100 per cent bars. In order to obtain these voltage swings, the following gain corrections are made to the components:

$$P_r = 0.71327 \ (R - Y) \text{ and } P_b = 0.56433 \ (B-Y)$$

Within waveform monitors, the colour difference signals may be offset by 350 mV as in Figure 2.28(b) to match the luma range for display purposes.

2.10 Motion portrayal and dynamic resolution

As the eye uses involuntary tracking at all times, the criterion for measuring the definition of moving-image portrayal systems has to be dynamic resolution, defined as the apparent resolution perceived by the viewer in an object moving within the limits of accurate eye tracking. The traditional metric of static resolution in film and television has to be abandoned as unrepresentative.

Figure 2.29(a) shows that when the moving eye tracks an object on the screen, the viewer is watching with respect to the optic flow axis, not the time axis, and these are not parallel when there is motion. The optic flow axis is defined as an imaginary axis in the spatio-temporal volume which joins the same points on objects in successive frames. Clearly, when many objects move independently there will be one optic flow axis for each.

The optic flow axis is identified by motion-compensated standards convertors to eliminate judder and also by MPEG compressors because the greatest similarity from one picture to the next is along that axis. The success of these devices is testimony to the importance of the theory.

Figure 2.29(b) shows that when the eye is tracking, successive pictures appear in different places with respect to the retina. In other words if an object is moving down the screen and followed by the eye, the raster is actually moving up with respect to the retina. Although the tracked object is stationary with respect to the retina and temporal frequencies are zero, the object is moving with respect to the sensor and the display and in those units high temporal frequencies will exist. If the motion of the object on the sensor is not correctly portrayed, dynamic resolution will suffer.

In real-life eye tracking, the motion of the background will be smooth, but in an image-portrayal system based on periodic presentation of frames, the background will be presented to the retina in a different position in each frame. The retina seperately perceives each impression of the background leading to an effect called background strobing.

The criterion for the selection of a display frame rate in an imaging system is sufficient reduction of background strobing. It is a complete myth that the display rate simply needs to exceed the critical flicker frequency. Manufacturers of graphics displays which use frame rates well in excess of those used in film and television are doing so for a valid reason: it gives better results! Note that the

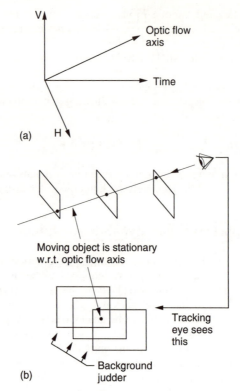

Figure 2.29 The optic flow axis (a) joins points on a moving object in successive pictures. (b) When a tracking eye follows a moving object on a screen, that screen will be seen in a different place at each picture. This is the origin of background strobing.

display rate and the transmission rate need not be the same in an advanced system.

Dynamic resolution analysis confirms that both interlaced television and conventionally projected cinema film are both seriously sub-optimal. In contrast, progressively scanned television systems have no such defects.

2.11 Progressive or interlaced scan?

Interlaced scanning is a crude compression technique which was developed empirically in the 1930s as a way of increasing the picture rate to reduce flicker without increasing the video bandwidth. Instead of transmitting entire frames, the lines of the frame are sorted into odd lines and even lines. Odd lines are transmitted in one field, even lines in the next. A pair of fields will interlace to produce a frame. Vertical detail such as an edge may only be present in one field of the pair and this results in frame rate flicker called 'interlace twitter'.

Figure 2.30(a) shows a dynamic resolution analysis of interlaced scanning. When there is no motion, the optic flow axis and the time axis are parallel and the apparent vertical sampling rate is the number of lines in a frame. However, when there is vertical motion, (b), the optic flow axis turns. In the case shown,

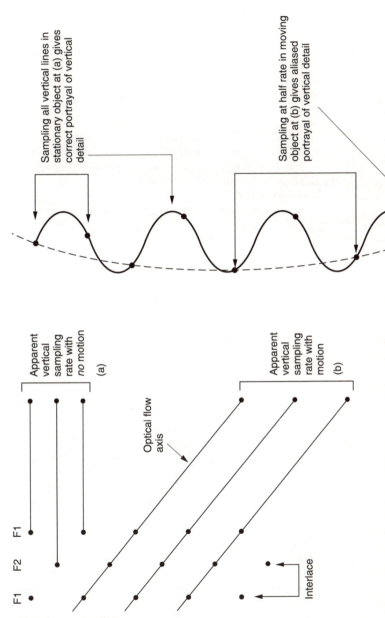

Figure 2.30 When an interlaced picture is stationary, viewing takes place along the time axis as shown in (a). When a vertical component of motion exists, viewing takes place along the optic flow axis. (b) The vertical sampling rate falls to one half its stationary value. (c) The halving in sampling rate causes high spatial frequencies to alias.

the sampling structure due to interlace results in the vertical sampling rate falling to one half of its stationary value.

Consequently interlace does exactly what would be expected from a half-bandwidth filter. It halves the vertical resolution when any motion with a vertical component occurs. In a practical television system, there is no anti-aliasing filter in the vertical axis and so when the vertical sampling rate of an interlaced system is halved by motion, high spatial freqiencies will alias or heterodyne causing annoying artifacts in the picture. This is easily demonstrated.

Figure 2.30(c) shows how a vertical spatial frequency well within the static resolution of the system aliases when motion occurs. In a progressive scan system this effect is absent and the dynamic resolution due to scanning can be the same as the static case.

This analysis also illustrates why interlaced television systems have to have horizontal raster lines. This is because in real life, horizontal motion is more common than vertical. It is easy to calculate the vertical image motion velocity needed to obtain the half-bandwidth speed of interlace, because it amounts to one raster line per field. In 525/60 (NTSC) there are about 500 active lines, so motion as slow as one picture height in 8 seconds will halve the dynamic resolution. In 625/50 (PAL) there are about 600 lines, so the half-bandwidth speed falls to one picture height in 12 seconds. This is why NTSC, with fewer lines and lower bandwidth, doesn't look as soft as it should compared to PAL, because it has better dynamic resolution.

The situation deteriorates rapidly if an attempt is made to use interlaced scanning in systems with a lot of lines. In 1250/50, the resolution is halved at a vertical speed of just one picture height in 24 seconds. In other words on real moving video a 1250/50 interlaced system has the same dynamic resolution as a 625/50 progressive system. By the same argument a 1080 *I* system has the same performance as a 480 *P* system.

2.12 Binary codes

For digital video use, the prime purpose of binary numbers is to express the values of the samples which represent the original analog video waveform. Figure 2.31 shows some binary numbers and their equivalent in decimal. The radix point has the same significance in binary: symbols to the right of it represent one half, one quarter and so on. Binary is convenient for electronic circuits, which do not get tired, but numbers expressed in binary become very long, and writing them is tedious and error-prone. The octal and hexadecimal notations are both used for writing binary since conversion is so simple. Figure 2.31 also shows that a binary number is split into groups of three or four digits starting at the least significant end, and the groups are individually converted to octal or hexadecimal digits. Since sixteen different symbols are required in hex. the letters A–F are used for the numbers above nine.

There will be a fixed number of bits in a PCM video sample, and this number determines the size of the quantizing range. In the eight-bit samples used in much digital video equipment, there are 256 different numbers. Each number represents a different analog signal voltage, and care must be taken during conversion to ensure that the signal does not go outside the convertor range, or it will be clipped. In Figure 2.32(a) it will be seen that in an eight-bit pure binary system, the number range goes from 00 hex, which represents the smallest

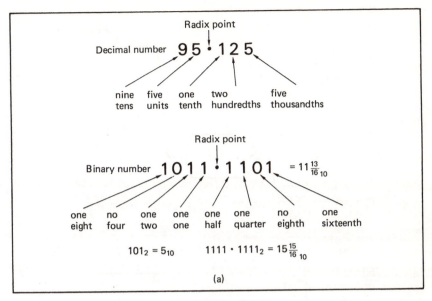

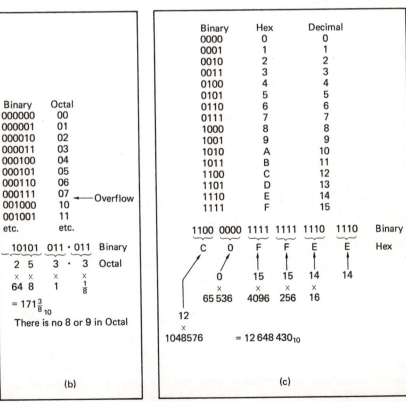

Figure 2.31 (a) Binary and decimal; (b) In octal, groups of three bits make one symbol 0–7. (c) In hex, groups of four bits make one symbol O–F. Note how much shorter the number is in hex.

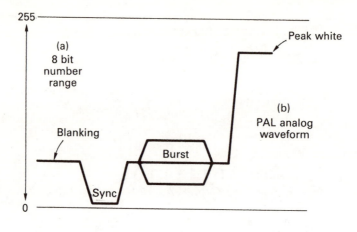

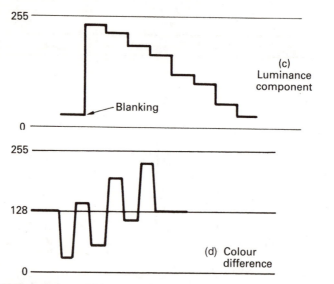

Figure 2.32 The unipolar quantizing range of an eight-bit pure binary system is shown at (a). The analog input must be shifted to fit into the quantizing range, as shown for PAL at (b). In component, sync pulses are not digitized, so the quantizing intervals can be smaller as at (c). An offset of half scale is used for colour difference signals (d).

voltage, through to FF hex, which represents the largest positive voltage. The video waveform must be accommodated within this voltage range, and (b) shows how this can be done for a PAL composite signal. A luminance signal is shown in (c). As component digital systems only handle the active line, the quantizing range is optimized to suit the gamut of the unblanked luminance. There is a small offset in order to handle slightly maladjusted inputs.

Colour difference signals are bipolar and so blanking is in the centre of the signal range. In order to accommodate colour difference signals in the quantizing range, the blanking voltage level of the analog waveform has been shifted as in Figure 2.32(d) so that the positive and negative voltages in a real video signal can be expressed by binary numbers which are only positive. This approach is called offset binary. Strictly speaking, both the composite and luminance signals are also offset binary because the blanking level is part-way up the quantizing scale.

Offset binary is perfectly acceptable where the signal has been digitized only for recording or transmission from one place to another, after which it will be converted directly back to analog. Under these conditions it is not necessary for the quantizing steps to be uniform, provided both ADC and DAC are constructed to the same standard. In practice, it is the requirements of signal processing in the digital domain which make both non-uniform quantizing and offset binary unsuitable.

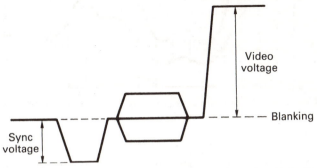

Figure 2.33 All video signal voltages are referred to blanking and must be added with respect to that level.

Figure 2.33 shows that analog video signal voltages are referred to blanking. The level of the signal is measured by how far the waveform deviates from blanking, and attenuation, gain and mixing all take place around blanking level. Digital vision mixing is achieved by adding sample values from two or more different sources, but unless all the quantizing intervals are of the same size and there is no offset, the sum of two sample values will not represent the sum of the two original analog voltages. Thus sample values which have been obtained by non-uniform or offset quantizing cannot readily be processed because the binary numbers are not proportional to the signal voltage.

If two offset binary sample streams are added together in an attempt to perform digital mixing, the result will be that the offsets are also added and this may lead to an overflow. Similarly, if an attempt is made to attenuate by, say, 6.02 dB by dividing all the sample values by two, Figure 2.34 shows that the offset is also divided and the waveform suffers a shifted baseline. This problem can be overcome with digital luminance signals simply by subtracting the offset from each sample before processing as this results in numbers truly proportional to the luminance voltage. This approach is not suitable for colour difference or composite signals because negative numbers would result when the analog voltage goes below blanking and pure binary coding cannot handle them. The problem with offset binary is that it works with reference to one end of the range. What is needed is a numbering system which operates symmetrically with reference to the centre of the range.

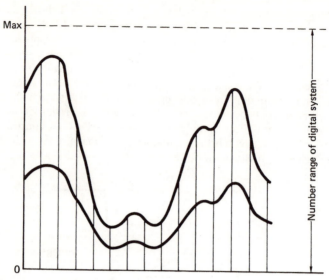

Figure 2.34 The result of an attempted attenuation in pure binary code is an offset. Pure binary cannot be used for digital video processing.

In the two's complement system, the upper half of the pure binary number range has been redefined to represent negative quantities. If a pure binary counter is constantly incremented and allowed to overflow, it will produce all the numbers in the range permitted by the number of available bits, and these are shown for a four-bit example drawn around the circle in Figure 2.35. As a circle has no real beginning, it is possible to consider it to start wherever it is convenient. In two's complement, the quantizing range represented by the circle of numbers does not start at zero, but starts on the diametrically opposite side of the circle. Zero is midrange, and all numbers with the MSB (most significant bit) set are considered negative. The MSB is thus the equivalent of a sign bit where 1 = minus. Two's complement notation differs from pure binary in that the most significant bit is inverted in order to achieve the half-circle rotation.

Figure 2.36 shows how a real ADC is configured to produce two's complement output. At (a) an analog offset voltage equal to one half the quantizing range is added to the bipolar analog signal in order to make it unipolar as at (b). The ADC produces positive only numbers at (c) which are proportional to the input voltage. The MSB is then inverted at (d) so that the all-zeros code moves to the centre of the quantizing range. The analog offset is often incorporated into the ADC as is the MSB inversion. Some convertors are designed to be used in either pure binary or two's complement mode. In this case the designer must arrange the appropriate DC conditions at the input. The MSB inversion may be selectable by an external logic level. In the digital video interface standards the colour difference signals use offset binary because the codes of all zeros and all ones are at the end of the range and can be reserved for synchronizing. A digital vision mixer simply inverts the MSB of each colour difference sample to convert it to two's complement.

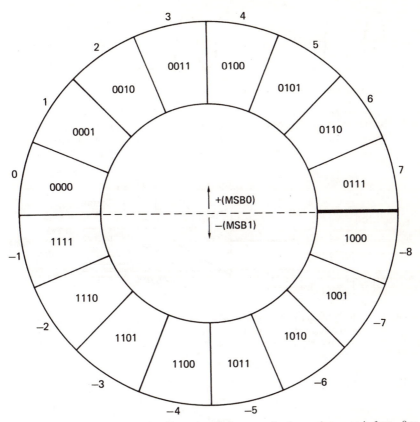

Figure 2.35 In this example of a four-bit two's complement code, the number range is from −8 to +7. Note that the MSB determines polarity.

The two's complement system allows two sample values to be added, or mixed in video parlance, and the result will be referred to the system midrange; this is analogous to adding analog signals in an operational amplifier.

Figure 2.37 illustrates how adding two's complement samples simulates a bipolar mixing process. The waveform of input A is depicted by solid black samples, and that of B by samples with a solid outline. The result of mixing is the linear sum of the two waveforms obtained by adding pairs of sample values. The dashed lines depict the output values. Beneath each set of samples is the calculation which will be seen to give the correct result. Note that the calculations are pure binary. No special arithmetic is needed to handle two's complement numbers.

It is sometimes necessary to phase reverse or invert a digital signal. The process of inversion in two's complement is simple. All bits of the sample value are inverted to form the one's complement, and one is added. This can be checked by mentally inverting some of the values in Figure 2.35. The inversion is transparent and performing a second inversion gives the original sample values.

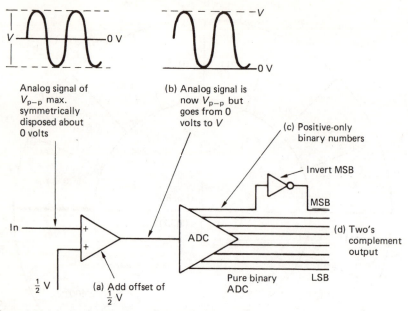

Figure 2.36 A two's complement ADC. At (a) an analog offset voltage equal to one-half the quantizing range is added to the bipolar analog signal in order to make it unipolar as at (b). The ADC produces positive only numbers at (c), but the MSB is then inverted at (d) to give a two's complement output.

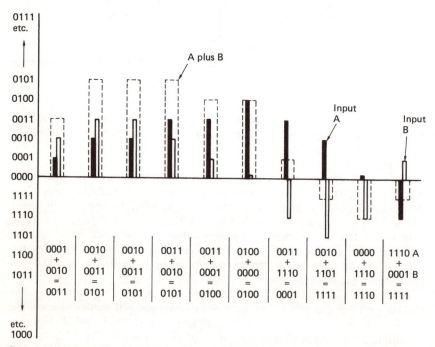

Figure 2.37 Using two's complement arithmetic, single values from two waveforms are added together with respect to midrange to give a correct mixing function.

Using inversion, signal subtraction can be performed using only adding logic. The inverted input is added to perform a subtraction, just as in the analog domain. This permits a significant saving in hardware complexity, since only carry logic is necessary and no borrow mechanism need be supported.

In summary, two's complement notation is the most appropriate scheme for bipolar signals, and allows simple mixing in conventional binary adders. It is in virtually universal use in digital video and audio processing.

Two's complement numbers can have a radix point and bits below it just as pure binary numbers can. It should, however, be noted that in two's complement, if a radix point exists, numbers to the right of it are added. For example, 1100.1 is not −4.5, it is − 4 + 0.5 = − 3.5.

2.13 Introduction to digital logic

However complex a digital process, it can be broken down into smaller stages until finally one finds that there are really only two basic types of element in use, and these can be combined in some way and supplied with a clock to implement

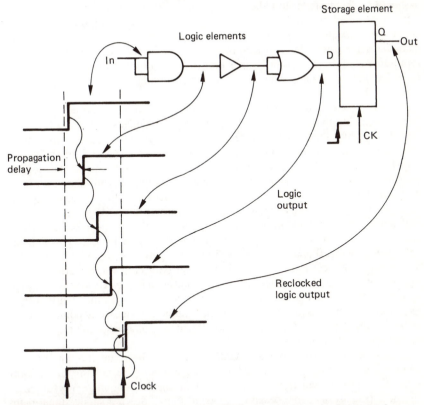

Figure 2.38 Logic elements have a finite propagation delay between input and output and cascading them delays the signal an arbitrary amount. Storage elements sample the input on a clock edge and can return a signal to near coincidence with the system clock. This is known as reclocking. Reclocking eliminates variations in propagation delay in logic elements.

virtually any process. Figure 2.38 shows that the first type is a *logic* element. This produces an output which is a logical function of the input with minimal delay. The second type is a *storage* element which samples the state of the input(s) when clocked and holds or delays that state. The strength of binary logic is that the signal has only two states, and considerable noise and distortion of the binary waveform can be tolerated before the state becomes uncertain. At every logic element, the signal is compared with a threshold, and can thus can pass through any number of stages without being degraded.

In addition, the use of a storage element at regular locations throughout logic circuits eliminates time variations or jitter. Figure 2.38 shows that if the inputs to a logic element change, the output will not change until the *propagation delay* of the element has elapsed. However, if the output of the logic element forms the input to a storage element, the output of that element will not change until the input is sampled *at the next clock edge*. In this way the signal edge is aligned to the system clock and the propagation delay of the logic becomes irrelevant. The process is known as reclocking.

The two states of the signal when measured with an oscilloscope are simply two voltages, usually referred to as high and low. As there are only two states, there can only be *true* or *false* meanings. The true state of the signal can be assigned by the designer to either voltage state. When a high voltage represents a true logic condition and a low voltage represents a false condition, the system is known as *positive logic*, or *high true* logic. This is the usual system, but sometimes the low voltage represents the true condition and the high voltage represents the false condition. This is known as *negative logic* or *low true* logic. Provided that everyone is aware of the logic convention in use, both work equally well.

In logic systems, all logical functions, however complex, can be configured from combinations of a few fundamental logic elements or *gates*. It is not profitable to spend too much time debating which are the truly fundamental ones, since most can be made from combinations of others. Figure 2.39 shows the important simple gates and their derivatives, and introduces the logical expressions to describe them, which can be compared with the truth-table notation. The figure also shows the important fact that when negative logic is used, the OR gate function interchanges with that of the AND gate.

If numerical quantities need to be conveyed down the two-state signal paths described here, then the only appropriate numbering system is binary, which has only two symbols, 0 and 1. Just as positive or negative logic could be used for the truth of a logical binary signal, it can also be used for a numerical binary signal. Normally, a high voltage level will represent a binary 1 and a low voltage will represent a binary 0, described as a 'high for a one' system. Clearly a 'low for a one' system is just as feasible. Decimal numbers have several columns, each of which represents a different power of ten; in binary the column position specifies the power of two.

Several binary digits or bits are needed to express the value of a binary video sample. These bits can be conveyed at the same time by several signals to form a parallel system, which is most convenient inside equipment or for short distances because it is inexpensive, or one at a time down a single signal path, which is more complex, but convenient for cables between pieces of equipment because the connectors require fewer pins. When a binary system is used to convey numbers in this way, it can be called a digital system.

Positive logic name	Boolean expression	Positive logic symbol	Positive logic truth table	Plain English
Inverter or NOT gate	$Q = \overline{A}$	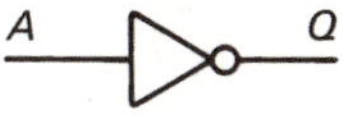	$\begin{array}{c\|c} A & Q \\ \hline 0 & 1 \\ 1 & 0 \end{array}$	Output is opposite of input
AND gate	$Q = A \cdot B$	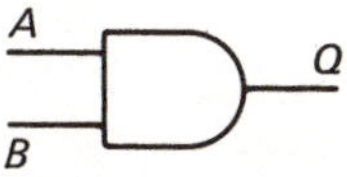	$\begin{array}{cc\|c} A & B & Q \\ \hline 0 & 0 & 0 \\ 0 & 1 & 0 \\ 1 & 0 & 0 \\ 1 & 1 & 1 \end{array}$	Output true when both inputs are true only
NAND (Not AND) gate	$Q = \overline{A \cdot B}$ $= \overline{A} + \overline{B}$	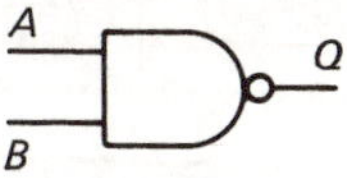	$\begin{array}{cc\|c} A & B & Q \\ \hline 0 & 0 & 1 \\ 0 & 1 & 1 \\ 1 & 0 & 1 \\ 1 & 1 & 0 \end{array}$	Output false when both inputs are true only
OR gate	$Q = A + B$	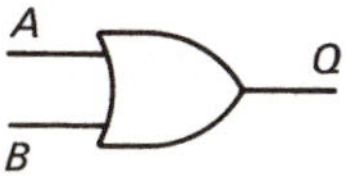	$\begin{array}{cc\|c} A & B & Q \\ \hline 0 & 0 & 0 \\ 0 & 1 & 1 \\ 1 & 0 & 1 \\ 1 & 1 & 1 \end{array}$	Output true if either or both inputs true
NOR (Not OR) gate	$Q = \overline{A + B}$ $= \overline{A} \cdot \overline{B}$	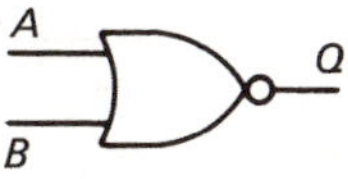	$\begin{array}{cc\|c} A & B & Q \\ \hline 0 & 0 & 1 \\ 0 & 1 & 0 \\ 1 & 0 & 0 \\ 1 & 1 & 0 \end{array}$	Output false if either or both inputs true
Exclusive OR (XOR) gate	$Q = A \oplus B$	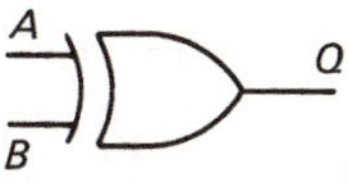	$\begin{array}{cc\|c} A & B & Q \\ \hline 0 & 0 & 0 \\ 0 & 1 & 1 \\ 1 & 0 & 1 \\ 1 & 1 & 0 \end{array}$	Output true if inputs are different

Figure 2.39 The basic logic gates compared.

The basic memory element in logic circuits is the latch, which is constructed from two gates as shown in Figure 2.40(a), and which can be set or reset. A more useful variant is the D-type latch shown at (b) which remembers the state of the input at the time a separate clock either changes state for an edge-triggered device, or after it goes false for a level-triggered device. D-type latches are commonly available with four or eight latches to the chip. A shift register can be made from a series of latches by connecting the Q output of one latch to the D input of the next and connecting all the clock inputs in parallel. Data are delayed by the number of stages in the register. Shift registers are also useful for converting between serial and parallel data transmissions.

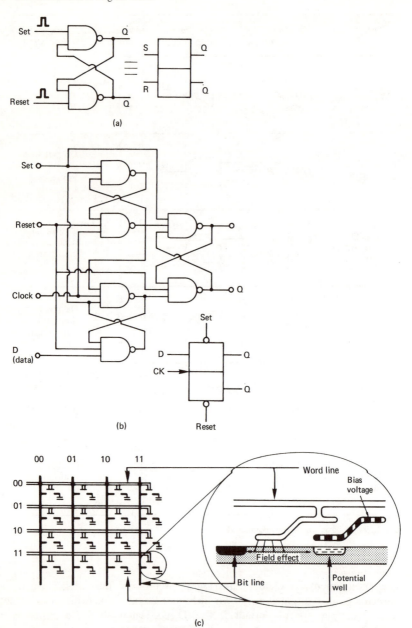

Figure 2.40 Digital semiconductor memory types. In (a), one data bit can be stored in a simple set-reset latch, which has little application because the D-type latch in (b) can store the state of the single data input when the clock occurs. These devices can be implemented with bipolar transistors or FETs, and are called static memories because they can store indefinitely. They consume a lot of power.

In (c), a bit is stored as the charge in a potential well in the substrate of a chip. It is accessed by connecting the bit line with the field effect from the word line. The single well where the two lines cross can then be written or read. These devices are called dynamic RAMs because the charge decays, and they must be read and rewritten (refreshed) periodically.

Where large numbers of bits are to be stored, cross-coupled latches are less suitable because they are more complicated to fabricate inside integrated circuits than dynamic memory, and consume more current.

In large random access memories (RAMs), the data bits are stored as the presence or absence of charge in a tiny capacitor as shown in Figure 2.40(c). The capacitor is formed by a metal electrode, insulated by a layer of silicon dioxide from a semiconductor substrate, hence the term MOS (metal oxide semi-conductor). The charge will suffer leakage, and the value would become indeterminate after a few milliseconds. Where the delay needed is less than this, decay is of no consequence, as data will be read out before they have had a chance to decay. Where longer delays are necessary, such memories must be refreshed periodically by reading the bit value and writing it back to the same place. Most modern MOS RAM chips have suitable circuitry built-in. Large RAMs store thousands of bits, and it is clearly impractical to have a connection to each one. Instead, the desired bit has to be addressed before it can be read or written. The size of the chip package restricts the number of pins available, so that large memories use the same address pins more than once. The bits are arranged internally as rows and columns, and the row address and the column address are specified sequentially on the same pins.

The circuitry necessary for adding pure binary or two's complement numbers is shown in Figure 2.41. Addition in binary requires two bits to be taken at a time from the same position in each word, starting at the least significant bit. Should both be ones, the output is zero, and there is a *carry-out* generated. Such a circuit is called a half adder, shown in Figure 2.41(a) and is suitable for the least significant bit of the calculation. All higher stages will require a circuit which can accept a carry input as well as two data inputs. This is known as a full adder (Figure 2.41(b)). Multibit full adders are available in chip form, and have carry-in and carry-out terminals to allow them to be cascaded to operate on long wordlengths. Such a device is also convenient for inverting a two's complement number, in conjunction with a set of invertors. The adder chip has one set of inputs grounded, and the carry-in permanently held true, such that it adds one to the one's complement number from the invertor.

When mixing by adding sample values, care has to be taken to ensure that if the sum of the two sample values exceeds the number range the result will be clipping rather than wraparound. In two's complement, the action necessary depends on the polarities of the two signals. Clearly, if one positive and one negative number are added, the result cannot exceed the number range. If two positive numbers are added, the symptom of positive overflow is that the most significant bit sets, causing an erroneous negative result, whereas a negative overflow results in the most significant bit clearing. The overflow control circuit will be designed to detect these two conditions, and override the adder output. If the MSB of both inputs is zero, the numbers are both positive, thus if the sum has the MSB set, the output is replaced with the maximum positive code (0111 . . .). If the MSB of both inputs is set, the numbers are both negative, and if the sum has no MSB set, the output is replaced with the maximum negative code (1000 . . .). These conditions can also be connected to warning indicators. Figure 2.41(c) shows this system in hardware. The resultant clipping on overload is sudden, and sometimes a PROM is included which translates values around and beyond maximum to soft-clipped values below or equal to maximum.

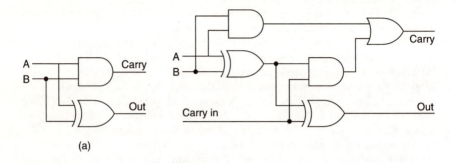

(a)

Data A	Bits B	Carry in	Out	Carry out
0	0	0	0	0
0	0	1	1	0
0	1	0	1	0
0	1	1	0	1
1	0	0	1	0
1	0	1	0	1
1	1	0	0	1
1	1	1	1	1

(b)

(c)

Figure 2.41 (a) Half adder; (b) full-adder circuit and truth table; (c) comparison of sign bits prevents wraparound on adder overflow by substituting clipping level.

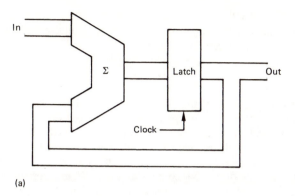

(a)

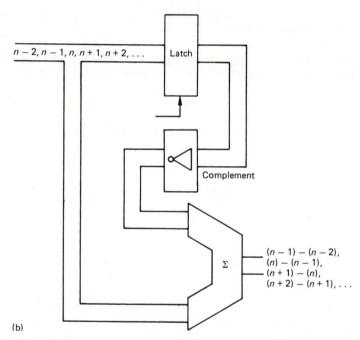

(b)

Figure 2.42 Two configurations which are common in processing. In (a) the feedback around the adder adds the previous sum to each input to perform accumulation or digital integration. In (b) an inverter allows the difference between successive inputs to be computed. This is differentiation.

A storage element can be combined with an adder to obtain a number of useful functional blocks which will crop up frequently in audio equipment. Figure 2.42(a) shows that a latch is connected in a feedback loop around an adder. The latch contents are added to the input each time it is clocked. The configuration is known as an accumulator in computation because it adds up or accumulates values fed into it. In filtering, it is known as an discrete time integrator. If the input is held at some constant value, the output increases by that amount on each clock. The output is thus a sampled ramp.

Figure 2.42(b) shows that the addition of an invertor allows the difference between successive inputs to be obtained. This is digital differentiation. The output is proportional to the slope of the input.

2.14 The computer

The computer is now a vital part of digital video systems, being used both for control purposes and to process video signals as data. In control, the computer finds applications in database management, automation, editing, and in electromechanical systems such as tape drives and robotic cassette handling. Now that processing speeds have advanced sufficiently, computers are able to manipulate certain types of digital video in real time. Where very complex calculations are needed, real-time operation may not be possible and instead the computation proceeds as fast as it can in a process called *rendering*. The rendered data are stored so that they can be viewed in real time from a storage medium when the rendering is complete.

The computer is a programmable device in that its operation is not determined by its construction alone, but instead by a series of *instructions* forming a *program*. The program is supplied to the computer one instruction at a time so that the desired sequence of events takes place.

Programming of this kind has been used for over a century in electro-mechanical devices, including automated knitting machines and street organs which are programmed by punched cards. However, the computer differs from these devices in that the program is not fixed, but can be modified by the computer itself. This possibility led to the creation of the term *software* to suggest a contrast to the constancy of hardware.

Computer instructions are binary numbers each of which is interpreted in a specific way. As these instructions don't differ from any other kind of data, they can be stored in RAM. The computer can change its own instructions by accessing the RAM. Most types of RAM are volatile, in that they lose data when power is removed. Clearly if a program is entirely stored in this way, the computer will not be able to recover fom a power failure. The solution is that a very simple starting or *bootstrap* program is stored in non-volatile ROM which will contain instructions that will bring in the main program from a storage system such as a disk drive after power is applied. As programs in ROM cannot be altered, they are sometimes referred to as *firmware* to indicate that they are classified between hardware and software.

Making a computer do useful work requires more than simply a program which performs the required computation. There is also a lot of mundane activity which does not differ significantly from one program to the next. This includes deciding which part of the RAM will be occupied by the program and which by the data, producing commands to the storage disk drive to read the input data from a file and write back the results. It would be very inefficient if all programs had to handle these processes themselves. Consequently the concept of an *operating system* was developed. This manages all the mundane decisions and creates an environment in which useful programs or *applications* can execute.

The ability of the computer to change its own instructions makes it very powerful, but it also makes it vulnerable to abuse. Programs exist which are deliberately written to do damage. These *viruses* are generally attached to plausible messages or data files and enter computers through storage media or communications paths.

There is also the possibility that programs contain logical errors such that in certain combinations of circumstances the wrong result is obtained. If this results in the unwitting modification of an instruction, the next time that instruction is accessed the computer will crash. In consumer-grade software, written for the vast personal computer market, this kind of thing is unfortunately accepted.

For critical applications, software must be *verified*. This is a process which can prove that a program can recover from absolutely every combination of circumstances and keep running properly. This is a non-trivial process, because the number of combinations of states a computer can get into is staggering. As a result most software is unverified.

It is of the utmost importance that networked computers which can suffer virus infection or computers running unverified software are never used in a life-support or critical application.

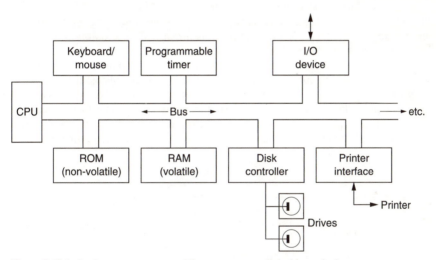

Figure 2.43 A simple computer system. All components are linked by a single data/address/control bus. Although cheap and flexible, such a bus can only make one connection at a time, so it is slow.

Figure 2.43 shows a simple computer system. The various parts are linked by a bus which allows binary numbers to be transferred from one place to another. This will generally use tri-state logic so that when one device is sending to another, all other devices present a high impedance to the bus.

The ROM stores the startup program, the RAM stores the operating system, applications programs and the data to be processed. The disk drive stores large quantities of data in a non-volatile form. The RAM only needs to be able to hold part of one program as other parts can be brought from the disk as required. A program executes by *fetching* one instruction at a time from the RAM to the processor along the bus.

The bus also allows keyboard/mouse inputs and outputs to the display and printer. Inputs and outputs are generally abbreviated to I/O. Finally a programmable timer will be present which acts as a kind of alarm clock for the processor.

2.15 The processor

The processor or CPU (central processing unit) is the heart of the system. Figure 2.44 shows a simple example of a CPU. The CPU has a bus interface which allows it to generate bus addresses and input or output data. Sequential instructions are stored in RAM at contiguously increasing locations so that a program can be executed by fetching instructions from a RAM address specified by the program counter (PC) to the instruction register in the CPU. As each instruction is completed, the PC is incremented so that it points to the next instruction. In this way the time taken to execute the instruction can vary.

The processor is notionally divided into data paths and control paths. Figure 2.44 shows the data path. The CPU contains a number of general-purpose registers or scratchpads which can be used to store partial results in complex calculations. Pairs of these registers can be addressed so that their contents go to the ALU (arithmetic logic unit). This performs various arithmetic (add, subtract, etc.) or logical (and, or, etc.) functions on the input data. The output of the ALU may be routed back to a register or output. By reversing this process it is possible to get data into the registers from the RAM. The ALU also outputs the conditions resulting from the calculation, which can control conditional instructions.

Which function the ALU performs and which registers are involved are determined by the instruction currently in the instruction register then is decoded in the control path. One pass through the ALU can be completed in one cycle of the processor's clock. Instructions vary in complexity as do the number of clock

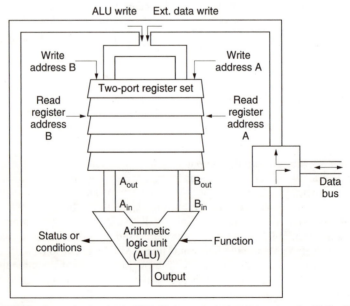

Figure 2.44 The data path of a simple CPU. Under control of an instruction, the ALU will perform some function on a pair of input values from the registers and store or output the result.

cycles needed to complete them. Incoming instructions are decoded and used to access a look-up table which converts them into *microinstructions*, one of which controls the CPU at each clock cycle.

2.16 Timebase correction

In Chapter 1 it was stated that a strength of digital technology is the ease with which delay can be provided. Accurate control of delay is the essence of timebase correction, necessary whenever the instantaneous time of arrival or rate from a data source does not match the destination. In digital video, the destination will almost always have perfectly regular timing, namely the sampling rate clock of the final DAC. Timebase correction consists of aligning jittery signals from storage media or transmission channels with that stable reference.

A further function of timebase correction is to reverse the time compression applied prior to recording or transmission. As was shown in section 1.7, digital recorders compress data into blocks to facilitate editing and error correction as well as to permit head switching between blocks in rotary-head machines. Owing to the spaces between blocks, data arrive in bursts on replay, but must be fed to the output convertors in an unbroken stream at the sampling rate.

In computer hard-disk drives, which are used in digital video workstations, time compression is also used, but a converse problem also arises. Data from the disk blocks arrive at a reasonably constant rate, but cannot necessarily be accepted at a steady rate by the logic because of contention for the use of buses and memory by the different parts of the system. In this case the data must be buffered by a relative of the timebase corrector which is usually referred to as a silo.

Although delay is easily implemented, it is not possible to advance a data stream. Most real machines cause instabilities balanced about the correct timing: the output jitters between too early and too late. Since the information cannot be advanced in the corrector, only delayed, the solution is to run the machine in advance of real time. In this case, correctly timed output signals will need a nominal delay to align them with reference timing. Early output signals will receive more delay, and late output signals will receive less delay.

Section 2.13 showed the principles of digital storage elements which can be used for delay purposes. The shift-register approach and the RAM approach to delay are very similar, as a shift register can be thought of as a memory whose address increases automatically when clocked. The data rate and the maximum delay determine the capacity of the RAM required. Figure 2.45 shows that the addressing of the RAM is by a counter that overflows endlessly from the end of the memory back to the beginning, giving the memory a ring-like structure. The write address is determined by the incoming data, and the read address is determined by the outgoing data. This means that the RAM has to be able to read and write at the same time. The switching between read and write involves not only a data multiplexer but also an address multiplexer. In general the arbitration between read and write will be done by signals from the stable side of the TBC as Figure 2.46 shows. In the replay case the stable clock will be on the read side. The stable side of the RAM will read a sample when it demands, and the writing will be locked out for that period. The input data cannot be interrupted in many applications, however, so a small buffer silo is installed before the memory, which fills up as the writing is locked out, and empties again as writing is

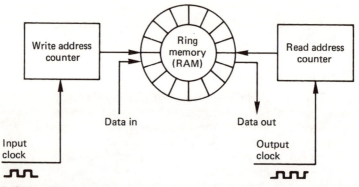

Figure 2.45 Most TBCs are implemented as a memory addressed by a counter which periodically overflows to give a ring structure. The memory allows the read and write sides to be asynchronous.

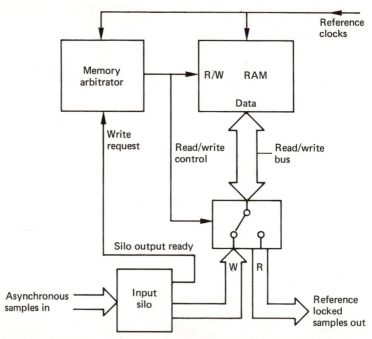

Figure 2.46 In a RAM-based TBC, the RAM is reference synchronous, and an arbitrator decides when it will read and when it will write. During reading, asynchronous input data back up in the input silo, asserting a write request to the arbitrator. Arbitrator will then cause a write cycle between read cycles.

permitted. Alternatively, the memory will be split into blocks as was shown in Chapter 1, such that when one block is reading a different block will be writing and the problem does not arise.

In most digital video applications, the sampling rate exceeds the rate at which economically available RAM chips can operate. The solution is to arrange several video samples into one longer word, known as a superword, and to construct the memory so that it stores superwords in parallel.

Figure 2.47 shows the operation of a FIFO chip, colloquially known as a silo because the data are tipped in at the top on delivery and drawn off at the bottom when needed. Each stage of the chip has a data register and a small amount of logic, including a data-valid or V bit. If the input register does not contain data, the first V bit will be reset, and this will cause the chip to assert 'input ready'. If data are presented at the input, and clocked into the first stage, the V bit will set, and the 'input ready' signal will become false. However, the logic associated with the next stage sees the V bit set in the top stage, and if its own V bit is clear, it will clock the data into its own register, set its own V bit, and clear the input V bit, causing 'input ready' to reassert, when another word can be fed in. This process then continues as the word moves down the silo, until it arrives at the last register in the chip. The V bit of the last stage becomes the 'output ready' signal,

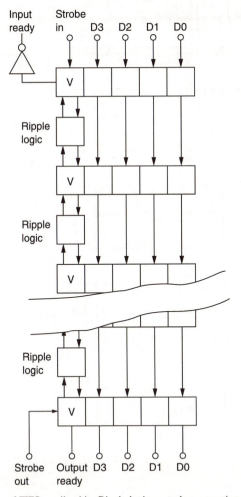

Figure 2.47 Structure of FIFO or silo chip. Ripple logic controls propagation of data down silo.

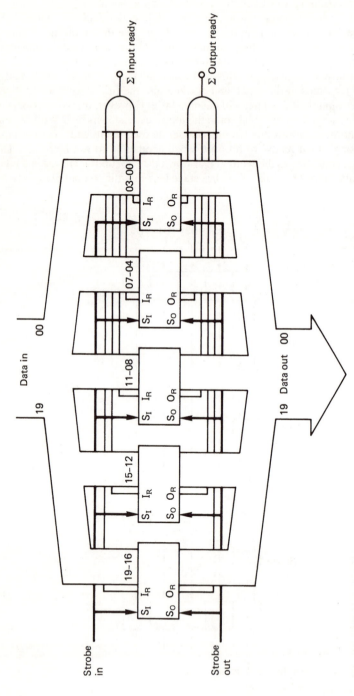

Figure 2.48 In this example, a twenty-bit wordlength silo is made from five parallel FIFO chips. The asynchronous ripple action of FIFOs means that it is necessary to 'AND' together the ready signals.

telling subsequent circuitry that there are data to be read. If this word is not read, the next word entered will ripple down to the stage above. Words thus stack up at the bottom of the silo. When a word is read out, an external signal must be provided which resets the bottom V bit. The 'output ready' signal now goes false, and the logic associated with the last stage now sees valid data above, and loads down the word when it will become ready again. The last register but one will now have no V bit set, and will see data above itself and bring that down. In this way a reset V bit propagates up the chip while the data ripple down, rather like a hole in a semiconductor going the opposite way to the electrons. Silo chips are usually available in four-bit wordlengths, but can easily be connected in parallel to form superwords. Silo chips are asynchronous, and paralleled chips will not necessarily all work at the same speed. This problem is easily overcome by 'anding' together all of the input-ready and output-ready signals and parallel-connecting the strobes. Figure 2.48 shows this mode of operation.

When used in a hard-disk system, a silo will allow data to and from the disk, which is turning at constant speed. When reading the disk, Figure 2.49(a) shows that the silo starts empty, and if there is bus contention, the silo will start to fill. Where the bus is free, the disk controller will attempt to empty the silo into the memory. The system can take advantage of the interblock gaps on the disk, containing headers, preambles and redundancy, for in these areas there are no

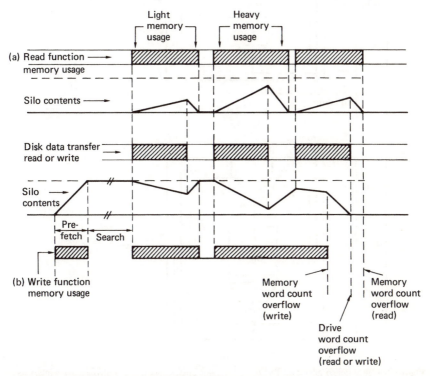

Figure 2.49 The silo contents during read functions (a) appear different from those during write functions (b). In (a), the control logic attempts to keep the silo as empty as possible; in (b) the logic prefills the silo and attempts to keep it full until the memory word count overflows.

data to transfer, and there is some breathing space to empty the silo before the next block. In practice the silo need not be empty at the start of every block, provided it never becomes full before the end of the transfer. If this happens some data are lost and the function must be aborted. The block containing the silo overflow will generally be reread on the next revolution. In sophisticated systems, the silo has a kind of dipstick, and can interrupt the CPU if the data get too deep. The CPU can then suspend some bus activity to allow the disk controller more time to empty the silo.

When the disk is to be written, as in Figure 2.49(b), a continuous data stream must be provided during each block, as the disk cannot stop. The silo will be pre-filled before the disk attempts to write, and the disk controller attempts to keep it full. In this case all will be well if the silo does not become empty before the

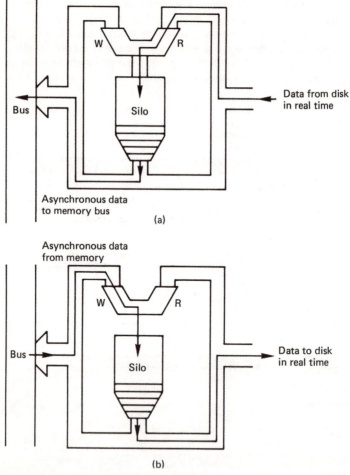

Figure 2.50 In order to guarantee that the drive can transfer data in real time at regular intervals (determined by disk speed and density) the silo provides buffering to the asynchronous operation of the memory access process. At (a) the silo is configured for a disk read. The same silo is used at (b) for a disk write.

end of the transfer. Figure 2.50 shows the silo of a typical disk controller with the multiplexers necessary to put it in the read data stream or the write data stream.

2.17 Multiplexing

Multiplexing is used where several signals are to be transmitted down the same channel. The channel bit rate must be the same as or greater than the sum of the source bit rates. Figure 2.51 shows that when multiplexing is used, the data from each source have to be time compressed. This is done by buffering source data in a memory at the multiplexer. It is written into the memory in real time as it arrives, but will be read from the memory with a clock which has a much higher rate. This means that the readout occurs in a smaller timespan. If, for example, the clock frequency is raised by a factor of ten, the data for a given signal will be transmitted in a tenth of the normal time, leaving time in the multiplex for nine more such signals.

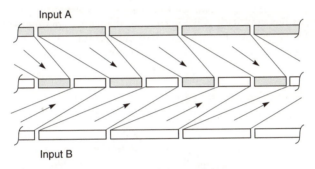

Figure 2.51 Multiplexing requires time compression on each input.

In the demultiplexer another buffer memory will be required. Only the data for the selected signal will be written into this memory at the bit rate of the multiplex. When the memory is read at the correct speed, the data will emerge with their original timebase.

In practice it is essential to have mechanisms to identify the separate signals to prevent them being mixed up and to convey the original signal clock frequency to the demultiplexer. In time-division multiplexing the timebase of the transmission is broken into equal slots, one for each signal. This makes it easy for the demultiplexer, but forces a rigid structure on all the signals such that they must all be locked to one another and have an unchanging bit rate. Packet multiplexing overcomes these limitations.

The multiplexer must switch between different time-compressed signals to create the bitstream and this is much easier to organize if each signal is in the form of data packets of constant size. Figure 2.52 shows a packet multiplexing system.

Each packet consists of two components: the header, which identifies the packet, and the payload, which is the data to be transmitted. The header will contain at least an identification code (ID) which is unique for each signal in the

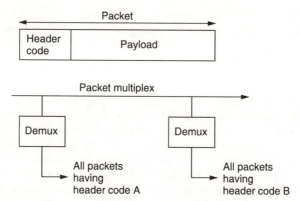

Figure 2.52 Packet multiplexing relies on headers to identify the packets.

multiplex. The demultiplexer checks the ID codes of all incoming packets and discards those which do not have the wanted ID.

In complex systems it is common to have a mechanism to check that packets are not lost or repeated. This is the purpose of the packet continuity count which is carried in the header. For packets carrying the same ID, the count should increase by one from one packet to the next. Upon reaching the maximum binary value, the count overflows and recommences.

2.18 Statistical multiplexing

Packet multiplexing has advantages over time-division multiplexing because it does not set the bit rate of each signal. A demultiplexer simply checks packet IDs and selects all packets with the wanted code. It will do this however frequently such packets arrive. Consequently it is practicable to have variable bit rate signals in a packet multiplex. The multiplexer has to ensure that the total bit rate does not exceed the rate of the channel, but that rate can be allocated arbitrarily between the various signals.

As a practical matter is is usually necessary to keep the bit rate of the multiplex constant. With variable rate inputs this is done by creating null packets which are generally called *stuffing* or *packing*. The headers of these packets contain an unique ID which the demultiplexer does not recognize and so these packets are discarded on arrival.

In an MPEG environment, statistical multiplexing can be extremely useful because it allows for the varying difficulty of real program material. In a multiplex of several television programs, it is unlikely that all the programs will encounter difficult material simultaneously. When one program encounters a detailed scene or frequent cuts which are hard to compress, more data rate can be allocated at the allowable expense of the remaining programs which are handling easy material.

2.19 Filters and transforms

One of the most important processes in digital video is filtering, and its parallel topic of transforms. Filters and transforms are relevant to sampling, displays, recording, transmission and compression systems.

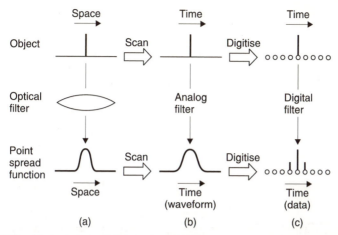

Figure 2.53 In optical systems an infinitely sharp line is reproduced as a point spread function (a) which is the impulse response of the optical path. Scanning either object or image produces an analog time-variant waveform (b). The scanned object waveform can be converted to the scanned image waveform with an electrical filter having an impulse response which is an analog of the point spread function. (c) The object and image may also be sampled or the object samples can be converted to the image samples by a filter with an analogous discrete impulse response.

Figure 2.53 shows an optical system of finite resolution. If an object containing an infinitely sharp line is presented to this system, the image will be a symmetrical intensity function known in optics as a *point spread function* which is a spatial impulse response. All images passing through the optical system are convolved with it.

Figure 2.53(b) shows that the object may be scanned by an analog system to produce a waveform. The image may also be scanned in this way. These waveforms are now temporal. However, the second waveform may be obtained in another way, using an analog filter in series with the first scanned waveform which has an equivalent impulse response. This filter must have linear phase, i.e. its impulse response must be symmetrical in order to replicate the point spread function.

Figure 2.53(c) shows that the object may also be sampled in which case all samples but one will have a value of zero. The image may also be sampled, and owing to the point spread function, there will now be a number of non-zero sample values. However, the image samples may also be obtained by passing the input sample into a digital filter having the appropriate impulse response. Note that it is possible to obtain the same result as (c) by passing the scanned waveform of (b) into an ADC and storing the samples in a memory.

It should be clear from Figure 2.53 why video signal paths need to have linear phase. In general, analog circuitry and filters tend not to have linear phase because they must be *causal* which means that the output can only occur after the input. Figure 2.54(a) shows a simple RC network and its impulse response. This is the familiar exponential decay due to the capacitor discharging through the resistor (in series with the source impedance which is assumed here to be negligible). The figure also shows the response to a squarewave at (b). With other waveforms the process is inevitably more complex.

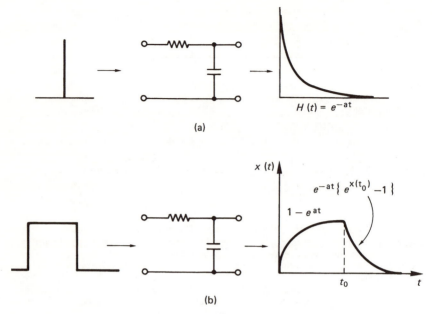

(a)

(b)

Figure 2.54 (a) The impulse response of a simple RC network is an exponential decay. This can used to calculate the response to a square wave, as in (b).

Filtering is unavoidable. Sometimes a process has a filtering effect which is undesirable, for example the limited frequency response of a CRT drive amplifier or loss of resolution in a lens, and we try to minimize it. On other occasions a filtering effect is specifically required. Analog or digital filters, and sometimes both, are required in ADCs, DACs, in the data channels of digital recorders and transmission systems and in DSP. Optical filters may also be necessary in imaging systems to convert between sampled and continuous images. Optical systems used in displays and in laser recorders also act as spatial filters.[1]

Figure 2.55 shows that impulse response testing tells a great deal about a filter. In a perfect filter, all frequencies should experience the same time delay. If some

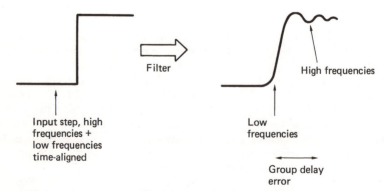

Figure 2.55 Group delay time-displaces signals as a function of frequency.

groups of frequencies experience a different delay from others, there is a group-delay error. As an impulse contains an infinite spectrum, a filter suffering from group-delay error will separate the different frequencies of an impulse along the time axis.

A pure delay will cause a phase shift proportional to frequency, and a filter with this characteristic is said to be phase-linear. The impulse response of a phase-linear filter is symmetrical. If a filter suffers from group-delay error it cannot be phase-linear. It is almost impossible to make a perfectly phase-linear analog filter, and many filters have a group-delay equalization stage following them which is often as complex as the filter itself. In the digital domain it is straightforward to make a phase-linear filter, and phase equalization becomes unnecessary.

Because of the sampled nature of the signal, whatever the response at low frequencies may be, all PCM channels act as low-pass filters because they cannot contain frequencies above the Nyquist limit of half the sampling frequency.

Transforms are a useful subject because they can help to understand processes which cause undesirable filtering or to design filters. The information itself may be subject to a transform. Transforming converts the information into another analog. The information is still there, but expressed with respect to temporal or spatial frequency rather than time or space. Instead of binary numbers representing the magnitude of samples, there are binary numbers representing the magnitude of frequency coefficients. What happens in the frequency domain must always be consistent with what happens in the time or space domains. Every combination of frequency and phase response has a corresponding impulse response in the time domain.

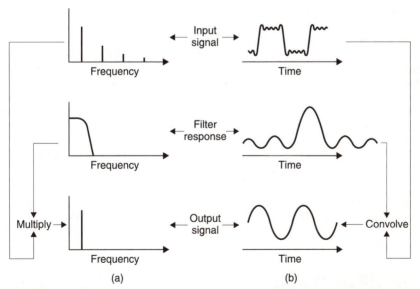

Figure 2.56 If a signal having a given spectrum is passed into a filter, multiplying the two spectra will give the output spectrum at (a). Equally transforming the filter frequency response will yield the impulse response of the filter. If this is convolved with the time domain waveform, the result will be the output waveform, whose transform is the output spectrum (b).

Figure 2.56 shows the relationship between the domains. On the left is the frequency domain. Here an input signal having a given spectrum is input to a filter having a given frequency response. The output spectrum will be the product of the two functions. If the functions are expressed logarithmically in deciBels, the product can be obtained by simple addition.

On the right, the time-domain output waveform represents the convolution of the impulse response with the input waveform. However, if the frequency transform of the output waveform is taken, it must be the same as the result obtained from the frequency response and the input spectrum. This is a useful result because it means that when image or audio sampling is considered, it will be possible to explain the process in both domains.

When a waveform is input to a system, the output waveform will be the convolution of the input waveform and the impulse response of the system. Convolution can be followed by reference to a graphic example in Figure 2.57. Where the impulse response is asymmetrical, the decaying tail occurs *after* the input. As a result it is necessary to reverse the impulse response in time so that it is mirrored prior to sweeping it through the input waveform. The output voltage is proportional to the shaded area shown where the two impulses overlap. If the impulse response is symmetrical, as would be the case with a linear phase filter, or in an optical system, the mirroring process is superfluous.

The same process can be performed in the sampled, or discrete time domain as shown in Figure 2.58. The impulse and the input are now a set of discrete samples which clearly must have the same sample spacing. The impulse response only has value where impulses coincide. Elsewhere it is zero. The impulse response is therefore stepped through the input one sample period at a time. At each step, the area is still proportional to the output, but as the time steps are of uniform width, the area is proportional to the impulse height and so the output is obtained by adding up the lengths of overlap. In mathematical terms, the output samples represent the convolution of the input and the impulse response by summing the coincident cross-products.

Filters can be described in two main classes, as shown in Figure 2.59, according to the nature of the impulse response. Finite-impulse response (FIR) filters are always stable and, as their name suggests, respond to an impulse once, as they have only a forward path. In the temporal domain, the time for which the filter responds to an input is finite, fixed and readily established. The same is therefore true about the distance over which a FIR filter responds in the spatial domain. FIR filters can be made perfectly phase-linear if a significant processing delay is accepted. Most filters used for image processing, sampling rate conversion and oversampling fall into this category.

Infinite-impulse response (IIR) filters respond to an impulse indefinitely and are not necessarily stable, as they have a return path from the output to the input. For this reason they are also called recursive filters. As the impulse response is not symmetrical, IIR filters are not phase-linear. Audio equalizers often employ recursive filters.

2.20 FIR filters

A FIR filter performs convolution of the input waveform with its own impulse response. It does this by graphically constructing the impulse response for every input sample and superimposing all these responses. It is first necessary to

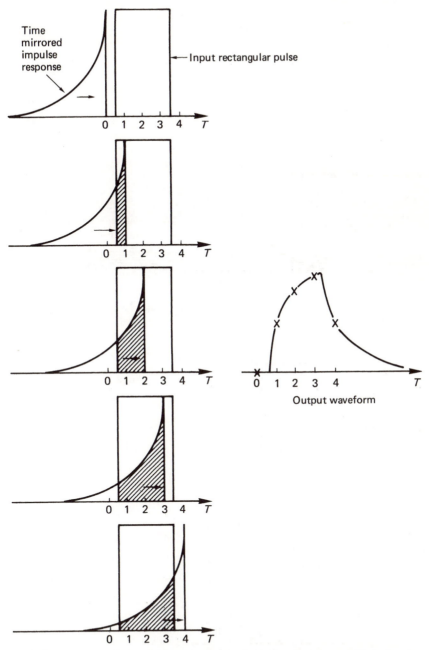

Figure 2.57 In the convolution of two continuous signals (the impulse response with the input), the impulse must be time reversed or mirrored. This is necessary because the impulse will be moved from left to right, and mirroring gives the impulse the correct time-domain response when it is moved past a fixed point. As the impulse response slides continuously through the input waveform, the area where the two overlap determines the instantaneous output amplitude. This is shown for five different times by the crosses on the output waveform.

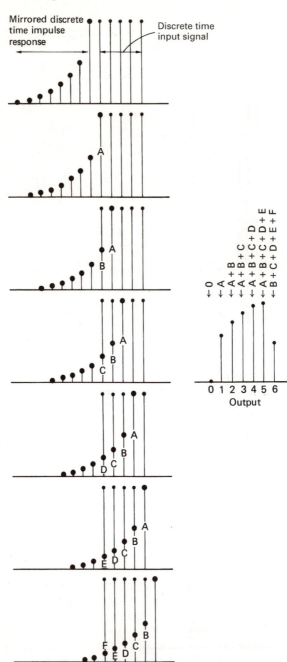

Figure 2.58 In discrete time convolution, the mirrored impulse response is stepped through the input one sample period at a time. At each step, the sum of the cross-products is used to form an output value. As the input in this example is a constant height pulse, the output is simply proportional to the sum of the coincident impulse response samples. This figure should be compared with Figure 2.57.

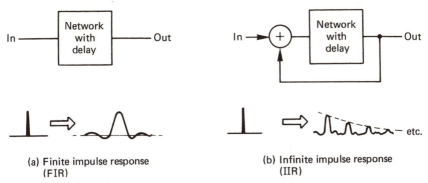

(a) Finite impulse response
(FIR)

(b) Infinite impulse response
(IIR)

Figure 2.59 An FIR filter (a) responds only once to an input, whereas the output of an IIR filter (b) continues indefinitely rather like a decaying echo.

establish the correct impulse response. Figure 2.60(a) shows an example of a low-pass filter which cuts off at $\frac{1}{4}$ of the sampling rate. The impulse response of an ideal low-pass filter is a $\sin x/x$ curve, where the time between the two central zero crossings is the reciprocal of the cut-off frequency. According to the mathematics, the waveform has always existed, and carries on for ever.

The peak value of the output coincides with the input impulse. This means that the filter cannot be causal, because the output has changed before the input is known. Thus in all practical applications it is necessary to truncate the extreme ends of the impulse response, which causes an aperture effect, and to introduce a time delay in the filter equal to half the duration of the truncated impulse in order to make the filter causal. As an input impulse is shifted through the series of registers in Figure 2.60(b), the impulse response is created, because at each point it is multiplied by a coefficient as in Figure 2.60(c).

These coefficients are simply the result of sampling and quantizing the desired impulse response. Clearly the sampling rate used to sample the impulse must be the same as the sampling rate for which the filter is being designed. In practice the coefficients are calculated, rather than attempting to sample an actual impulse response. The coefficient wordlength will be a compromise between cost and performance. Because the input sample shifts across the system registers to create the shape of the impulse response, the configuration is also known as a transversal filter. In operation with real sample streams, there will be several consecutive sample values in the filter registers at any time in order to convolve the input with the impulse response.

Simply truncating the impulse response causes an abrupt transition from input samples which matter and those which do not. Truncating the filter superimposes a rectangular shape on the time-domain impulse response. In the frequency domain the rectangular shape transforms to a $\sin x/x$ characteristic which is superimposed on the desired frequency response as a ripple. One consequence of this is known as Gibb's phenomenon; a tendency for the response to peak just before the cut-off frequency.[2,3] As a result, the length of the impulse which must be considered will depend not only on the frequency response, but also on the amount of ripple which can be tolerated. If the relevant period of the impulse is measured in sample periods, the result will be the number of points or multiplications needed in the filter. Figure 2.61 compares the performance of

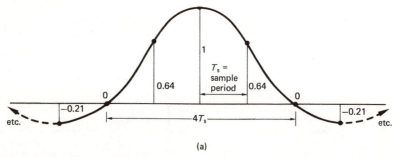

(a)

Figure 2.60(a) The impulse response of an LPF is a sin*x*/*x* curve which stretches from −∞ to +∞ in time. The ends of the response must be neglected, and a delay introduced to make the filter causal.

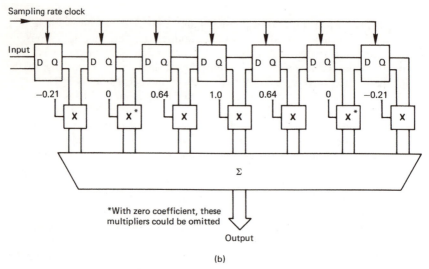

(b)

Figure 2.60(b) The structure of an FIR LPF. Input samples shift across the register and at each point are multiplied by different coefficients.

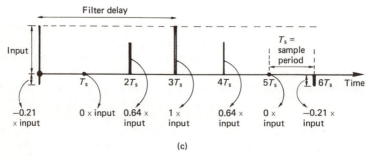

(c)

Figure 2.60(c) When a single unit sample shifts across the circuit of Figure 2.60(b), the impulse response is created at the output as the impulse is multiplied by each coefficient in turn.

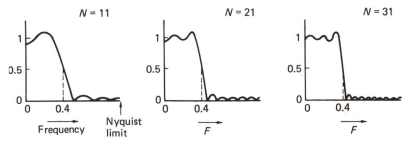

Figure 2.61 The truncation of the impulse in an FIR filter caused by the use of a finite number of points (N) results in ripple in the response. Shown here are three different numbers of points for the same impulse response. The filter is an LPF which rolls off at 0.4 of the fundamental interval. (Courtesy *Philips Technical Review*.)

filters with different numbers of points. A high-quality digital audio FIR filter may need in excess of 100 points.

Rather than simply truncate the impulse response in time, it is better to make a smooth transition from samples which do not count to those that do. This can be done by multiplying the coefficients in the filter by a window function which peaks in the centre of the impulse. In the example of Figure 2.62, a low-pass FIR filter is shown which is intended to allow downsampling by a factor of two. The key feature is that the stopband must have begun before one half of the output sampling rate. This is most readily achieved using a Hamming window because it was designed empirically to have a flat stopband so that good aliasing attenuation is possible. The width of the transition band determines the number of significant sample periods embraced by the impulse. The Hamming window doubles the width of the transition band. This determines in turn both the number of points in the filter and the filter delay. For the purposes of illustration, the number of points is much smaller than would normally be the case.

As the impulse is symmetrical, the delay will be half the impulse period. The impulse response is a sin$x$/x function, and this has been calculated in the figure. The equation for the Hamming window function is shown with the window values which result. The sin$x$/x response is next multiplied by the Hamming window function to give the windowed impulse response shown.

If the coefficients are not quantized finely enough, it will be as if they had been calculated inaccurately, and the performance of the filter will be less than expected. Figure 2.63 shows an example of quantizing coefficients. Conversely, raising the wordlength of the coefficients increases cost.

The FIR structure is inherently phase-linear because it is easy to make the impulse response absolutely symmetrical. The individual samples in a digital system do not know in isolation what frequency they represent, and they can only pass through the filter at a rate determined by the clock. Because of this inherent phase-linearity, a FIR filter can be designed for a specific impulse response, and the frequency response will follow.

The frequency response of the filter can be changed at will by changing the coefficients. A programmable filter only requires a series of PROMs to supply the coefficients; the address supplied to the PROMs will select the response. The frequency response of a digital filter will also change if the clock rate is changed, so it is often less ambiguous to specify a frequency of interest in a digital filter

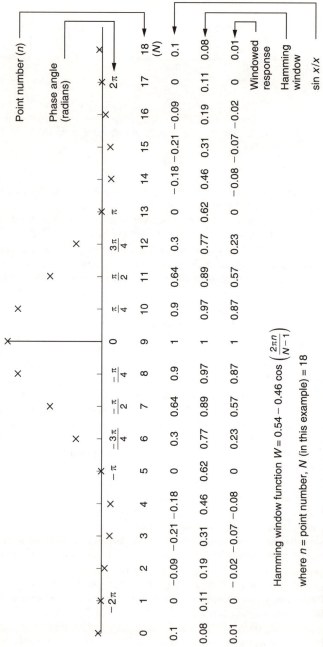

Point number (n)	0	1	2	3	4	5	6	7	8	9	10	11	12	13	14	15	16	17	18 (N)
Phase angle (radians)	-2π					$-\pi$	$-\frac{3\pi}{4}$	$-\frac{\pi}{2}$	$-\frac{\pi}{4}$	0	$\frac{\pi}{4}$	$\frac{\pi}{2}$	$\frac{3\pi}{4}$	π				2π	
$\sin x/x$	0.1	0	-0.09	-0.21	-0.18	0	0.3	0.64	0.9	1	0.9	0.64	0.3	0	-0.18	-0.21	-0.09	0	0.1
Hamming window	0.08	0.11	0.19	0.31	0.46	0.62	0.77	0.89	0.97	1	0.97	0.89	0.77	0.62	0.46	0.31	0.19	0.11	0.08
Windowed response	0.01	0	-0.02	-0.07	-0.08	0	0.23	0.57	0.87	1	0.87	0.57	0.23	0	-0.08	-0.07	-0.02	0	0.01

Hamming window function $W = 0.54 - 0.46 \cos\left(\dfrac{2\pi n}{N-1}\right)$

where n = point number, N (in this example) = 18

Figure 2.62 A downsampling filter using the Hamming window.

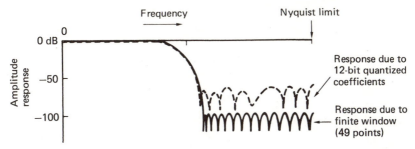

Figure 2.63 Frequency response of a 49-point transversal filter with infinite precision (solid line) shows ripple due to finite window size. Quantizing coefficients to twelve bits reduces attenuation in the stopband. (Responses courtesy *Philips Technical Review*.)

in terms of a fraction of the fundamental interval rather than in absolute terms. The configuration shown in Figure 2.60 serves to illustrate the principle. The units used on the diagrams are sample periods and the response is proportional to these periods or spacings, and so it is not necessary to use actual figures.

Where the impulse response is symmetrical, it is often possible to reduce the number of multiplications, because the same product can be used twice, at equal distances before and after the centre of the window. This is known as folding the filter. A folded filter is shown in Figure 2.64.

2.21 Sampling-rate conversion

Sampling-rate conversion or interpolation is an important enabling technology on which a large number of practical digital video devices are based. In digital video, the sampling rate takes on many guises. When analog video is sampled in real time, the sampling rate is temporal, but where pixels form a static array, the sampling rate is a spatial frequency.

Some of the applications of interpolation are set out here:

1 Video standards convertors need to change two of the sampling rates of the signal they handle, namely the temporal frame rate and the vertical line spacing, which is in fact a spatial sampling frequency. In some low-bit rate video applications such as Internet video, the frame rate may deliberately be reduced. The display will have to increase it again to avoid flicker.
2 To take advantage of oversampling convertors, an increase in sampling rate is necessary for DACs and a reduction in sampling rate is necessary for ADCs. In oversampling the factors by which the rates are changed are simpler than in other applications.
3 In image processing, a large number of different standard pixel array sizes exists. Changing between these formats may be necessary in order to view an incoming image on an avilable display. This technique is generally known as *resizing* and is essentially a two-dimensional sampling rate conversion. The rate in this case is the spatial frequency of the pixels.

There are three basic but related categories of rate conversion, as shown in Figure 2.65. The most straightforward (a) changes the rate by an integer ratio, up

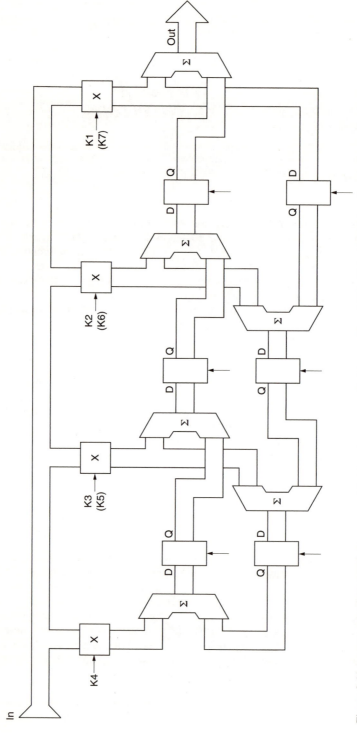

Figure 2.64 A seven-point folded filter for a symmetrical impulse response. In this case K1 and K7 will be identical, and so the input sample can be multiplied once, and the product fed into the output shift system in two different places. The centre coefficient K4 appears once. In an even-numbered filter the centre coefficient would also be used twice.

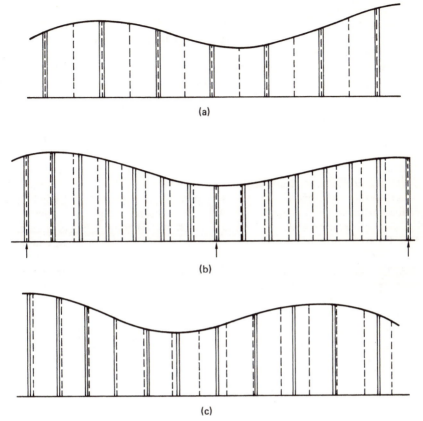

Figure 2.65 Categories of rate conversion. (a) Integer-ratio conversion, where the lower-rate samples are always coincident with those of the higher rate. There are a small number of phases needed. (b) Fractional-ratio conversion, where sample coincidence is periodic. A larger number of phases is required. Example here is conversion from 50.4 kHz to 44.1 kHz (8/7). (c) Variable-ratio conversion, where there is no fixed relationship, and a large number of phases are required.

or down. The timing of the system is thus simplified because all samples (input and output) are present on edges of the higher-rate sampling clock. Such a system is generally adopted for oversampling convertors; the exact sampling rate immediately adjacent to the analog domain is not critical, and will be chosen to make the filters easier to implement.

Next in order of difficulty is the category shown at (b) where the rate is changed by the ratio of two small integers. Samples in the input periodically time-align with the output. Such devices can be used for converting between the various rates of ITU–601.

The most complex rate-conversion category is where there is no simple relationship between input and output sampling rates, and in fact they may vary. This situation, shown at (c), is known as variable-ratio conversion. The temporal or spatial relationship of input and output samples is arbitrary. This problem will be met in effects machines which zoom or rotate images.

The technique of integer-ratio conversion is used in conjunction with oversampling convertors in digital video and audio and in motion-estimation and compression systems where sub-sampled or reduced resolution versions of an input image are required.

In considering how interpolators work it should be recalled that all sampled systems have finite bandwidth and need a reconstruction filter to remove the frequencies above the baseband due to sampling. After reconstruction, one infinitely short digital sample ideally represents a $\sin x/x$ pulse whose central peak width is determined by the response of the reconstruction filter, and whose amplitude is proportional to the sample value. This implies that, in reality, one sample value has meaning over a considerable timespan, rather than just at the sample instant. This will be detailed in Chapter 3. Were this not true, it would be impossible to build an interpolator.

Performing the steps of rate increase separately is inefficient. The bandwidth of the information is unchanged when the sampling rate is increased; therefore the original input samples will pass through the filter unchanged, and it is superfluous to compute them. The combination of the two processes into an interpolating filter minimizes the amount of computation.

As the purpose of the system is purely to increase the sampling rate, the filter must be as transparent as possible, and this implies that a linear-phase configuration is mandatory, suggesting the use of an FIR structure. Figure 2.66 shows that the theoretical impulse response of such a filter is a $\sin x/x$ curve which has zero value at the position of adjacent input samples. In practice this impulse cannot be implemented because it is infinite. The impulse response used will be truncated and windowed as described earlier. To simplify this discussion, assume that a $\sin x/x$ impulse is to be used. There is a strong parallel with the operation

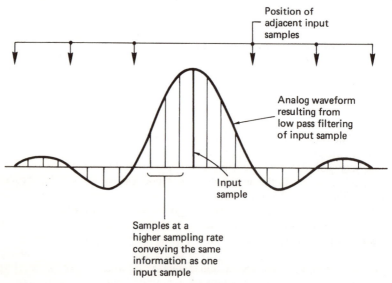

Figure 2.66 A single sample results in a $\sin x/x$ waveform after filtering in the analog domain. At a new, higher, sampling rate, the same waveform after filtering will be obtained if the numerous samples of differing size shown here are used. It follows that the values of these new samples can be calculated from the input samples in the digital domain in an FIR filter.

of a DAC where the analog voltage is returned to the time-continuous state by summing the analog impulses due to each sample. In a digital interpolating filter, this process is duplicated.[4]

If the sampling rate is to be doubled, new samples must be interpolated exactly half-way between existing samples. The necessary impulse response is shown in Figure 2.67; it can be sampled at the *output* sample period and quantized to form

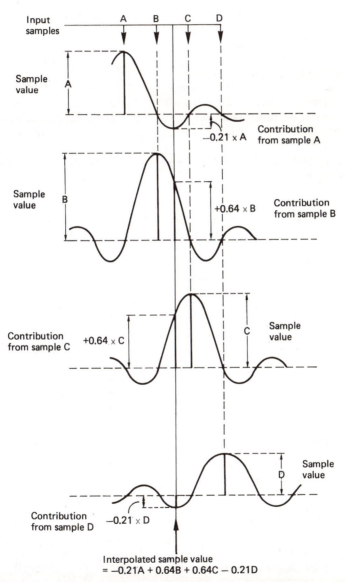

Figure 2.67 A two times oversampling interpolator. To compute an intermediate sample, the input samples are imagined to be sin*x*/*x* impulses, and the contributions from each at the point of interest can be calculated. In practice, rather more samples on either side need to be taken into account.

coefficients. If a single input sample is multiplied by each of these coefficients in turn, the impulse response of that sample at the new sampling rate will be obtained. Note that every other coefficient is zero, which confirms that no computation is necessary on the existing samples; they are just transferred to the output. The intermediate sample is then computed by adding together the impulse responses of every input sample in the window. The figure shows how this mechanism operates. If the sampling rate is to be increased by a factor of four, three sample values must be interpolated between existing input samples. It is then necessary to sample the impulse response at one-quarter the period of input samples to obtain three sets of coefficients which will be used in turn. In hardware-implemented filters, the input sample which is passed straight to the output is transferred by using a fourth filter phase where all coefficients are zero except the central one, which is unity.

Fractional ratio conversion allows interchange between different images having different pixel array sizes. Fractional ratios also occur in the vertical axis of standards convertors. Figure 2.65 showed that when the two sampling rates have a simple fractional relationship m/n, there is a periodicity in the relationship between samples in the two streams. It is possible to have a system clock running at the least-common multiple frequency which will divide by different integers to give each sampling rate.[5]

In a variable-ratio interpolator, values will exist for the points at which input samples were made, but it is necessary to compute what the sample values would have been at absolutely any point between available samples. The general concept of the interpolator is the same as for the fractional-ratio convertor, except that an infinite number of filter phases is ideally necessary. Since a realizable filter will have a finite number of phases, it is necessary to study the degradation this causes. The desired continuous temporal or spatial axis of the interpolator is quantized by the phase spacing, and a sample value needed at a particular point will be replaced by a value for the nearest available filter phase. The number of phases in the filter therefore determines the accuracy of the interpolation. The effects of calculating a value for the wrong point are identical to those of sampling with clock jitter, in that an error occurs proportional to the slope of the signal. The result is program-modulated noise. The higher the noise specification, the greater the desired time accuracy and the greater the number of phases required. The number of phases is equal to the number of sets of coefficients available, and should not be confused with the number of points in the filter, which is equal to the number of coefficients in a set (and the number of multiplications needed to calculate one output value).

The sampling jitter accuracy necessary for eight-bit working is measured in picoseconds. This implies that something like 32 filter phases will be required for adequate performance in an eight-bit sampling-rate convertor.

2.22 Transforms and duality

The duality of transforms provides an interesting insight into what is happening in common processes. Fourier analysis holds that any periodic waveform can be reproduced by adding together an arbitrary number of harmonically related sinusoids of various amplitudes and phases. Figure 2.68 shows how a square wave can be built up of harmonics. The spectrum can be drawn by plotting the amplitude of the harmonics against frequency. It will be seen that this gives a

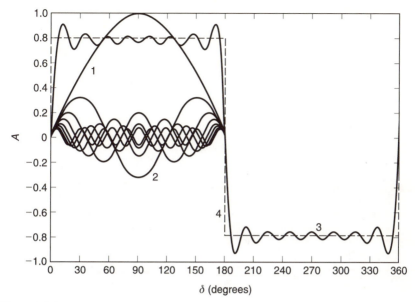

Figure 2.68 Fourier analysis of a square wave into fundamental and harmonics. A, amplitude; δ, phase of fundamental wave in degrees; 1, first harmonic (fundamental); 2, odd harmonics 3–15; 3, sum of harmonics 1–15; 4, ideal square wave.

spectrum which is a decaying wave. It passes through zero at all even multiples of the fundamental. The shape of the spectrum is a sin$x$/x curve. If a square wave has a sin$x$/x spectrum, it follows that a filter with a rectangular impulse response will have a sin$x$/x spectrum.

A low-pass filter has a rectangular spectrum, and this has a sin$x$/x impulse response. These characteristics are known as a transform pair. In transform pairs, if one domain has one shape of the pair, the other domain will have the other shape. Figure 2.69 shows a number of transform pairs.

At (a) a square wave has a sin$x$/x spectrum and a sin$x$/x impulse has a square spectrum. In general the product of equivalent parameters on either side of a transform remains constant, so that if one increases, the other must fall. If (a) shows a filter with a wider bandwidth, having a narrow impulse response, then (b) shows a filter of narrower bandwidth which has a wide impulse response. This is duality in action. The limiting case of this behaviour is where one parameter becomes zero, the other goes to infinity. At (c) a time-domain pulse of infinitely short duration has a flat spectrum. Thus a flat waveform, i.e. DC, has only zero in its spectrum. The impulse response of the optics of a laser disk (d) has a $\sin^2 x/x^2$ intensity function, and this is responsible for the triangular falling frequency response of the pickup. The lens is a rectangular aperture, but as there is no such thing as negative light, a sin$x$/x impulse response is impossible. The squaring process is consistent with a positive-only impulse response. Interestingly the transform of a Gaussian response in still Gaussian.

Duality also holds for sampled systems. A sampling process is periodic in the time domain. This results in a spectrum which is periodic in the frequency domain. If the time between the samples is reduced, the bandwidth of the system

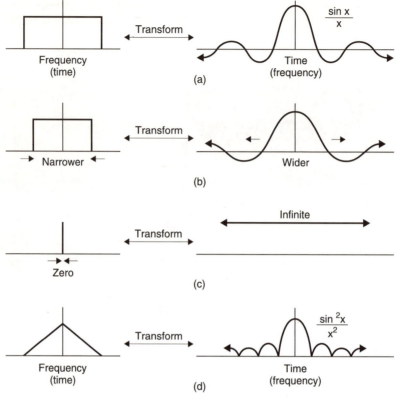

Figure 2.69 Transform pairs. At (a) the dual of a rectangle is a sin$x$/x function. If one is time domain, the other is frequency domain. At (b), narrowing one domain widens the other. The limiting case of this is (c). Transform of the sin$x$/x squared function is triangular (d).

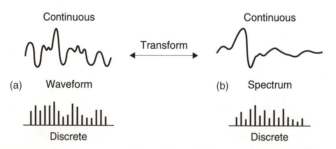

Figure 2.70 Continuous time signal (a) has continuous spectrum. Discrete time signal (b) has discrete spectrum.

rises. Figure 2.70(a) shows that a continuous time signal has a continuous spectrum whereas at (b) the frequency transform of a sampled signal is also discrete. In other words sampled signals can only be analysed into a finite number of frequencies. The more accurate the frequency analysis has to be, the more samples are needed in the block. Making the block longer reduces the ability to locate a transient in time. This is the Heisenberg inequality, which is the limiting case of duality, because when infinite accuracy is achieved in one domain, there is no accuracy at all in the other.

2.23 The Fourier transform

Figure 2.68 showed that if the amplitude and phase of each frequency component is known, linearly adding the resultant components in an inverse transform results in the original waveform. In digital systems the waveform is expressed as a number of discrete samples. As a result the Fourier transform analyses the signal into an equal number of discrete frequencies. This is known as a discrete Fourier transform or DFT in which the number of frequency coefficients is equal to the number of input samples. The fast Fourier transform is no more than an efficient way of computing the DFT.[6]

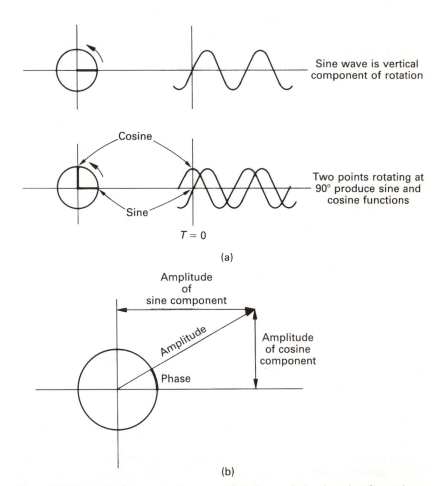

Figure 2.71 The origin of sine and cosine waves is to take a particular viewpoint of a rotation. Any phase can be synthesized by adding proportions of sine and cosine waves.

It will be evident from Figure 2.68 that the knowledge of the phase of the frequency component is vital, as changing the phase of any component will seriously alter the reconstructed waveform. Thus the DFT must accurately analyse the phase of the signal components.

There are a number of ways of expressing phase. Figure 2.71 shows a point which is rotating about a fixed axis at constant speed. Looked at from the side, the point oscillates up and down at constant frequency. The waveform of that motion is a sine wave, and that is what we would see if the rotating point were to translate along its axis whilst we continued to look from the side.

One way of defining the phase of a waveform is to specify the angle through which the point has rotated at time zero $(T = 0)$. If a second point is made to revolve at 90° to the first, it would produce a cosine wave when translated. It is possible to produce a waveform having arbitrary phase by adding together the sine and cosine wave in various proportions and polarities. For example, adding the sine and cosine waves in equal proportion results in a waveform lagging the sine wave by 45°.

Figure 2.71 shows that the proportions necessary are respectively the sine and the cosine of the phase angle. Thus the two methods of describing phase can be readily interchanged.

The discrete Fourier transform spectrum-analyses a string of samples by searching separately for each discrete target frequency. It does this by multiplying the input waveform by a sine wave, known as the basis function, having the target frequency and adding up or integrating the products. Figure 2.72(a) shows that multiplying by basis functions gives a non-zero integral when the input frequency is the same, whereas (b) shows that with a different input frequency (in fact all other different frequencies) the integral is zero showing that no component of the target frequency exists. Thus from a real waveform containing many frequencies all frequencies except the target frequency are excluded. The magnitude of the integral is proportional to the amplitude of the target component.

Figure 2.72(c) shows that the target frequency will not be detected if it is phase shifted 90° as the product of quadrature waveforms is always zero. Thus the discrete Fourier transform must make a further search for the target frequency using a cosine basis function. It follows from the arguments above that the relative proportions of the sine and cosine integrals reveal the phase of the input component. Thus each discrete frequency in the spectrum must be the result of a pair of quadrature searches.

Searching for one frequency at a time as above will result in a DFT, but only after considerable computation. However, a lot of the calculations are repeated many times over in different searches. The fast Fourier transform gives the same result with less computation by logically gathering together all the places where the same calculation is needed and making the calculation once.

2.24 The discrete cosine transform (DCT)

The DCT is a special case of a discrete Fourier transform in which the sine components of the coefficients have been eliminated leaving a single number. This is actually quite easy. Figure 2.73(a) shows a block of input samples to a transform process. By repeating the samples in a time-reversed order and performing a discrete Fourier transform on the double-length sample set a DCT

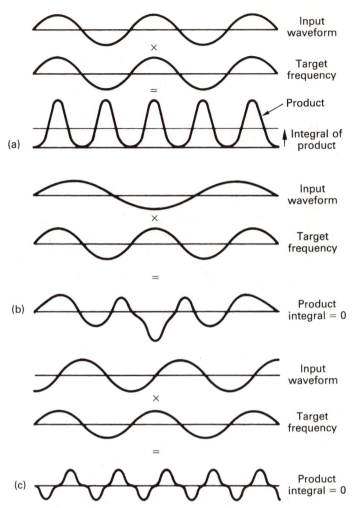

Figure 2.72 The input waveform is multiplied by the target frequency and the result is averaged or integrated. At (a) the target frequency is present and a large integral results. With another input frequency the integral is zero as at (b). The correct frequency will also result in a zero integral shown at (c) if it is at 90° to the phase of the search frequency. This is overcome by making two searches in quadrature.

is obtained. The effect of mirroring the input waveform is to turn it into an even function whose sine coefficients are all zero. The result can be understood by considering the effect of individually transforming the input block and the reversed block.

Figure 2.73(b) shows that the phase of all the components of one block are in the opposite sense to those in the other. This means that when the components are added to give the transform of the double length block all the sine components cancel out, leaving only the cosine coefficients, hence the name of the transform.[6] In practice the sine component calculation is eliminated. Another advantage is that doubling the block length by mirroring doubles the frequency

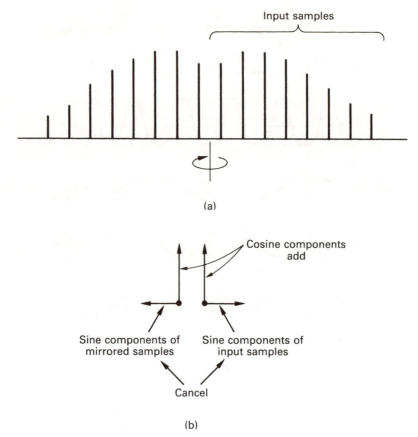

Figure 2.73 The DCT is obtained by mirroring the input block as shown at (a) prior to an FFT. The mirroring cancels out the sine components as at (b), leaving only cosine coefficients.

resolution, so that twice as many useful coefficients are produced. In fact a DCT produces as many useful coefficients as input samples.

For image processing two-dimensional transforms are needed. In this case for every horizontal frequency, a search is made for all possible vertical frequencies. A two-dimensional DCT is shown in Figure 2.74. The DCT is separable in that the two-dimensional DCT can be obtained by computing in each dimension separately. Fast DCT algorithms are available.[7]

Figure 2.75 shows how a two-dimensional DCT is calculated by multiplying each pixel in the input block by terms which represent sampled cosine waves of various spatial frequencies. A given DCT coefficient is obtained when the result of multiplying every input pixel in the block is summed. Although most compression systems, including JPEG and MPEG, use square DCT blocks, this is not a necessity and rectangular DCT blocks are possible and are used in, for example, Digital Betacam and DVC.

The DCT is primarily used in MPEG-2 because it converts the input waveform into a form where redundancy can be easily detected and removed. More details of the DCT can be found in Chapter 5.

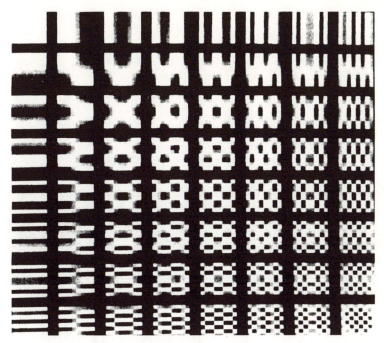

Figure 2.74 The discrete cosine transform breaks up an image area into discrete frequencies in two dimensions. The lowest frequency can be seen here at the top left corner. Horizontal frequency increases to the right and vertical frequency increases downwards.

2.25 Modulo-*n* arithmetic

Conventional arithmetic which is in everyday use relates to the real world of counting actual objects, and to obtain correct answers the concepts of borrow and carry are necessary in the calculations.

There is an alternative type of arithmetic which has no borrow or carry which is known as modulo arithmetic. In modulo-*n* no number can exceed *n*. If it does, *n* or whole multiples of *n* are subtracted until it does not. Thus 25 *mod*ulo-16 is 9 and 12 modulo-5 is 2. The count shown in Figure 2.35 is from a four-bit device which overflows when it reaches 1111 because the carry out is ignored. If a number of clock pulses *m* are applied from the zero state, the state of the counter will be given by *m* Mod.16. Thus modulo arithmetic is appropriate to systems in which there is a fixed wordlength and this means that the range of values the system can have is restricted by that wordlength. A number range which is restricted in this way is called a finite field.

Modulo-2 is a numbering scheme which is used frequently in digital processes. Figure 2.76 shows that in modulo-2 the conventional addition and subtraction are replaced by the XOR function such that:

A + B Mod.2 = A XOR B

When multi-bit values are added Mod.2, each column is computed quite independently of any other. This makes Mod.2 circuitry very fast in operation as

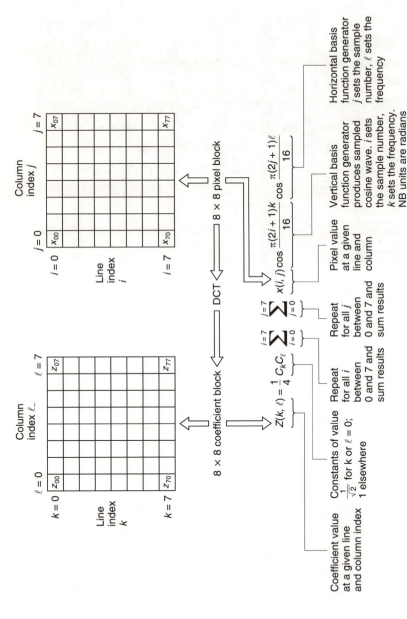

Figure 2.75 A two-dimensional DCT is calculated as shown here. Starting with an input pixel block one calculation is necessary to find a value for each coefficient. After 64 calculations using different basis functions the coefficient block is complete.

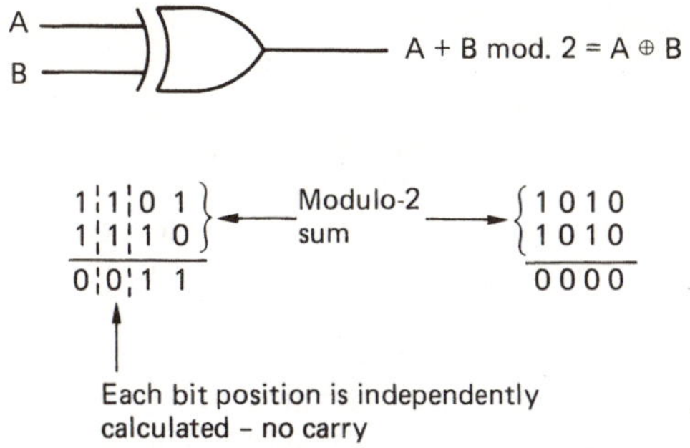

$$A + B \text{ mod. } 2 = A \oplus B$$

$$
\begin{array}{l}
1\,\vert 1\,\vert 0\ \ 1 \\
1\,\vert 1\,\vert 1\ \ 0 \\
\hline
0\,\vert 0\,\vert 1\ \ 1
\end{array}
\left.\vphantom{\begin{array}{l}1\\1\end{array}}\right\}
\xleftarrow{\ \ \begin{array}{c}\text{Modulo-2}\\ \text{sum}\end{array}\ \ }
\left\{
\begin{array}{l}
1\ 0\ 1\ 0 \\
1\ 0\ 1\ 0 \\
\hline
0\ 0\ 0\ 0
\end{array}
\right.
$$

Each bit position is independently
calculated – no carry

Figure 2.76 In modulo-2 calculations, there can be no carry or borrow operations and conventional addition and subtraction become identical. The XOR gate is a modulo-2 adder.

it is not necessary to wait for the carries from lower-order bits to ripple up to the high-order bits.

Modulo-2 arithmetic is not the same as conventional arithmetic and takes some getting used to. For example, adding something to itself in Mod.2 always gives the answer zero.

2.26 The Galois field

Figure 2.77 shows a simple circuit consisting of three D-type latches which are clocked simultaneously. They are connected in series to form a shift register. At (a) a feedback connection has been taken from the output to the input and the

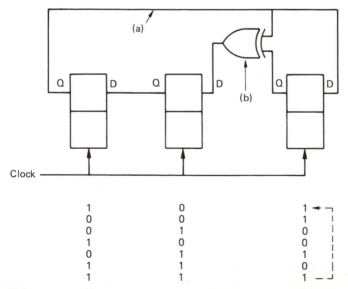

Figure 2.77 The circuit shown is a twisted-ring counter which has an unusual feedback arrangement. Clocking the counter causes it to pass through a series of non-sequential values. See text for details.

result is a ring counter where the bits contained will recirculate endlessly. At (b) one XOR gate is added so that the output is fed back to more than one stage. The result is known as a twisted-ring counter and it has some interesting properties. Whenever the circuit is clocked, the left-hand bit moves to the right-hand latch, the centre bit moves to the left-hand latch and the centre latch becomes the XOR of the two outer latches. The figure shows that whatever the starting condition of the three bits in the latches, the same state will always be reached again after seven clocks, except if zero is used. The states of the latches form an endless ring of non-sequential numbers called a Galois field after the French mathematical prodigy Evariste Galois who discovered them. The states of the circuit form a maximum length sequence because there are as many states as are permitted by the wordlength. As the states of the sequence have many of the characteristics of random numbers, yet are repeatable, the result can also be called a pseudo-random sequence (prs). As the all-zeros case is disallowed, the length of a maximum length sequence generated by a register of m bits cannot exceed $(2m^{-1})$ states. The Galois field, however, includes the zero term. It is useful to explore the bizarre mathematics of Galois fields which use modulo-2 arithmetic. Familiarity with such manipulations is helpful when studying the error correction, particularly the Reed–Solomon codes used in recorders and treated in Chapter 6. They will also be found in processes which require pseudo-random numbers such as digital dither, considered in Chapter 3, and randomized channel codes used in, for example, DVB and discussed in Chapter 9.

The circuit of Figure 2.77 can be considered as a counter and the four points shown will then be representing different powers of 2 from the MSB on the left to the LSB on the right. The feedback connection from the MSB to the other stages means that whenever the MSB becomes 1, two other powers are also forced to one so that the code of 1011 is generated.

Each state of the circuit can be described by combinations of powers of x, such as

$$x^2 = 100$$

$$x = 010$$

$$x^2 + x = 110, \text{ etc.}$$

The fact that three bits have the same state because they are connected together is represented by the Mod.2 equation:

$$x^3 + x + 1 = 0$$

Let $x = a$, which is a primitive element. Now

$$a^3 + a + 1 = 0 \qquad\qquad (2.1)$$

In modulo-2

$$a + a = a^2 + a^2 = 0$$

$$a = x = 010$$

$$a^2 = x^2 = 100$$

$a^3 = a + 1 = 011$ from (2.1)

$a^4 = a \times a^3 = a(a + 1) = a^2 + a = 110$

$a^5 = a^2 + a + 1 = 111$

$a^6 = a \times a^5 = a(a^2 + a + 1)$

$= a^3 + a^2 + a = a + 1 + a^2 + a$

$= a^2 + 1 = 101$

$a^7 = a(a^2 + 1) = a^3 + a$

$= a + 1 + a = 1 = 001$

In this way it can be seen that the complete set of elements of the Galois field can be expressed by successive powers of the primitive element. Note that the twisted-ring circuit of Figure 2.77 simply raises a to higher and higher powers as it is clocked. Thus the seemingly complex multibit changes caused by a single clock of the register become simple to calculate using the correct primitive and the appropriate power.

The numbers produced by the twisted-ring counter are not random; they are completely predictable if the equation is known. However, the sequences produced are sufficiently similar to random numbers that in many cases they will be useful. They are thus referred to as pseudo-random sequences. The feedback connection is chosen such that the expression it implements will not factorize. Otherwise a maximum-length sequence could not be generated because the circuit might sequence around one or other of the factors depending on the initial condition. A useful analogy is to compare the operation of a pair of meshed gears. If the gears have a number of teeth which is relatively prime, many revolutions are necessary to make the same pair of teeth touch again. If the number of teeth have a common multiple, far fewer turns are needed.

2.27 The phase-locked loop

All digital video systems need to be clocked at the appropriate rate in order to function properly. Whilst a clock may be obtained from a fixed-frequency oscillator such as a crystal, many operations in video require *genlocking* or synchronizing the clock to an external source. The phase-locked loop excels at this job, and many others, particularly in connection with recording and transmission.

In phase-locked loops, the oscillator can run at a range of frequencies according to the voltage applied to a control terminal. This is called a voltage-controlled oscillator or VCO. Figure 2.78 shows that the VCO is driven by a phase error measured between the output and some reference. The error changes the control voltage in such a way that the error is reduced, such that the output eventually has the same frequency as the reference. A low-pass filter is fitted in the control voltage path to prevent the loop becoming unstable. If a divider is placed between the VCO and the phase comparator, as in the figure, the VCO frequency can be made to be a multiple of the reference. This also has the effect of making the loop more heavily damped, so that it is less likely to change frequency if the input is irregular.

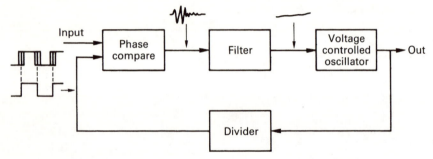

Figure 2.78 A phase-locked loop requires these components as a minimum. The filter in the control voltage serves to reduce clock jitter.

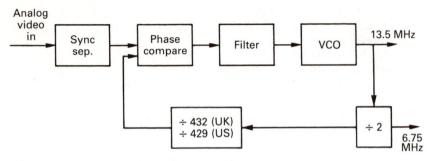

Figure 2.79 In order to obtain 13.5 MHz from input syncs, a PLL with an appropriate division ratio is required.

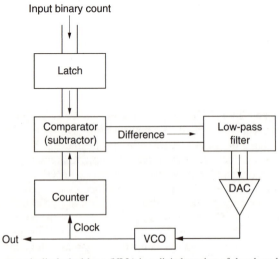

Figure 2.80 The numerically locked loop (NLL) is a digital version of the phase-locked loop.

In digital video, the frequency multiplication of a phase-locked loop is extremely useful. Figure 2.79 shows how the 13.5 MHz clock of component digital video is obtained from the sync pulses of an analog reference by such a multiplication process.

Figure 2.80 shows the NLL or numerically locked loop. This is similar to a phase-locked loop, except that the two phases concerned are represented by the state of a binary number. The NLL is useful to generate a remote clock from a master. The state of a clock count in the master is periodically transmitted to the NLL which will re-create the same clock frequency. The technique is used in MPEG transport streams.

References

1. Ray, S.F., *Applied Photographic Optics*, Chapter 17, Oxford: Focal Press (1988)
2. van den Enden, A.W.M. and Verhoeckx, N.A.M., Digital signal processing: theoretical background. *Philips Tech. Rev.* **42**, 110–144, (1985)
3. McClellan, J.H., Parks, T.W. and Rabiner, L.R., A computer program for designing optimum FIR linear-phase digital filters. *IEEE Trans. Audio and Electroacoustics*, **AU–21**, 506–526 (1973)
4. Crochiere, R.E. and Rabiner, L.R., Interpolation and decimation of digital signals – a tutorial review. *Proc. IEEE*, **69**, 300–331 (1981)
5. Rabiner, L.R., Digital techniques for changing the sampling rate of a signal. In B. Blesser, B. Locanthi and T.G. Stockham Jr (eds), *Digitic Audio*, pp. 79–89, New York: Audio Engineering Society (1982)
6. Kraniauskas, P., *Transforms in Signals and Systems*, Chapter 6, Wokingham: Addison-Wesley (1992)
7. Ahmed, N., Natarajan, T. and Rao, K., Discrete Cosine Transform. *IEEE Trans. Computers*, **C–23**, 90–93 (1974)

Chapter 3

Conversion

3.1 Introduction to conversion

The most useful and therefore common signal representation is pulse code modulation or PCM which was introduced in Chapter 1. The input is a continuous-time, continuous-voltage video waveform, and this is converted into a discrete-time, discrete-voltage format by a combination of sampling and quantizing. As these two processes are independent they can be performed in either order. Figure 3.1(a) shows an analog sampler preceding a quantizer, whereas (b) shows an asynchronous quantizer preceding a digital sampler.

The independence of sampling and quantizing allows each to be discussed quite separately in some detail, prior to combining the processes for a full understanding of conversion.

3.2 Sampling and aliasing

Sampling can take place in space or time and in several dimensions at once. Figure 3.2 shows that in *temporal sampling* the frequency of the signal to be sampled and the sampling rate F_s are measured in Hertz (Hz). In still images there is no temporal change and Figure 3.2 also shows that the sampling is *spatial*. The sampling rate is now a spatial frequency. The absolute unit of spatial frequency is cycles per metre, although for imaging purposes cycles-per-mm is more practical. Spatial and temporal frequencies are related by the process of scanning as given by:

Temporal frequency = Spatial frequency × scanning velocity

Figure 3.3 shows that if the 1024 pixels along one line of an SVGA monitor were scanned in one tenth of a millisecond, the sampling clock frequency would be 10.24 MHz.

The sampling process originates with a pulse train which is shown in Figure 3.4(a) to be of constant amplitude and period. This pulse train can be temporal or spatial. The information to be sampled amplitude-modulates the pulse train in much the same way as the carrier is modulated in an AM radio transmitter. It is important to avoid overmodulating the pulse train as shown in (b) and this is achieved by suitably biasing the information waveform as at (c).

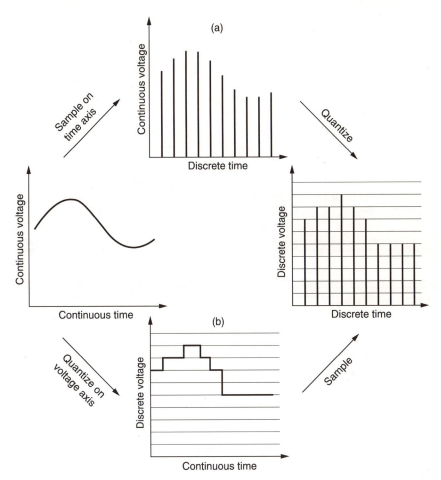

Figure 3.1 Since sampling and quantizing are orthogonal, the order in which they are performed is not important. In (a) sampling is performed first and the samples are quantized. This is common in audio convertors. In (b) the analog input is quantized into an asynchronous binary code. Sampling takes place when this code is latched on sampling clock edges. This approach is universal in video convertors.

In the same way that AM radio produces sidebands or identical images above and below the carrier, sampling also produces sidebands although the carrier is now a pulse train and has an infinite series of harmonics as shown in Figure 3.5(a). The sidebands repeat above and below each harmonic of the sampling rate as shown in (b). As the spectrum of the baseband signal is simply repeated, sampling need not lose any information.

The sampled signal can be returned to the continuous domain simply by passing it into a low-pass filter which prevents the images from passing, hence the term 'anti-image filter'. If considered in the time domain it can be called a reconstruction filter. It can also be considered as a spatial filter if a sampled still image is being returned to a continuous image. Such a filter will be two-dimensional.

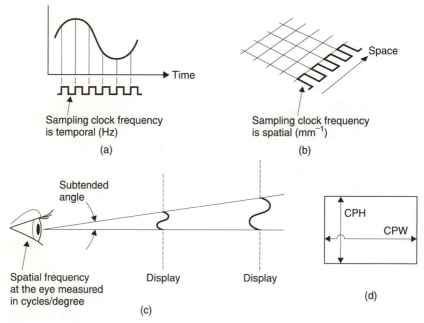

Figure 3.2 (a) Electrical waveforms are sampled temporally at a sampling rate measured in Hz. (b) Image information must be sampled spatially, but there is no single unit of spatial sampling frequency. (c) The acuity of the eye is measured as a subtended angle, and here two different displays of different resolutions give the same result at the eye because they are at a different distance. (d) Size-independent units such as cycles per picture height will also be found.

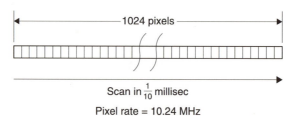

Figure 3.3 The connection between image resolution and pixel rate is the scanning speed. Scanning the above line in 1/10 ms produces a pixel rate of 10.24 MHz.

If the input has excessive bandwidth, the sidebands will overlap (Figure 3.5(c)) and the result is aliasing, where certain output frequencies are not the same as their input frequencies but instead become difference frequencies (Figure 3.5(d)). It will be seen that aliasing does not occur when the input bandwidth is equal to or less than half the sampling rate, and this derives the most fundamental rule of sampling, which is that the sampling rate must be at least twice the input bandwidth.

Nyquist[1] is generally credited with being the first to point this out (1928), although the mathematical proofs were given independently by Shannon[2,3] and Kotelnikov. It subsequently transpired that Whittaker[4] beat them all to it, although his work was not widely known at the time. One half of the sampling frequency is often called the Nyquist frequency.

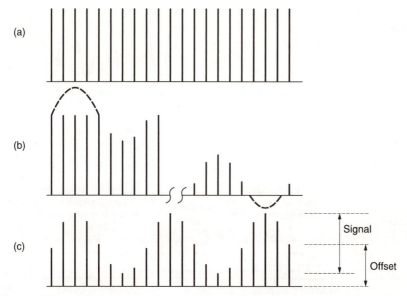

Figure 3.4 The sampling process requires a constant-amplitude pulse train as shown in (a). This is amplitude modulated by the waveform to be sampled. If the input waveform has excessive amplitude or incorrect level, the pulse train clips as shown in (b). For a bipolar waveform, the greatest signal level is possible when an offset of half the pulse amplitude is used to centre the waveform as shown in (c).

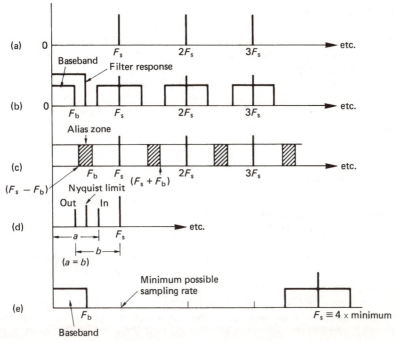

Figure 3.5 (a) Spectrum of sampling pulses. (b) Spectrum of samples. (c) Aliasing due to sideband overlap. (d) Beat-frequency production. (e) 4 × oversampling.

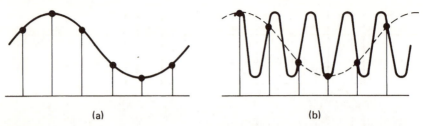

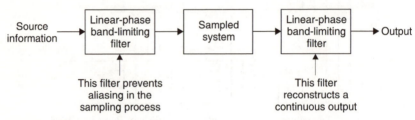

(a)   (b)

Figure 3.6 In (a), the sampling is adequate to reconstruct the original signal. In (b) the sampling rate is inadequate, and reconstruction produces the wrong waveform (dotted). Aliasing has taken place.

```
Source          Linear-phase         Sampled          Linear-phase
information  →  band-limiting  →     system      →    band-limiting  →  Output
                filter                                filter

                This filter prevents                 This filter
                aliasing in the                       reconstructs a
                sampling process                      continuous output
```

Figure 3.7 Sampling systems depend completely on the use of band limiting filters before and after the sampling stage. Implementing these filters rigorously is non-trivial.

Aliasing can be described equally well in the time domain. In Figure 3.6(a) the sampling rate is obviously adequate to describe the waveform, but at (b) it is inadequate and aliasing has occurred. Where there is no control over the spectrum of input signals it becomes necessary to have a low-pass filter at the input which prevents frequencies of more than half the sampling rate from reaching the sampling stage.

Figure 3.7 shows that all practical sampling systems consist of a pair of filters, the anti-aliasing filter before the sampling process and the reconstruction filter after it. It should be clear that the results obtained will be strongly affected by the quality of these filters which may be spatial or temporal according to the application.

3.3 Reconstruction

Perfect reconstruction was theoretically demonstrated by Shannon as shown in Figure 3.8. The input must be band limited by an ideal linear-phase low-pass filter with a rectangular frequency response and a bandwidth of one-half the sampling frequency. The samples must be taken at an instant with no averaging of the waveform. These instantaneous samples can then be passed through a second, identical filter which will perfectly reconstruct that part of the input waveform which was within the passband.

It was shown in Chapter 2 that the impulse response of a linear-phase ideal low-pass filter is a $\sin x/x$ waveform, and this is repeated in Figure 3.9(a). Such a waveform passes through zero volts periodically. If the cut-off frequency of the filter is one-half of the sampling rate, the impulse passes through zero *at the sites of all other samples*. It can be seen from Figure 3.9(b) that at the output of such

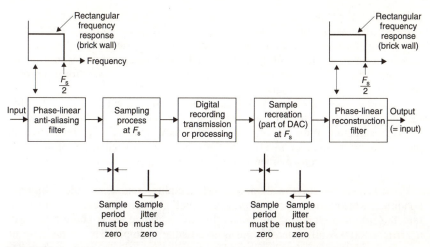

Figure 3.8 Shannon's concept of perfect reconstruction requires the hypothetical approach shown here. The anti-aliasing and reconstruction filters must have linear phase and rectangular frequency response. The sample period must be infinitely short and the sample clock must be perfectly regular. Then the output and input waveforms will be identical if the sampling frequency is twice the input bandwidth (or more).

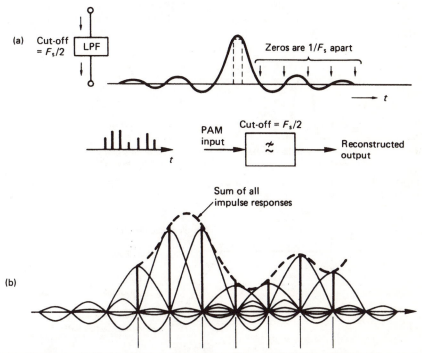

Figure 3.9 An ideal low-pass filter has an impulse response shown in (a). The impulse passes through zero at intervals equal to the sampling period. When convolved with a pulse train at the sampling rate, as shown in (b), the voltage at each sample instant is due to that sample alone as the impulses from all other samples pass through zero there.

a filter, the voltage at the centre of a sample is due to that sample alone, since the value of *all* other samples is zero at that instant. In other words the continuous output waveform must pass through the tops of the input samples. In between the sample instants, the output of the filter is the sum of the contributions from many impulses (theoretically an infinite number), causing the waveform to pass smoothly from sample to sample.

It is a consequence of the band-limiting of the original anti-aliasing filter that the filtered analog waveform could only take one path between the samples. As the reconstruction filter has the same frequency response, the reconstructed output waveform must be identical to the original band-limited waveform prior to sampling. A rigorous mathematical proof of reconstruction can be found in Porat[5] or Betts.[6]

Perfect reconstruction with a Nyquist sampling rate is an ideal limiting condition which can only be approached, but it forms a useful performance target. Zero duration pulses are impossible and the ideal linear-phase filter with a vertical 'brick-wall' cut-off slope is impossible to implement. In the case of temporal sampling, as the slope tends to vertical, the delay caused by the filter goes to infinity. In the case of spatial sampling, sharp cut optical filters are impossible to build. Figure 3.10 shows that the spatial impulse response of an ideal lens is a symmetrical intensity function. Note that the function is positive only as the expression for intensity contains a squaring process. The negative excursions of the sin$x$/x curve can be handled in an analog or digital filter by negative voltages or numbers, but in optics there is no negative light. The restriction to positive only impulse response limits the sharpness of optical filters.

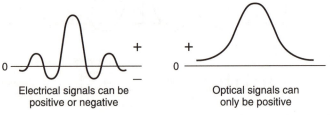

Electrical signals can be
positive or negative

Optical signals can
only be positive

Figure 3.10 In optical systems the spatial impulse response cannot have negative excursions and so ideal filters in optics are more difficult to make.

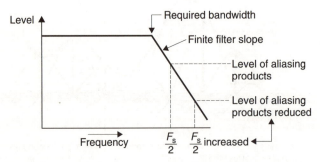

Figure 3.11 With finite slope filters, aliasing is always possible, but it can be set at an arbitrarily low level by raising the sampling rate.

In practice, real filters with finite slopes can still be used as shown in Figure 3.11. The cut-off slope begins at the edge of the required pass band, and because the slope is not vertical, aliasing will always occur, but the sampling rate can be raised to drive aliasing products to an arbitrarily low level. The perfect reconstruction process still works, but the system is a little less efficient in information terms because the sampling rate has to be raised. There is no absolute factor by which the sampling rate must be raised. A figure of 10 per cent is typical in temporal sampling, although it depends upon the filters which are available and the level of aliasing products that are acceptable.

3.4 Aperture effect

The reconstruction process of Figure 3.9 only operates exactly as shown if the impulses are of negligible duration. In many processes this is not the case, and many real devices keep the analog signal constant for a substantial part of or even the whole period. The result is a waveform which is more like a staircase than a pulse train. The case where the pulses have been extended in width to become equal to the sample period is known as a zero-order hold system and has a 100 per cent aperture ratio.

Whereas pulses of negligible width have a uniform spectrum, which is flat within the baseband, pulses of 100 per cent aperture ratio have a $\sin x/x$ spectrum which is shown in Figure 3.12. The frequency response falls to a null at the sampling rate, and as a result is about 4 dB down at the edge of the baseband. If the pulse width is stable, the reduction of high frequencies is constant and predictable, and an appropriate equalization circuit can render the overall response flat once more. An alternative is to use resampling which is shown in Figure 3.13. Resampling passes the zero-order hold waveform through a further synchronous sampling stage which consists of an analog switch that closes briefly in the centre of each sample period. The output of the switch will be pulses which are narrower than the original. If, for example, the aperture ratio is reduced to 50 per cent of the sample period, the first frequency response null is

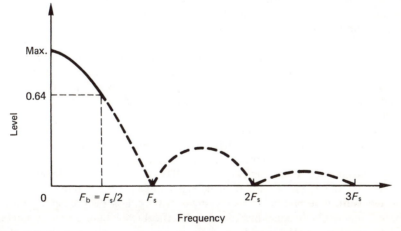

Figure 3.12 Frequency response with 100 per cent aperture nulls at multiples of sampling rate. Area of interest is up to half sampling rate.

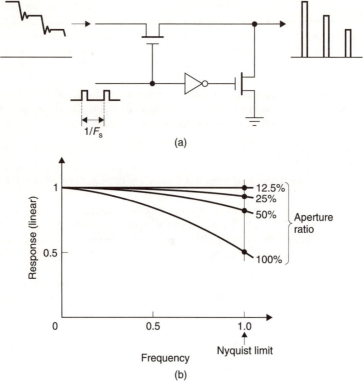

Figure 3.13 (a) Resampling circuit eliminates transients and reduces aperture ratio. (b) Response of various aperture ratios.

now at twice the sampling rate, and the loss at the edge of the audio band is reduced. As the figure shows, the frequency response becomes flatter as the aperture ratio falls. The process should not be carried too far, as with very small aperture ratios there is little energy in the pulses and noise can be a problem. A practical limit is around 12.5 per cent where the frequency response is virtually ideal.

The aperture effect will show up in many aspects of television. Lenses have finite MTF (modulation transfer function), such that a very small object becomes spread in the image. The image sensor will also have a finite aperture function. In tube cameras, the beam will have a finite radius, and will not necessarily have a uniform energy distribution across its diameter. In CCD cameras, the sensor is split into elements which may almost touch in some cases. The element integrates light falling on its surface, and so will have a rectangular aperture. In both cases there will be a roll-off of higher spatial frequencies.

In conventional tube cameras and CRTs the horizontal dimension is continuous, whereas the vertical dimension is sampled. The aperture effect means that the vertical resolution in real systems will be less than sampling theory permits, and to obtain equal horizontal and vertical resolutions a greater number of lines is necessary. The magnitude of the increase is described by the so-called Kell factor,[7] although the term 'factor' is a misnomer since it can have a range of values depending on the apertures in use and the methods used to measure resolution. In digital video, sampling takes place in horizontal and vertical dimensions, and the Kell parameter becomes unnecessary. The outputs of digital systems will, however, be displayed on raster scan CRTs, and the Kell parameter of the display will then be effectively in series with the other system constraints.

The temporal aperture effect varies according to the equipment used. Tube cameras have a long integration time and thus a wide temporal aperture. Whilst this reduces temporal aliasing, it causes smear on moving objects. CCD cameras do not suffer from lag and as a result their temporal response is better. Some CCD cameras deliberately have a short temporal aperture as the time axis is resampled by a shutter. The intention is to reduce smear, hence the popularity of such devices for sporting events, but there will be more aliasing on certain subjects.

The eye has a temporal aperture effect which is known as persistence of vision, and the phosphors of CRTs continue to emit light after the electron beam has passed. These produce further temporal aperture effects in series with those in the camera.

Current liquid crystal displays do not generate light, but act as a modulator to a separate light source. Their temporal response is rather slow, but there is a possibility to resample by pulsing the light source.

3.5 Two-dimensional sampling

Analog video samples in the time domain and vertically, whereas a two-dimensional still image such as a photograph must be sampled horizontally and vertically. In both cases a two-dimensional spectrum will result, one vertical/ temporal and one vertical/horizontal.

Figure 3.14(a) shows a square matrix of sampling sites which has an identical spatial sampling frequency both vertically and horizontally. The corresponding spectrum is shown in (b). The baseband spectrum is in the centre of the diagram, and the repeating sampling sideband spectrum extends vertically and horizontally. The star-shaped spectrum results from viewing an image of a man-made object such as a building containing primarily horizontal and vertical elements. A more natural scene such as foliage would result in a more circular or elliptical spectrum.

In order to return to the baseband image, the sidebands must be filtered out with a two-dimensional spatial filter. The shape of the two-dimensional frequency response shown in Figure 3.14(c) is known as a Brillouin zone.

Figure 3.14(d) shows an alternative sampling site matrix known as quincunx sampling because of the similarity to the pattern of five dots on a dice. The resultant spectrum has the same characteristic pattern as shown in (e). The corresponding Brillouin zones are shown in (f). Quincunx sampling offers a better compromise between diagonal and horizontal/vertical resolution but is complex to implement.

It is highly desirable to prevent spatial aliasing, since the result is visually irritating. In tube cameras the spatial aliasing will be in the vertical dimension only, since the horizontal dimension is continuously scanned. Such cameras seldom attempt to prevent vertical aliasing. CCD sensors can, however, alias in both horizontal and vertical dimensions, and so an anti-aliasing optical filter is

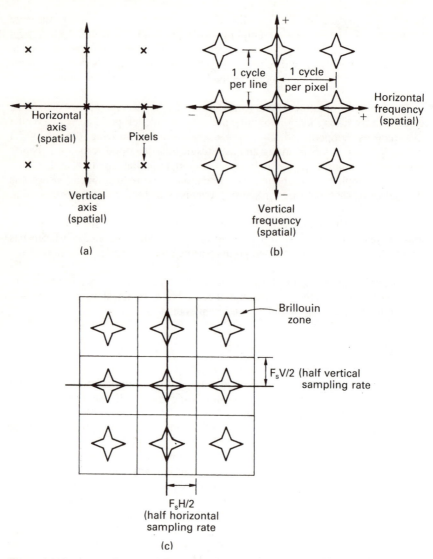

Figure 3.14 Image-sampling spectra. The rectangular array of (a) has a spectrum shown at (b) having a rectangular repeating structure. Filtering to return to the baseband requires a two-dimensional filter whose response lies within the Brillouin zone shown at (c).

generally fitted between the lens and the sensor. This takes the form of a plate which diffuses the image formed by the lens. Such a device can never have a sharp cut-off nor will the aperture be rectangular. The aperture of the anti-aliasing plate is in series with the aperture effect of the CCD elements, and the combination of the two effectively prevents spatial aliasing, and generally gives a good balance between horizontal and vertical resolution, allowing the picture a natural appearance.

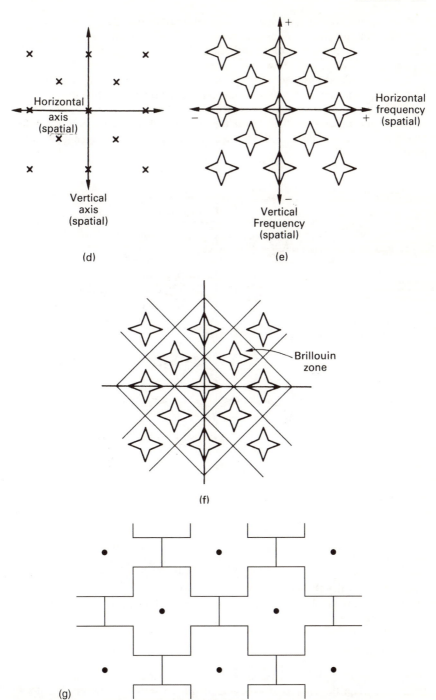

(d)

(e)

(f)

(g)

Figure 3.14 (*Continued*) Quincunx sampling is shown at (d) to have a similar spectral structure (e). An appropriate Brillouin zone is required as at (f). (g) An alternative Brillouin zone for quincunx sampling.

With a conventional approach, there are effectively two choices. If aliasing is permitted, the theoretical information rate of the system can be approached. If aliasing is prevented, realizable anti-aliasing filters cannot sharp cut, and the information conveyed is below system capacity.

These considerations also apply at the television display. The display must filter out spatial frequencies above one half the sampling rate. In a conventional CRT this means that a vertical optical filter should be fitted in front of the screen to render the raster invisible. Again the aperture of a simply realizable filter would attenuate too much of the wanted spectrum, and so the technique is not used.

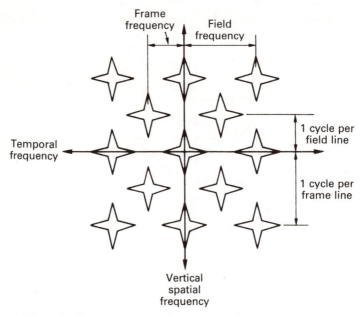

Figure 3.15 The vertical/temporal spectrum of monochrome video due to interlace.

Figure 3.15 shows the spectrum of analog monochrome video (or of an analog component). The use of interlace has a similar effect on the vertical/temporal spectrum as the use of quincunx sampling on the vertical/horizontal spectrum. The concept of the Brillouin zone cannot really be applied to reconstruction in the spatial/temporal domains. This is because the necessary temporal filtering will be absent in real systems.

3.6 Choice of sampling rate

If the reason for digitizing a video signal is simply to convey it from one place to another, then the choice of sampling frequency can be determined only by sampling theory and available filters. If, however, processing of the video in the digital domain is contemplated, the choice becomes smaller. In order to produce a two-dimensional array of samples which form rows and vertical columns, the sampling rate has to be an integer multiple of the line rate. This allows for the vertical picture processing. Whilst the bandwidth needed by 525/59.94 video is less than that of 625/50, and a lower sampling rate might be used, practicality dictated the choice of a standard sampling rate for video components.

This was the goal of CCIR Recommendation 601, which combined the 625/50 input of EBU Docs Tech. 3246 and 3247 and the 525/59.94 input of SMPTE RP 125. The result is not one sampling rate, but a family of rates based upon the magic frequency of 13.5 MHz. Using this frequency as a sampling rate produces 858 samples in the line period of 525/59.94 and 864 samples in the line period of 625/50. For lower bandwidths, the rate can be divided by three quarters, one half or one quarter to give sampling rates of 10.125, 6.75 and 3.375 MHz respectively. If the lowest frequency is considered to be 1, then the highest is 4. For maximum quality *RGB* working, then three parallel, identical sample streams would be required, which would be denoted by 4:4:4. Colour difference signals intended for post-production, where a wider colour difference bandwidth is needed, require 4:2:2 sampling for luminance, *R–Y* and *B–Y* respectively. 4:2:2 has the advantage that an integer number of colour difference samples also exist in both line standards.

Figure 3.16 shows the spatial arrangement given by 4:2:2 sampling. Luminance samples appear at half the spacing of colour difference samples, and half of the luminance samples are in the same physical position as a pair of colour difference samples, these being called co-sited samples.

Where the signal is likely to be broadcast as PAL or NTSC, a standard of 4:1:1 is acceptable, since this still delivers a colour difference bandwidth in excess of 1 MHz. Where data rate is at a premium, 3:1:1 can be used, and can still offer just about enough bandwidth for 525 lines. This would not be enough for 625-line working, but would be acceptable for ENG applications. The problem with the factors three and one is that they do not offer a columnar sampling structure, and

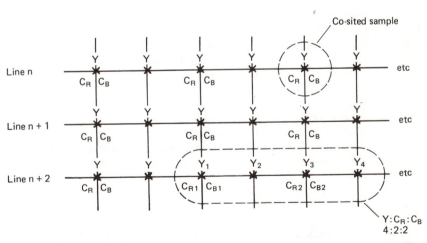

Figure 3.16 In CCIR-601 sampling mode 4:2:2, the line synchronous sampling rate of 13.5 MHz results in samples having the same position in successive lines, so that vertical columns are generated. The sampling rates of the colour difference signals C_R, C_B are one-half of that of luminance, i.e. 6.75 MHz, so that there are alternate Y only samples and co-sited samples which describe Y, C_R and C_B. In a run of four samples, there will be four Y samples, two C_R samples and two C_B samples, hence 4:2:2.

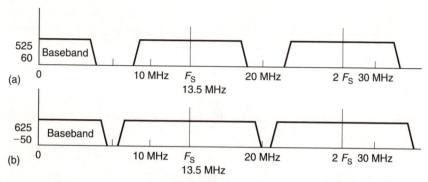

Figure 3.17 Spectra of video sampled at 13.5 MHz. In (a) the baseband 525/60 signal at left becomes the sidebands of the sampling rate and its harmonics. In (b) the same process for the 625/50 signal results in a smaller gap between baseband and sideband because of the wider bandwidth of the 625 system. The same sampling rate for both standards results in a great deal of commonality between 50 Hz and 60 Hz equipment.

so are not appropriate for processing systems. Figure 3.17(a) shows the one-dimensional spectrum which results from sampling 525/59.94 video at 13.5 MHz, and (b) shows the result for 625/50 video.

The traditional TV screen has an aspect ratio of 4:3, whereas digital TV broadcasts have adopted an aspect ratio of 16:9. Expressing 4:3 as 12:9 makes it clear that the 16:9 picture is 16/12 or 4/3 times as wide. There are two ways of handling 16:9 pictures in the digital domain. One is to retain the standard sampling rate of 13.5 MHz, which results in the horizontal resolution falling to 3/4 of its previous value, the other is to increase the sampling rate in proportion to the screen width. This results in a luminance sampling rate of 13.5 × 4/3 MHz or 18.0 MHz.

When composite video is to be digitized, the input will be a single waveform having spectrally interleaved luminance and chroma. Any sampling rate which allows sufficient bandwidth would convey composite video from one point to another, indeed 13.5 MHz has been successfully used to sample PAL and NTSC. However, if simple processing in the digital domain is contemplated, there will be less choice.

In many cases it will be necessary to decode the composite signal which will require some kind of digital filter. Whilst it is possible to construct filters with any desired response, it is a fact that a digital filter whose response is simply related to the sampling rate will be much less complex to implement. This is the reasoning which has led to the near-universal use of four times subcarrier sampling rate.

3.7 Jitter

The instants at which samples are taken in an ADC and the instants at which DACs make conversions must be evenly spaced, otherwise unwanted signals can be added to the video. Figure 3.18 shows the effect of sampling clock jitter on a sloping waveform. Samples are taken at the wrong times. When these samples have passed through a system, the timebase correction stage prior to the DAC will remove the jitter, and the result is shown at (b). The magnitude of the

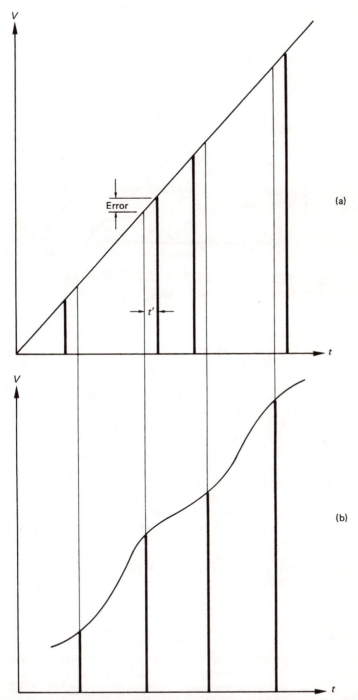

Figure 3.18 The effect of sampling timing jitter on noise. At (a) a sloping signal sampled with jitter has error proportional to the slope. When jitter is removed by reclocking, the result at (b) is noise.

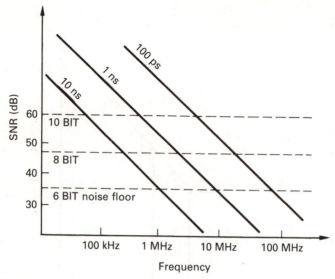

Figure 3.19 The effect of sampling clock jitter on signal-to-noise ratio at various frequencies, compared with the theoretical noise floors with different wordlengths.

unwanted signal is proportional to the slope of the waveform and so the amount of jitter which can be tolerated falls at 6 dB per octave. As the resolution of the system is increased by the use of longer sample wordlength, tolerance to jitter is further reduced. The nature of the unwanted signal depends on the spectrum of the jitter. If the jitter is random, the effect is noise-like and relatively benign unless the amplitude is excessive. Figure 3.19 shows the effect of differing amounts of random jitter with respect to the noise floor of various wordlengths. Note that even small amounts of jitter can degrade a ten-bit convertor to the performance of a good eight-bit unit. There is thus no point in upgrading to higher-resolution convertors if the clock stability of the system is insufficient to allow their performance to be realized.

The allowable jitter is measured in picoseconds, and clearly steps must be taken to eliminate it by design. Convertor clocks must be generated from clean power supplies which are well decoupled from the power used by the logic because a convertor clock must have a signal-to-noise ratio of the same order as that of the signal. Otherwise noise on the clock causes jitter which in turn causes noise in the video. The same effect will be found in digital audio signals, which are perhaps more critical.

3.8 Quantizing

Quantizing is the process of expressing some infinitely variable quantity by discrete or stepped values. It turns up in a remarkable number of everyday guises. Figure 3.20 shows that an inclined ramp enables infinitely variable height to be achieved, whereas a step-ladder allows only discrete heights to be had. A step-ladder quantizes height. When accountants round off sums of money to the nearest pound or dollar they are quantizing. Time passes continuously, but the

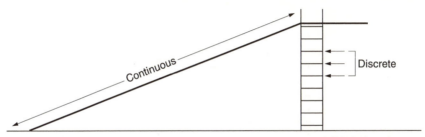

Figure 3.20 An analog parameter is continuous whereas a quantized parameter is restricted to certain values. Here the sloping side of a ramp can be used to obtain any height whereas a ladder only allows discrete heights.

display on a digital clock changes suddenly every minute because the clock is quantizing time.

In video and audio the values to be quantized are infinitely variable voltages from an analog source. Strict quantizing is a process which operates in the voltage domain only. For the purpose of studying the quantizing of a single sample, time is assumed to stand still. This is achieved in practice either by the use of a track/hold circuit or the adoption of a quantizer technology such as a flash convertor which operates before the sampling stage.

Figure 3.21(a) shows that the process of quantizing divides the voltage range up into quantizing intervals Q, also referred to as steps S. In applications such as telephony these may advantageously be of differing size, but for digital audio the quantizing intervals are made as identical as possible. If this is done, the binary numbers which result are truly proportional to the original analog voltage, and the digital equivalents of mixing and gain changing can be performed by adding and multiplying sample values. If the quantizing intervals are unequal this cannot be done. When all quantizing intervals are the same, the term 'uniform quantizing' is used. The term linear quantizing will be found, but this is, like military intelligence, a contradiction in terms.

The term LSB (least significant bit) will also be found in place of quantizing interval in some treatments, but this is a poor term because quantizing works in the voltage domain. A bit is not a unit of voltage and can only have two values. In studying quantizing voltages within a quantizing interval will be discussed, but there is no such thing as a fraction of a bit.

Whatever the exact voltage of the input signal, the quantizer will locate the quantizing interval in which it lies. In what may be considered a separate step, the quantizing interval is then allocated a code value which is typically some form of binary number. The information sent is the number of the quantizing interval in which the input voltage lay. Whereabouts that voltage lay within the interval is not conveyed, and this mechanism puts a limit on the accuracy of the quantizer. When the number of the quantizing interval is converted back to the analog domain, it will result in a voltage at the centre of the quantizing interval as this minimizes the magnitude of the error between input and output. The number range is limited by the wordlength of the binary numbers used. In an eight-bit system, 256 different quantizing intervals exist, although in digital video the codes at the extreme ends of the range are reserved for synchronizing.

It is possible to draw a transfer function for such an ideal quantizer followed by an ideal DAC, and this is also shown in Figure 3.21. A transfer function is

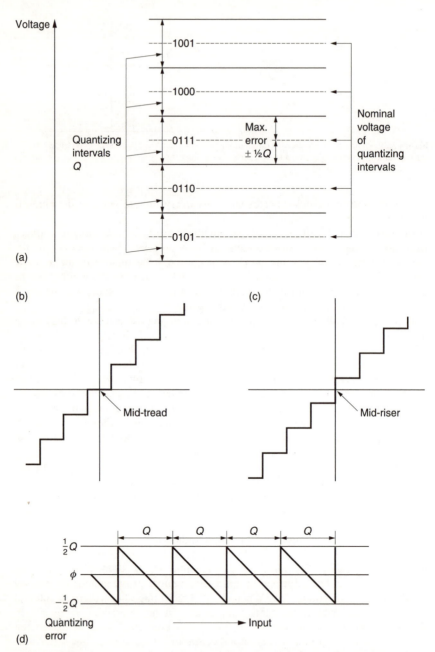

Figure 3.21 Quantizing assigns discrete numbers to variable voltages. All voltages within the same quantizing interval are assigned the same number which causes a DAC to produce the voltage at the centre of the intervals shown by the dashed lines in (a). This is the characteristic of the mid-tread quantizer shown in (b). An alternative system is the mid-riser system shown in (c). Here 0 Volts analog falls between two codes and there is no code for zero. Such quantizing cannot be used prior to signal processing because the number is no longer proportional to the voltage. Quantizing error cannot exceed $\pm\frac{1}{2}Q$ as shown in (d).

simply a graph of the output with respect to the input. In electronics, when the term 'linearity' is used, this generally means the overall straightness of the transfer function. Linearity is a goal in video and audio convertors, yet it will be seen that an ideal quantizer is anything but linear.

Figure 3.21(b) shows the transfer function is somewhat like a staircase, and blanking level is half-way up a quantizing interval, or on the centre of a tread. This is the so-called mid-tread quantizer which is universally used in video and audio.

Quantizing causes a voltage error in the sample which is given by the difference between the actual staircase transfer function and the ideal straight line. This is shown in Figure 3.21(d) to be a sawtooth-like function which is periodic in Q. The amplitude cannot exceed $\pm \frac{1}{2}Q$ peak-to-peak unless the input is so large that clipping occurs.

Quantizing error can also be studied in the time domain where it is better to avoid complicating matters with the aperture effect of the DAC. For this reason it is assumed here that output samples are of negligible duration. Then impulses from the DAC can be compared with the original analog waveform and the difference will be impulses representing the quantizing error waveform. This has been done in Figure 3.22. The horizontal lines in the drawing are the boundaries between the quantizing intervals, and the curve is the input waveform. The vertical bars are the quantized samples which reach to the centre of the quantizing interval. The quantizing error waveform shown at (b) can be thought of as an unwanted signal which the quantizing process adds to the perfect original. If a very small input signal remains within one quantizing interval, the quantizing error *is* the signal.

As the transfer function is non-linear, ideal quantizing can cause distortion. As a result, practical digital video equipment deliberately uses non-ideal quantizers to achieve linearity.

As the magnitude of the quantizing error is limited, its effect can be minimized by making the signal larger. This will require more quantizing intervals and more bits to express them. The number of quantizing intervals multiplied by their size gives the quantizing range of the convertor. A signal outside the range will be clipped. Provided that clipping is avoided, the larger the signal, the less will be the effect of the quantizing error.

Where the input signal exercises the whole quantizing range and has a complex waveform (such as from a contrasty, detailed scene), successive samples will have widely varying numerical values and the quantizing error on a given sample will be independent of that on others. In this case the size of the quantizing error will be distributed with equal probability between the limits. Figure 3.22(c)) shows the resultant uniform probability density. In this case the unwanted signal added by quantizing is an additive broadband noise uncorrelated with the signal, and it is appropriate in this case to call it quantizing noise. This is not quite the same as thermal noise which has a Gaussian (bell-shaped) probability shown in Figure 3.22(d). The difference is of no consequence as in the large signal case the noise is masked by the signal. Under these conditions, a meaningful signal-to-noise ratio can be calculated by taking the ratio between the largest signal amplitude which can be accommodated without clipping and the error amplitude. By way of example, an eight-bit system will offer very nearly 50 dB unweighted SNR.

Whilst the above result is true for a large complex input waveform, treatments which then assume that quantizing error is *always* noise give results which are at

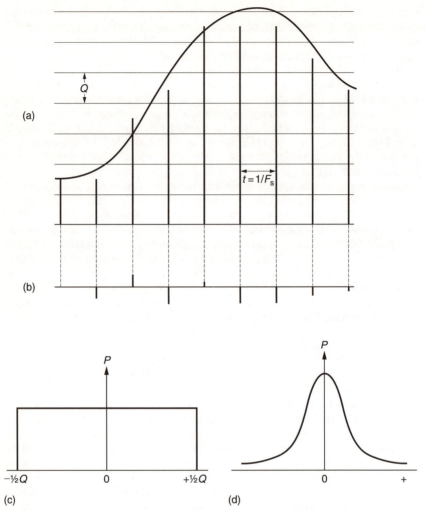

Figure 3.22 In (a) an arbitrary signal is represented to finite accuracy by PAM needles whose peaks are at the centre of the quantizing intervals. The errors caused can be thought of as an unwanted signal (b) added to the original. In (c) the amplitude of a quantizing error needle will be from $-\frac{1}{2}Q$ to $+\frac{1}{2}Q$ with equal probability. Note, however, that white noise in analog circuits generally has Gaussian amplitude distribution, shown in (d).

variance with reality. The expression above is only valid if the probability density of the quantizing error is uniform. Unfortunately at low depths of modulations, and particularly with flat fields or simple pictures, this is not the case.

At low modulation depth, quantizing error ceases to be random, and becomes a function of the input waveform and the quantizing structure as Figure 3.22 shows. Once an unwanted signal becomes a deterministic function of the wanted signal, it has to be classed as a distortion rather than a noise. Distortion can also be predicted from the non-linearity, or staircase nature, of the transfer function. With a large signal, there are so many steps involved that we must stand well back, and a staircase with 256 steps appears to be a slope. With a small signal there are few steps and they can no longer be ignored.

The effect can be visualized readily by considering a television camera viewing a uniformly painted wall. The geometry of the lighting and the coverage of the lens means that the brightness is not absolutely uniform, but falls slightly at the ends of the TV lines. After quantizing, the gently sloping waveform is replaced by one which stays at a constant quantizing level for many sampling periods and then suddenly jumps to the next quantizing level. The picture then consists of areas of constant brightness with steps between, resembling nothing more than a contour map, hence the use of the term *contouring* to describe the effect.

Needless to say, the occurrence of contouring precludes the use of an ideal quantizer for high-quality work. There is little point in studying the adverse effects further as they should be and can be eliminated completely in practical equipment by the use of dither. The importance of correctly dithering a quantizer cannot be emphasized enough, since failure to dither irrevocably distorts the converted signal: there can be no process which will subsequently remove that distortion.

3.9 Introduction to dither

At high signal levels, quantizing error is effectively noise. As the depth of modulation falls, the quantizing error of an ideal quantizer becomes more strongly correlated with the signal and the result is distortion, visible as contouring. If the quantizing error can be decorrelated from the input in some way, the system can remain linear but noisy. Dither performs the job of decorrelation by making the action of the quantizer unpredictable and gives the system a noise floor like an analog system.[8,9]

All practical digital video systems use non-subtractive dither where the dither signal is added prior to quantization and no attempt is made to remove it at the DAC.[10] The introduction of dither prior to a conventional quantizer inevitably causes a slight reduction in the signal-to-noise ratio attainable, but this reduction is a small price to pay for the elimination of non-linearities.

The ideal (noiseless) quantizer of Figure 3.21 has fixed quantizing intervals and must always produce the same quantizing error from the same signal. In Figure 3.23 it can be seen that an ideal quantizer can be dithered by linearly adding a controlled level of noise either to the input signal or to the reference voltage which is used to derive the quantizing intervals. There are several ways of considering how dither works, all of which are equally valid.

The addition of dither means that successive samples effectively find the quantizing intervals in different places on the voltage scale. The quantizing error becomes a function of the dither, rather than a predictable function of the input signal. The quantizing error is not eliminated, but the subjectively unacceptable distortion is converted into a broadband noise which is more benign to the viewer.

Some alternative ways of looking at dither are shown in Figure 3.24. Consider the situation where a low-level input signal is changing slowly within a quantizing interval. Without dither, the same numerical code is output for a number of sample periods, and the variations within the interval are lost. Dither has the effect of forcing the quantizer to switch between two or more states. The

(a)

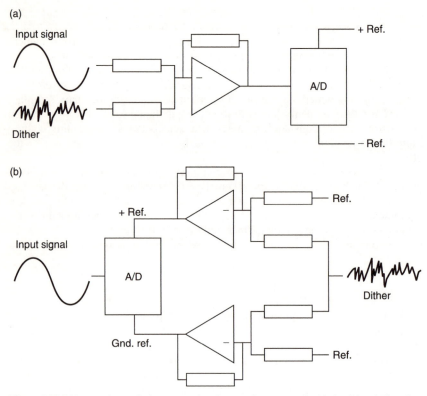

(b)

Figure 3.23 Dither can be applied to a quantizer in one of two ways. In (a) the dither is linearly added to the analog input signal, whereas in (b) it is added to the reference voltages of the quantizer.

higher the voltage of the input signal within a given interval, the more probable it becomes that the output code will take on the next higher value. The lower the input voltage within the interval, the more probable it is that the output code will take the next lower value. The dither has resulted in a form of duty cycle modulation, and the resolution of the system has been extended indefinitely instead of being limited by the size of the steps.

Dither can also be understood by considering what it does to the transfer function of the quantizer. This is normally a perfect staircase, but in the presence of dither it is smeared horizontally until with a certain amplitude the average transfer function becomes straight.

3.10 Requantizing and digital dither

Recent ADC technology allows the resolution of video samples to be raised from eight bits to ten or even twelve bits. The situation then arises that an existing eight-bit device such as a digital VTR needs to be connected to the output of an ADC with greater wordlength. The words need to be shortened in some way.

When a sample value is attenuated, the extra low-order bits which come into existence below the radix point preserve the resolution of the signal and the

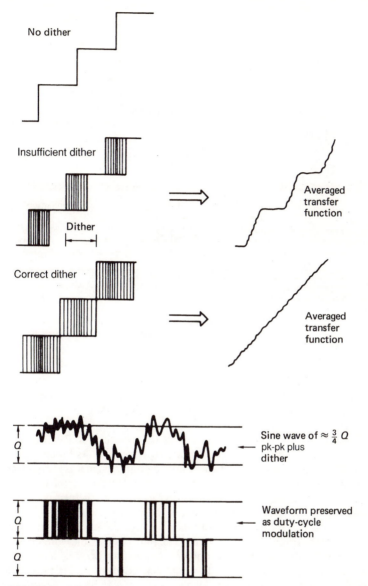

No dither

Insufficient dither

Dither

⇒

Averaged transfer function

Correct dither

⇒

Averaged transfer function

Q

Sine wave of $\approx \frac{3}{4} Q$ pk-pk plus dither

Q

Q

Waveform preserved as duty-cycle modulation

Figure 3.24 Wideband dither of the appropriate level linearizes the transfer function to produce noise instead of distortion. This can be confirmed by spectral analysis. In the voltage domain, dither causes frequent switching between codes and preserves resolution in the duty cycle of the switching.

dither in the least significant bit(s) which linearizes the system. The same word extension will occur in any process involving multiplication, such as digital filtering. It will subsequently be necessary to shorten the wordlength. Low-order bits must be removed in order to reduce the resolution whilst keeping the signal magnitude the same. Even if the original conversion was correctly dithered, the

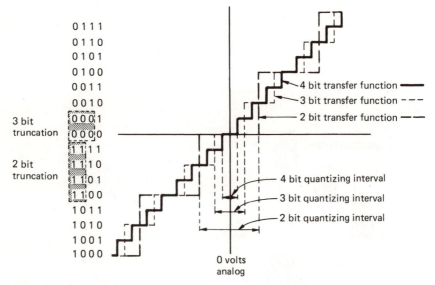

0 1 1 1
0 1 1 0
0 1 0 1
0 1 0 0
0 0 1 1
0 0 1 0

3 bit
truncation

0 0 0 1
0 0 0 0
1 1 1 1

2 bit
truncation

1 1 1 0
1 1 0 1
1 1 0 0
1 0 1 1
1 0 1 0
1 0 0 1
1 0 0 0

4 bit transfer function ——
3 bit transfer function - - -
2 bit transfer function ——

4 bit quantizing interval
3 bit quantizing interval
2 bit quantizing interval

0 volts
analog

Figure 3.25 Shortening the wordlength of a sample reduces the number of codes which can describe the voltage of the waveform. This makes the quantizing steps bigger hence the term 'requantizing'. It can be seen that simple truncation or omission of the bits does not give analogous behaviour. Rounding is necessary to give the same result as if the larger steps had been used in the original conversion.

random element in the low-order bits will now be some way below the end of the intended word. If the word is simply truncated by discarding the unwanted low-order bits or rounded to the nearest integer the linearizing effect of the original dither will be lost.

Shortening the wordlength of a sample reduces the number of quantizing intervals available without changing the signal amplitude. As Figure 3.25 shows, the quantizing intervals become larger and the original signal is *requantized* with the new interval structure. This will introduce requantizing distortion having the same characteristics as quantizing distortion in an ADC. It then is obvious that when shortening the wordlength of a ten-bit convertor to eight bits, the two low-order bits must be removed in a way that displays the same overall quantizing structure as if the original convertor had been only of eight-bit wordlength. It will be seen from Figure 3.25 that truncation cannot be used because it does not meet the above requirement but results in signal-dependent offsets because it always rounds in the same direction. Proper numerical rounding is essential in video applications because it accurately simulates analog quantizing to the new interval size. Unfortunately the ten-bit convertor will have a dither amplitude appropriate to quantizing intervals one quarter the size of an eight-bit unit and the result will be highly non-linear.

In practice, the wordlength of samples must be shortened in such a way that the requantizing error is converted to noise rather than distortion. One technique which meets this requirement is to use digital dithering[11] prior to rounding. This is directly equivalent to the analog dithering in an ADC.

Digital dither is a pseudo-random sequence of numbers. If it is required to simulate the analog dither signal of Figures 3.23 and 3.24, then it is obvious that

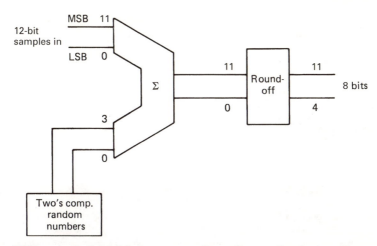

Figure 3.26 In a simple digital dithering system, two's complement values from a random number generator are added to low-order bits of the input. The dithered values are then rounded up or down according to the value of the bits to be removed. The dither linearizes the requantizing.

the noise must be bipolar so that it can have an average voltage of zero. Two's complement coding must be used for the dither values.

Figure 3.26 shows a simple digital dithering system (i.e. one without noise shaping) for shortening sample wordlength. The output of a two's complement pseudo-random sequence generator (see Chapter 2) of appropriate wordlength is added to input samples prior to rounding. The most significant of the bits to be discarded is examined in order to determine whether the bits to be removed sum to more or less than half a quantizing interval. The dithered sample is either rounded down, i.e. the unwanted bits are simply discarded, or rounded up, i.e. the unwanted bits are discarded but one is added to the value of the new short word. The rounding process is no longer deterministic because of the added dither which provides a linearizing random component.

If this process is compared with that of Figure 3.23 it will be seen that the principles of analog and digital dither are identical; the processes simply take place in different domains using two's complement numbers which are rounded or voltages that are quantized as appropriate. In fact quantization of an analog dithered waveform is identical to the hypothetical case of rounding after bipolar digital dither where the number of bits to be removed is infinite, and remains identical for practical purposes when as few as eight bits are to be removed. Analog dither may actually be generated from bipolar digital dither (which is no more than random numbers with certain properties) using a DAC.

3.11 Basic digital-to-analog conversion

This direction of conversion will be discussed first, since ADCs often use embedded DACs in feedback loops.

The purpose of a digital-to-analog convertor is to take numerical values and reproduce the continuous waveform that they represent. Figure 3.27 shows the

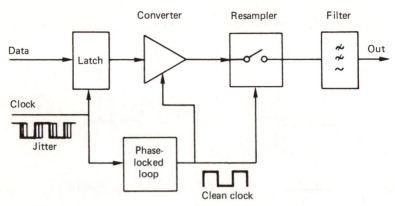

Figure 3.27 The components of a conventional convertor. A jitter-free clock drives the voltage conversion, whose output may be resampled prior to reconstruction.

major elements of a conventional conversion subsystem, i.e. one in which oversampling is not employed. The jitter in the clock needs to be removed with a VCO or VCXO. Sample values are buffered in a latch and fed to the convertor element which operates on each cycle of the clean clock. The output is then a voltage proportional to the number for at least a part of the sample period. A resampling stage may be found next, in order to remove switching transients, reduce the aperture ratio or allow the use of a convertor which takes a substantial part of the sample period to operate. The resampled waveform is then presented to a reconstruction filter which rejects frequencies above the audio band.

This section is primarily concerned with the implementation of the convertor element. The most common way of achieving this conversion is to control binary-weighted currents and sum them in a virtual earth. Figure 3.28 shows the classical $R-2R$ DAC structure. This is relatively simple to construct, but the

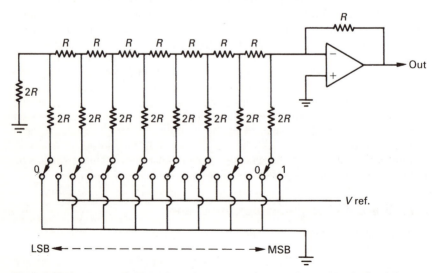

Figure 3.28 The classical $R-2R$ DAC requires precise resistance values and 'perfect' switches.

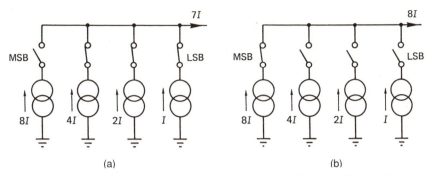

Figure 3.29 At (a) current flow with an input of 0111 is shown. At (b) current flow with input code one greater.

resistors have to be extremely accurate. To see why this is so, consider the example of Figure 3.29. At (a) the binary code is about to have a major overflow, and all the low-order currents are flowing. At (b), the binary input has increased by one, and only the most significant current flows. This current must equal the sum of all the others plus one. The accuracy must be such that the step size is within the required limits. In this eight-bit example, if the step size needs to be a rather casual 10 per cent accurate, the necessary accuracy is only one part in 2560, but for a ten-bit system it would become one part in 10 240. This degree of accuracy is difficult to achieve and maintain in the presence of ageing and temperature change.

3.12 Basic analog-to-digital conversion

The general principle of a quantizer is that different quantized voltages are compared with the unknown analog input until the closest quantized voltage is found. The code corresponding to this becomes the output. The comparisons can be made in turn with the minimal amount of hardware, or simultaneously with more hardware.

The flash convertor is probably the simplest technique available for PCM video conversion. The principle is shown in Figure 3.30. The threshold voltage of every quantizing interval is provided by a resistor chain which is fed by a reference voltage. This reference voltage can be varied to determine the sensitivity of the input. There is one voltage comparator connected to every reference voltage, and the other input of all the comparators is connected to the analog input. A comparator can be considered to be a one-bit ADC. The input voltage determines how many of the comparators will have a true output. As one comparator is necessary for each quantizing interval, then, for example, in an eight-bit system there will be 255 binary comparator outputs, and it is necessary to use a priority encoder to convert these to a binary code.

The quantizing stage is asynchronous; comparators change state as and when the variations in the input waveform result in a reference voltage being crossed. Sampling takes place when the comparator outputs are clocked into a subsequent latch. This is an example of quantizing before sampling as was illustrated in Figure 3.1. Although the device is simple in principle, it contains a lot of circuitry and can only be practicably implemented on a chip. The analog signal has to

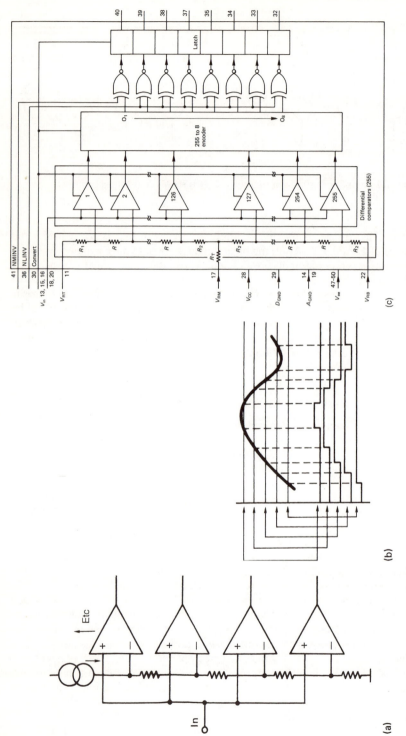

Figure 3.30 The flash convertor. In (a) each quantizing interval has its own comparator, resulting in waveforms of (b). A priority encoder is necessary to convert the comparator outputs to a binary code. Shown in (c) is a typical eight-bit flash convertor primarily intended for video applications. (Courtesy TRW).

drive many inputs which results in a significant parallel capacitance, and a low-impedance driver is essential to avoid restricting the slewing rate of the input. The extreme speed of a flash convertor is a distinct advantage in oversampling. Because computation of all bits is performed simultaneously, no track/hold circuit is required, and droop is eliminated. Figure 3.30(c) shows a flash convertor chip. Note the resistor ladder and the comparators followed by the priority encoder. The MSB can be selectively inverted so that the device can be used either in offset binary or two's complement mode.

3.13 Oversampling

Oversampling means using a sampling rate which is greater (generally substantially greater) than the Nyquist rate. Neither sampling theory nor quantizing theory *require* oversampling to be used to obtain a given signal quality, but Nyquist rate conversion places extremely high demands on component accuracy when a convertor is implemented. Oversampling allows a given signal quality to be reached without requiring very close tolerance, and therefore expensive, components.

Figure 3.31 shows the main advantages of oversampling. At (a) it will be seen that the use of a sampling rate considerably above the Nyquist rate allows the anti-aliasing and reconstruction filters to be realized with a much more gentle cut-off slope. There is then less likelihood of phase linearity and ripple problems in the passband.

Figure 3.31(b) shows that information in an analog signal is two-dimensional and can be depicted as an area which is the product of bandwidth and the linearly expressed signal-to-noise ratio. The figure also shows that the same amount of information can be conveyed down a channel with a SNR of half as much (6 dB

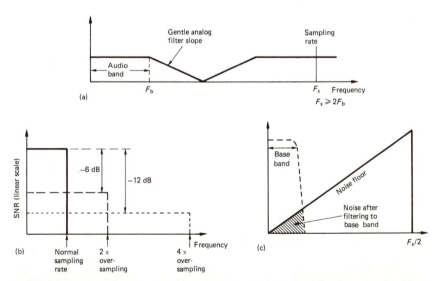

Figure 3.31 Oversampling has a number of advantages. In (a) it allows the slope of analog filters to be relaxed. In (b) it allows the resolution of convertors to be extended. In (c) *a noise-shaped* convertor allows a disproportionate improvement in resolution.

less) if the bandwidth used is doubled, with 12 dB less SNR if bandwidth is quadrupled, and so on, provided that the modulation scheme used is perfect.

The information in an analog signal can be conveyed using some analog modulation scheme in any combination of bandwidth and SNR which yields the appropriate channel capacity. If bandwidth is replaced by sampling rate and SNR is replaced by a function of wordlength, the same must be true for a digital signal as it is no more than a numerical analog. Thus raising the sampling rate potentially allows the wordlength of each sample to be reduced without information loss.

Information theory predicts that if a signal is spread over a much wider bandwidth by some modulation technique, the SNR of the demodulated signal can be higher than that of the channel it passes through, and this is also the case in digital systems. The concept is illustrated in Figure 3.32. At (a) four-bit samples are delivered at sampling rate F. As four bits have sixteen combinations, the information rate is $16F$. At (b) the same information rate is obtained with three-bit samples by raising the sampling rate to $2F$ and at (c) two-bit samples having four combinations require to be delivered at a rate of $4F$. Whilst the information rate has been maintained, it will be noticed that the bit-rate of (c) is twice that of (a). The reason for this is shown in Figure 3.33. A single binary digit can only have two states; thus it can only convey two pieces of information, perhaps 'yes' or 'no'. Two binary digits together can have four states, and can thus convey four pieces of information, perhaps 'spring summer autumn or winter', which is two pieces of information per bit. Three binary digits grouped together can have eight combinations, and convey eight pieces of information, perhaps 'doh re mi fah so lah te or doh', which is nearly three pieces of information per digit. Clearly the further this principle is taken, the greater the benefit. In a sixteen-bit system, each bit is worth 4K pieces of information. It is always more efficient, in information-capacity terms, to use the combinations of long binary words than to send single bits for every piece of information. The greatest efficiency is reached when the longest words are sent at the slowest rate

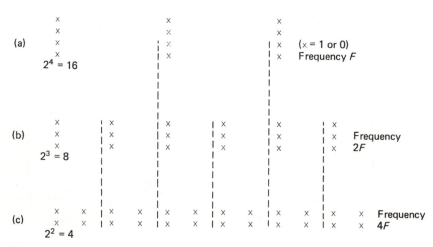

Figure 3.32 Information rate can be held constant when frequency doubles by removing one bit from each word. In all cases here it is $16F$. Note bit rate of (c) is double that of (a). Data storage in oversampled form is inefficient.

	0 = No 1 = Yes	00 = Spring 01 = Summer 10 = Autumn 11 = Winter	000 do 001 re 010 mi 011 fa 100 so 101 la 110 te 111 do	0000 0 0001 1 0010 2 0011 3 0100 4 0101 5 0110 6 0111 7 1000 8 1001 9 1010 A 1011 B 1100 C 1101 D 1110 E 1111 F	0000 FFFF Digital audio sample values
No of bits	1	2	3	4	16
Information per word	2	4	8	16	65 536
Information per bit	2	2	≈3	4	4096

Figure 3.33 The amount of information per bit increases disproportionately as wordlength increases. It is always more efficient to use the longest words possible at the lowest word rate. It will be evident that sixteen-bit PCM is 2048 times as efficient as delta modulation. Oversampled data are also inefficient for storage.

which must be the Nyquist rate. This is one reason why PCM recording is more common than delta modulation, despite the simplicity of implementation of the latter type of convertor. PCM simply makes more efficient use of the capacity of the binary channel.

As a result, oversampling is confined to convertor technology where it gives specific advantages in implementation. The storage or transmission system will usually employ PCM, where the sampling rate is a little more than twice the input bandwidth. Figure 3.34 shows a digital VTR using oversampling convertors. The ADC runs at n times the Nyquist rate, but once in the digital domain the rate needs to be reduced in a type of digital filter called a *decimator*. The output of this is conventional Nyquist rate PCM, according to the tape format, which is then recorded. On replay the sampling rate is raised once more in a further type of digital filter called an *interpolator*. The system now has the best of both worlds: using oversampling in the convertors overcomes the shortcomings of analog anti-aliasing and reconstruction filters and the wordlength of the convertor elements is reduced making them easier to construct; the recording is made with Nyquist rate PCM which minimizes tape consumption.

Oversampling is a method of overcoming practical implementation problems by replacing a single critical element or bottleneck by a number of elements whose overall performance is what counts. As Hauser[12] properly observed, oversampling tends to overlap the operations which are quite distinct in a conventional convertor. In earlier sections of this chapter, the vital subjects of filtering, sampling, quantizing and dither have been treated almost independently. Figure 3.35(a) shows that it is possible to construct an ADC of predictable

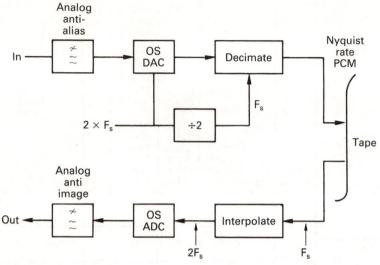

Figure 3.34 An oversampling DVTR. The convertors run faster than sampling theory suggests to ease analog filter design. Sampling-rate reduction allows efficient PCM recording on tape.

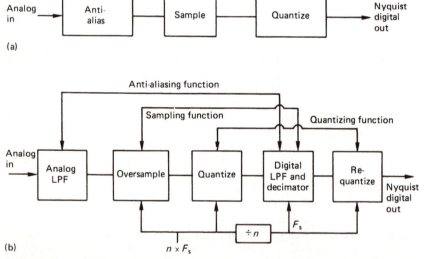

Figure 3.35 A conventional ADC performs each step in an identifiable location as in (a). With oversampling, many of the steps are distributed as shown in (b).

performance by taking a suitable anti-aliasing filter, a sampler, a dither source and a quantizer and assembling them like building bricks. The bricks are effectively in series and so the performance of each stage can only limit the overall performance. In contrast, Figure 3.35(b) shows that with oversampling the overlap of operations allows different processes to augment one another allowing a synergy which is absent in the conventional approach.

If the oversampling factor is n, the analog input must be bandwidth limited to $n.F_s/2$ by the analog anti-aliasing filter. This unit need only have flat frequency response and phase linearity within the audio band. Analog dither of an amplitude compatible with the quantizing interval size is added prior to sampling at $n.F_s$ and quantizing.

Next, the anti-aliasing function is completed in the digital domain by a low-pass filter which cuts off at $F_s/2$. Using an appropriate architecture this filter can be absolutely phase linear and implemented to arbitrary accuracy. The filter can be considered to be the demodulator of Figure 3.31 where the SNR improves as the bandwidth is reduced. The wordlength can be expected to increase. The multiplications taking place within the filter extend the wordlength considerably more than the bandwidth reduction alone would indicate. The analog filter serves only to prevent aliasing into the baseband at the oversampling rate; the signal spectrum is determined with greater precision by the digital filter.

With the information spectrum now Nyquist limited, the sampling process is completed when the rate is reduced in the decimator. One sample in n is retained. The excess wordlength extension due to the anti-aliasing filter arithmetic must then be removed. Digital dither is added, completing the dither process, and the quantizing process is completed by requantizing the dithered samples to the appropriate wordlength which will be greater than the wordlength of the first quantizer. Alternatively noise shaping may be employed.

Figure 3.36(a) shows the building-brick approach of a conventional DAC. The Nyquist rate samples are converted to analog voltages and then a steep-cut analog low-pass filter is needed to reject the sidebands of the sampled spectrum. Figure 3.36(b) shows the oversampling approach. The sampling rate is raised in an interpolator which contains a low-pass filter that restricts the baseband spectrum to the audio bandwidth shown. A large frequency gap now

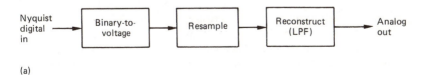

(a)

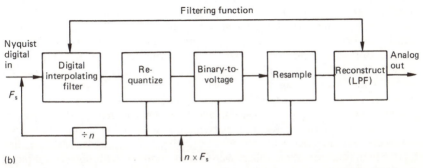

(b)

Figure 3.36 A conventional DAC in (a) is compared with the oversampling implementation in (b).

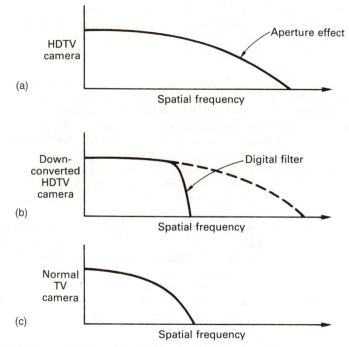

Figure 3.37 Using an HDTV camera with downconversion is a form of oversampling and gives better results than a normal camera because the aperture effect is overcome.

exists between the baseband and the lower sideband. The multiplications in the interpolator extend the wordlength considerably and this must be reduced within the capacity of the DAC element by the addition of digital dither prior to requantizing.

Oversampling may also be used to considerable benefit in other dimensions. Figure 3.37 shows how spatial oversampling can be used to increase the resolution of an imaging system. Assuming a 720×400 pixel system, Figure 3.37(a) shows that aperture effect would result in an early roll-off of the MTF. Instead a 1440×800 pixel sensor is used, having a response shown at (b). This outputs four times as much data, but if these data are passed into a two-dimensional low-pass filter which decimates by a factor of two in each axis, the original bit rate will be obtained once more. This will be a digital filter which can have arbitrarily accurate peformance, including a flat passband and steep cut-off slope. The combination of the aperture effect of the 1440×800 pixel camera and the LPF gives a spatial frequency response which is shown in (c). This is better than could be achieved with a 720×400 camera. The improvement in subjective quality is quite noticeable in practice.

Oversampling can also be used in the time domain in order to reduce or eliminate display flicker. A different type of standards convertor is necessary which doubles the input field rate by interpolation. The standards convertor must use motion compensation otherwise moving objects will not be correctly positioned in intermediate fields and will suffer from judder. Motion compensation is considered in Chapter 4.

3.14 Gamma in the digital domain

As was explained in section 2.2, the use of gamma makes the transfer function between luminance and the analog video voltage non-linear. This is done because the eye is more sensitive to noise at low brightness. The use of gamma allows the receiver to compress the video signal at low brightness and with it any transmission noise. In the digital domain transmission noise is eliminated, but instead the conversion process introduces quantizing noise. Consequently gamma is retained in the digital domain.

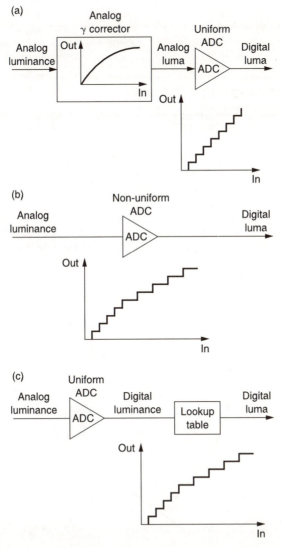

Figure 3.38 (a) Analog γ correction prior to ADC. (b) Non-uniform quantizer gives direct γ conversion. (c) Digital γ correction using look-up table

Figure 3.38 shows that digital luma can be considered in several equivalent ways. At (a) a linear analog luminance signal is passed through a gamma corrector to create luma and this is then quantized uniformly. At (b) the linear analog luminance signal is fed directly to a non-uniform quantizer. At (c) the linear analog luminance signal is uniformly quantized to produce digital luminance. This is converted to digital luma by a digital process having a non-linear transfer function.

Whilst the three techniques shown give the same result, (a) is the simplest, (b) requires a special ADC with gamma spaced quantizing steps, and (c) requires a high-resolution ADC of perhaps fourteen to sixteen bits because it works in the linear luminance domain where noise is highly visible. Technique (c) is used in digital processing cameras where long wordlength is common practice.

As digital luma with eight-bit resolution gives the same subjective performance as digital luminance with fourteen-bit resolution it will be seen that gamma can also be considered to be an effective perceptive compression technique.

3.15 Colour in the digital domain

Colour cameras and most graphics computers produce three signals, or components, R, G and B which are essentially monochrome video signals representing an image in each primary colour. Figure 3.39 shows that the three primaries are spaced apart in the chromaticity diagram and the only colours which can be generated fall within the resultant triangle.

RGB signals are only strictly compatible if the colour primaries assumed in the source are present in the display. If there is a primary difference the reproduced colours are different. Clearly, broadcast television must have a standard set of primaries. The EBU television systems have only ever had one set of primaries. NTSC began with one set and then adopted another because the phosphors were brighter. Computer displays have any number of standards because initially all

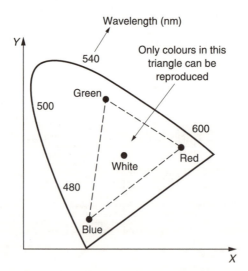

Figure 3.39 Additive mixing colour systems can only reproduce colours within a triangle where the primaries lie on each vertex.

computer colour was false. Now that computer displays are going to be used for television it will be necessary for them to adopt standard phosphors or to use colorimetric transcoding on the signals.

Fortunately the human visual system is quite accommodating. The colour of daylight changes throughout the day, so everything changes colour with it. Humans, however, accommodate to that. We tend to see the colour we expect rather than the actual colour. The colour reproduction of printing and photographic media is pretty appalling, but so is human colour memory, and so it's acceptable.

On the other hand, our ability to discriminate between colours presented simultaneously is razor-sharp, hence the difficulty car repairers have in getting paint to match.

RGB and *Y* signals are incompatible, yet when colour television was introduced it was a practical necessity that it should be possible to display colour signals on a monochrome display and vice versa.

Creating or transcoding a luma signal from *RGB* is relatively easy. The spectral response of the eye has a peak in the green region. Green objects will produce a larger stimulus than red objects of the same brightness, with blue objects producing the least stimulus. A luma signal can be obtained by adding *R*, *G* and *B* together, not in equal amounts, but in a sum which is weighted by the relative response of the human visual system. Thus:

$$Y = 0.299R + 0.587G + 0.114B$$

Note that the factors add up to one. If *Y* is derived in this way, a monochrome display will show nearly the same result as if a monochrome camera had been used in the first place. The results are not identical because of the non-linearities introduced by gamma correction.

As colour pictures require three signals, it is possible to send *Y* along with two colour difference signals. There are three possible colour difference signals, $R-Y$, $B-Y$ and $G-Y$. As the green signal makes the greatest contribution to *Y*, then the amplitude of $G-Y$ would be the smallest and would be most susceptible to noise. Thus $R-Y$ and $B-Y$ are used in practice as Figure 3.40 shows. In the digital domain $R-Y$ is known as C_r and $B-Y$ is known as C_b.

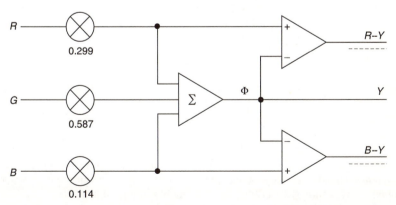

Figure 3.40 Colour components are converted to colour difference signals by the transcoding shown here.

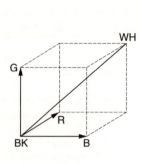

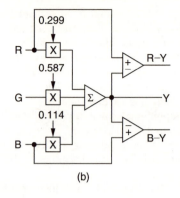

(a)   (b)

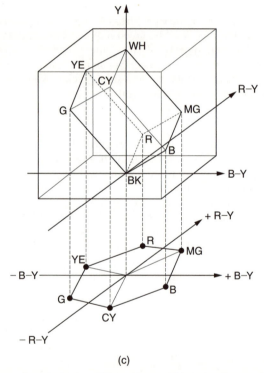

(c)

Figure 3.41 *RGB* transformed to colour difference space. This is done because *R–Y* and *B–Y* can be sent with reduced bandwidth. (a) *RGB* cube. WH–BK axis is diagonal. All locations within cube are legal, (b) *RGB* to colour difference transform. (c) *RGB* cube mapped into colour difference space is no longer a cube. Projection down creates conventional vectorscope display.

Whilst signals such as *Y*, *R*, *G* and *B* are unipolar or positive only, colour difference signals are bipolar and may meaningfully take on negative values. Figure 3.41(a) shows the colour space available in eight-bit *RGB*.

Figure 3.41(c) shows the *RGB* cube mapped into eight-bit colour difference space so that it is no longer a cube. Now the grey axis goes straight up the middle because greys correspond to both C_r and C_b being zero. To visualize colour difference space, imagine looking down along the grey axis. This makes the black and white corners coincide in the centre. The remaining six corners of the legal colour difference space now correspond to the six boxes on a component vectorscope. Although there are still 16 million combinations, many of these are now illegal. For example, as black or white are approached, the colour differences must fall to zero.

From an information theory standpoint, colour difference space is redundant. With some tedious geometry, it can be shown that less than a quarter of the codes are legal. The luminance resolution remains the same, but there is about half as much information in each colour axis. This due to the colour difference signals being bipolar. If the signal resolution has to be maintained, eight-bit *RGB* should be transformed to a longer wordlength in the colour difference domain, nine bits being adequate. At this stage the colour difference transform doesn't seem efficient because twenty-four-bit *RGB* converts to twenty-six-bit Y, C_r, C_b.

In most cases the loss of colour resolution is invisible to the eye, and eight-bit resolution is retained. The results of the transform computation must be digitally dithered to avoid posterizing.

The inverse transform to obtain *RGB* again at the display is straightforward. R and B are readily obtained by adding Y to the two colour difference signals. G is obtained by rearranging the expression for Y above such that:

$$G = \frac{Y - 0.3R - 0.11B}{0.59}$$

If a monochrome source having only a Y output is supplied to a colour display, C_r and C_b will be zero. It is reasonably obvious that if there are no colour difference signals the colour signals cannot be different from one another and $R = G = B$. As a result the colour display can produce only a neutral picture.

The use of colour difference signals is essential for compatibility in both directions between colour and monochrome, but it has a further advantage that follows from the way in which the eye works. In order to produce the highest resolution in the fovea, the eye will use signals from all types of cone, regardless of colour. In order to determine colour the stimuli from three cones must be compared.

There is evidence that the nervous system uses some form of colour difference processing to make this possible. As a result the full acuity of the human eye is available only in monochrome. Detail in colour changes cannot be resolved so well. A further factor is that the lens in the human eye is not achromatic and this means that the ends of the spectrum are not well focused. This is particularly noticeable on blue.

In this case there is no point is expending valuable bandwidth sending high-resolution colour signals. Colour difference working allows the luminance to be sent separately at full bandwidth. This determines the subjective sharpness of the picture. The colour difference information can be sent with considerably reduced resolution, as little as one quarter that of luminance, and the human eye is unable to tell.

The acuity of human vision is axisymmetric. In other words, detail can be resolved equally at all angles. When the human visual system assesses the

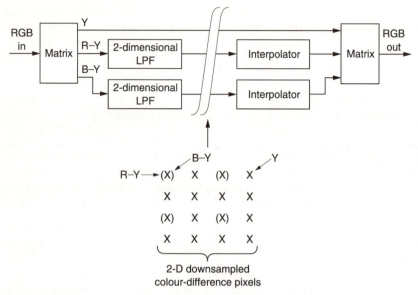

Figure 3.42 Ideal two-dimensionally downsampled colour-difference system. Colour resolution is half of luma resolution, but the eye cannot tell the difference.

sharpness of a TV picture, it will measure the quality of the worst axis and the extra information on the better axis is wasted. Consequently the most efficient row-and-column image sampling arrangement is the so-called 'square pixel'. Now pixels are dimensionless and so this is meaningless. However, it is understood to mean that the horizontal and vertical spacing between pixels is the same. Thus it is the sampling grid which is square, rather than the pixel.

The square pixel is optimal for luminance and also for colour difference signals. Figure 3.42 shows the ideal. The colour sampling is co-sited with the luminance sampling but the colour sample spacing is twice that of luminance. The colour difference signals after matrixing from *RGB* have to be low-pass filtered in two dimensions prior to downsampling in order to prevent aliasing of HF detail. At the display, the downsampled colour data have to be interpolated in two dimensions to produce colour information in every pixel. In an oversampling display the colour interpolation can be combined with the display upsampling stage.

Co-siting the colour and luminance pixels means that the transmitted colour values are displayed unchanged. Only the interpolated values need to be calculated. This minimizes generation loss in the filtering. Downsampling the colour by a factor of two in both axes means that the colour data are reduced to one quarter of their original amount. When viewed by a human this is essentially a lossless process.

References

1. Nyquist, H., Certain topics in telegraph transmission theory. *AIEE Trans.*, 617–644 (1928)
2. Shannon, C.E., A mathematical theory of communication. *Bell Syst. Tech. J.*, **27**, 379 (1948)
3. Jerri, A.J., The Shannon sampling theorem – its various extensions and applications: a tutorial review. *Proc. IEEE*, **65**, 1565–1596 (1977)

4. Whittaker, E.T., On the functions which are represented by the expansions of the interpolation theory. *Proc. R. Soc. Edinburgh*, 181–194 (1915)
5. Porat, B., *A Course in Digital Signal Processing*, New York: John Wiley (1997)
6. Betts, J.A., *Signal Processing Modulation and Noise*, Chapter 6, Sevenoaks: Hodder and Stoughton (1970)
7. Kell, R., Bedford, A. and Trainer, An experimental television system, Part 2. *Proc. IRE*, **22**, 1246–1265 (1934)
8. Goodall, W. M., Television by pulse code modulation. *Bell System Tech. Journal*, **30**, 33–49 (1951)
9. Roberts, L. G., Picture coding using pseudo-random noise. *IRE Trans. Inform. Theory*, **IT-8**, 145–154 (1962)
10. Vanderkooy, J. and Lipshitz, S.P., Resolution below the least significant bit in digital systems with dither. *J. Audio Eng. Soc.*, **32**, 106–113 (1984)
11. Vanderkooy, J. and Lipshitz, S.P., Digital dither. Presented at 81st Audio Eng. Soc. Conv. (Los Angeles 1986), Preprint 2412 (C–8)
12. Hauser, M.W., Principles of oversampling A/D conversion. *J. Audio Eng. Soc.*, **39**, 3–26 (1991)

Chapter 4

Digital video processing

The conversion process expresses the analog input as a numerical code. Within the digital domain, all signal processing must be performed by arithmetic manipulation of the code values. Whilst this was at one time done in dedicated logic circuits, the advancing speed of computers means that increasingly the processes explained here will be performed under software control. However, the principles remain the same.

4.1 A simple digital vision mixer

During production, the samples or pixels representing video images need to be mixed with others. Effects such as dissolves and soft-edged wipes can only be achieved if the level of each input signal can be controlled independently. Level is controlled in the digital domain by multiplying every pixel value by a coefficient. If that coefficient is less than one, attenuation will result; if it is greater than one, amplification can be obtained.

Multiplication in binary circuits is difficult. It can be performed by repeated adding, but this is slow. In fast multiplication, one of the inputs will be simultaneously multiplied by one, two, four, etc., by hard-wired bit shifting. Figure 4.1 shows that the other input bits will determine which of these powers will be added to produce the final sum, and which will be neglected. If multiplying by five, the process is the same as multiplying by four, multiplying by one, and adding the two products. This is achieved by adding the input to itself shifted two places. As the wordlength of such a device increases, the complexity increases exponentially, so this is a natural application for an integrated circuit. It is probably true that digital video would not have been viable without such chips.

In a digital mixer, the gain coefficients may originate in hand-operated faders, just as in analog. Analog faders may be retained and used to produce a varying voltage which is converted to a digital code or gain coefficient in an ADC, but it is also possible to obtain coefficients directly in digital faders. Digital faders are a form of displacement transducer known as an encoder in which the mechanical position of the control is converted directly to a digital code. The position of other controls, such as jog wheels on VTRs or editors, will also need to be digitized. Encoders can be linear or rotary, and absolute or relative. In an absolute encoder, the position of the knob determines the output directly. In a

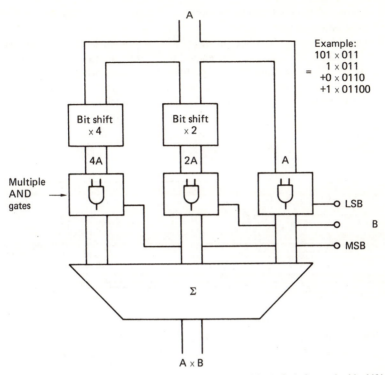

Example:
101 × 011
= 1 × 011
+0 × 0110
+1 × 01100

Figure 4.1 Structure of fast multiplier: the input A is multiplied by 1, 2, 4, 8, etc., by bit shifting. The digits of the B input then determine which multiples of A should be added together by enabling AND gates between the shifters and the adder. For long wordlengths, the number of gates required becomes enormous, and the device is best implemented in a chip.

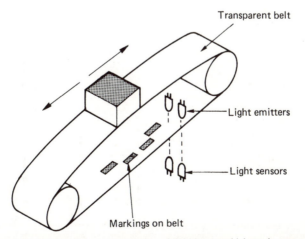

Figure 4.2 An absolute linear fader uses a number of light beams which are interrupted in various combinations according to the position of a grating. A Gray code shown in Figure 4.3 must be used to prevent false codes.

relative control, the knob can be moved to increase or decrease the output, but its
absolute position is meaningless.

Figure 4.2 shows an absolute linear encoder. A grating is moved with respect
to several light beams, one for each bit of the coefficient required. The
interruption of the beams by the grating determines which photocells are
illuminated. It is not possible to use a pure binary pattern on the grating because
this results in transient false codes due to mechanical tolerances. Figure 4.3

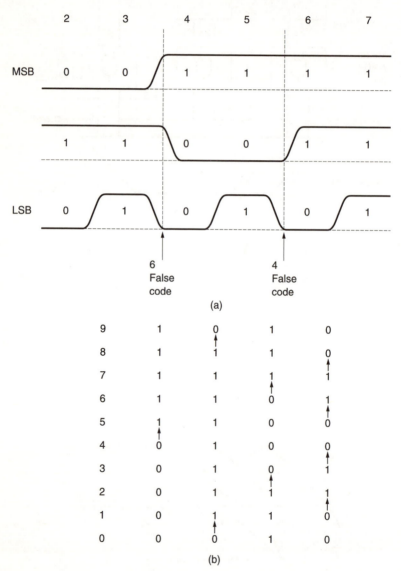

Figure 4.3 (a) Binary cannot be used for position encoders because mechanical tolerances cause
false codes to be produced. (b) In Gray code, only one bit (arrowed) changes in between
positions, so no false codes can be generated.

shows some examples of these false codes. For example, on moving the fader from 3 to 4, the MSB goes true slightly before the middle bit goes false. This results in a momentary value of $4 + 2 = 6$ between 3 and 4. The solution is to use a code in which only one bit ever changes in going from one value to the next. One such code is the Gray code, which was devised to overcome timing hazards in relay logic but is now used extensively in encoders. Gray code can be converted to binary in a suitable PROM or gate array or in software.

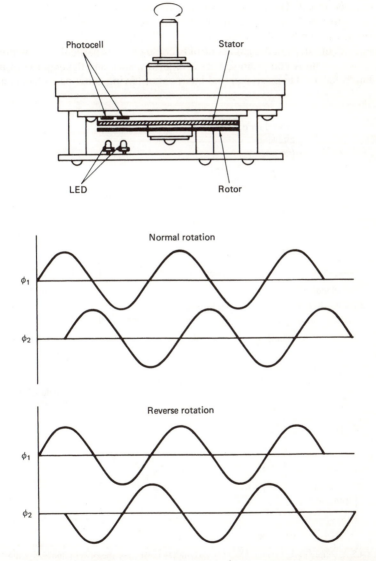

Figure 4.4 The fixed and rotating gratings produce moiré fringes which are detected by two light paths as quadrature sinusoids. The relative phase determines the direction, and the frequency is proportional to speed of rotation.

Figure 4.4 shows a rotary incremental encoder. This produces a sequence of pulses whose number is proportional to the angle through which it has been turned. The rotor carries a radial grating over its entire perimeter. This turns over a second fixed radial grating whose bars are not parallel to those of the first grating. The resultant moiré fringes travel inward or outward depending on the direction of rotation. Two suitably positioned light beams falling on photocells will produce outputs in quadrature. The relative phase determines the direction and the frequency is proportional to speed. The encoder outputs can be connected to a counter whose contents will increase or decrease according to the direction the rotor is turned. The counter provides the coefficient output.

The wordlength of the gain coefficients requires some thought as they determine the number of discrete gains available. If the coefficient wordlength is inadequate, the steps in the gain control become obvious particularly towards the end of a fadeout. A compromise between performance and the expense of high-resolution faders is to insert a digital interpolator having a low-pass characteristic

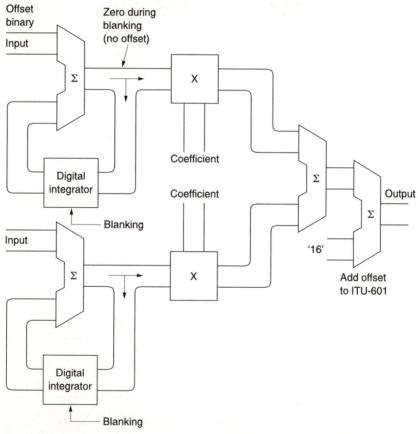

Figure 4.5 A simple digital mixer. Offset binary inputs must have the offset removed. A digital integrator will produce a counter-offset which is subtracted from every input sample. This will increase or reduce until the output of the subtractor is zero during blanking. The offset must be added back after processing if an ITU-601 output is required.

between the fader and the gain control stage. This will compute intermediate gains to higher resolution than the coarse fader scale so that the steps cannot be discerned.

The luminance path of a simple component digital mixer is shown in Figure 4.5. The ITU-601 digital input uses offset binary in that it has a nominal black level of 16_{10} in an eight-bit pixel, and a subtraction has to be made in order that fading will take place with respect to black. On a perfectly converted signal, subtracting 16 would achieve this, but on a signal subject to a DC offset it would not. Since the digital active line is slightly longer than the analog active line, the first sample should be blanking level, and this will be the value to subtract to obtain pure binary luminance with respect to black. This is the digital equivalent of black level clamping. The two inputs are then multiplied by their respective coefficients, and added together to achieve the mix. Peak limiting will be required as in section 2.13, and then, if the output is to be to ITU-601, 16_{10} must be added to each sample value to establish the correct offset. In some video applications, a crossfade will be needed, and a

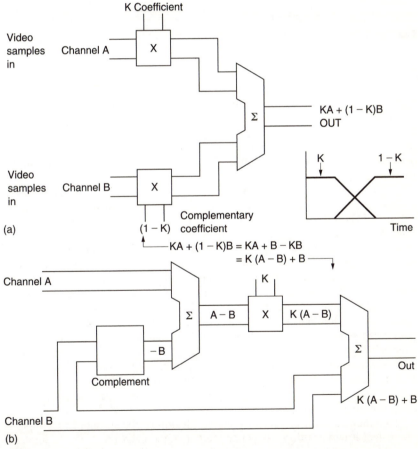

Figure 4.6 Crossfade at (a) requires two multipliers. Reconfiguration at (b) requires only one multiplier.

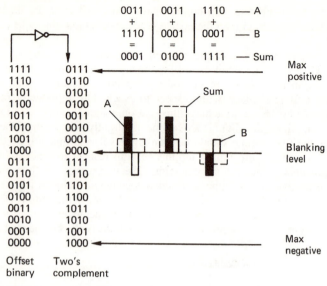

Figure 4.7 Offset binary colour difference values are converted to two's complement by reversing the state of the first bit. Two's complement values A and B will then add around blanking level.

rearrangement of the crossfading equation allows one multiplier to be used instead of two, as shown in Figure 4.6.

The colour difference signals are offset binary with an offset of 128_{10}, and again it is necessary to normalize these with respect to blanking level so that proper fading can be carried out. Since colour difference signals can be positive or negative, this process results in two's complement samples. Figure 4.7 shows some examples.

In this form, the samples can be added with respect to blanking level. Following addition, a limiting stage is used as before, and then, if it is desired to return to ITU-601 standard, the MSB must be inverted once more in order to convert from two's complement to offset binary.

In practice the same multiplier can be used to process luminance and colour difference signals. Since these will be arriving time multiplexed at 27 MHz, it is only necessary to ensure that the correct coefficients are provided at the right time. Figure 4.8 shows an example of part of a slow fade. As the co-sited samples C_B, Y and C_R enter, all are multiplied by the same coefficient K_n, but the next sample will be luminance only, so this will be multiplied by K_{n+1}. The next set of co-sited samples will be multiplied by K_{n+2} and so on. Clearly, coefficients must be provided which change at 13.5 MHz. The sampling rate of the two inputs must be exactly the same, and in the same phase, or the circuit will not be able to add on a sample-by-sample basis. If the two inputs have come from different sources, they must be synchronized by the same master clock, and/or timebase correction must be provided on the inputs.

Some thought must be given to the wordlength of the system. If a sample is attenuated, it will develop bits which are below the radix point. For example, if an eight-bit sample is attenuated by 24 dB, the sample value will be shifted four places down. Extra bits must be available within the mixer to accommodate this

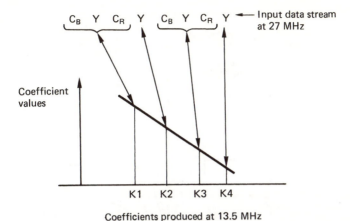

Coefficients produced at 13.5 MHz

Figure 4.8 When using one multiplier to fade both luminance and colour difference in a 27 MHz multiplex 4:2:2 system, one coefficient will be used three times on the co-sited samples, whereas the next coefficient will only be used for a single luminance sample.

shift. Digital vision mixers may have an internal wordlength of sixteen bits or more. When several attenuated sources are added together to produce the final mix, the result will be a sixteen-bit sample stream. As the output will generally need to be of the same format as the input, the wordlength must be shortened.

Shortening the wordlength of samples effectively makes the quantizing intervals larger and Chapter 3 showed that this can be called requantizing. This must be done using digital dithering to avoid artifacts.

4.2 Keying

Keying is the process where one video signal can be cut into another to replace part of the picture with a different image. One application of keying is where a switcher can wipe from one input to another using one of a variety of different patterns. Figure 4.9 shows that an analog switcher performs such an effect by generating a binary switching waveform in a pattern generator. Video switching between inputs actually takes place during the active line. In most analog switchers, the switching waveform is digitally generated, then fed to a D-A convertor, whereas in a digital switcher, the pattern generator outputs become the coefficients supplied to the crossfader, which is sometimes referred to as a cutter. The switching edge must be positioned to an accuracy of a few nanoseconds, much less than the spacing of the pixels, otherwise slow wipes will not appear to move smoothly, and diagonal wipes will have stepped edges, a phenomenon known as ratcheting.

Positioning the switch point to sub-pixel accuracy is not particularly difficult, as Figure 4.10 shows. A suitable series of coefficients can position the effective crossover point anywhere. The finite slope of the coefficients results in a brief crossfade from one video signal to the other. This soft-keying gives a much more realistic effect than binary switchers, which often give a 'cut out with scissors' appearance. In some machines the slope of the crossfade can be adjusted to achieve the desired degree of softness.

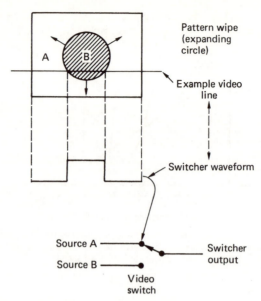

Pattern wipe
(expanding
circle)

Example video
line

Switcher waveform

Source A

Source B

Switcher
output

Video
switch

Figure 4.9 In a video switcher a pattern generator produces a switching waveform which changes from line to line and from frame to frame to allow moving pattern wipes between sources.

Another application of keying is to derive the switching signal by processing video from a camera in some way. By analysing colour difference signals, it is possible to determine where in a picture a particular colour occurs. When a key signal is generated in this way, the process is known as chroma keying, which is the electronic equivalent of matting in film.

In a 4:2:2 component system, It will be necessary to provide coefficients to the luminance crossfader at 13.5 MHz. Chroma samples only occur at half this frequency, so it is necessary to provide a chroma interpolator to artificially raise the chroma sampling rate. For chroma keying a simple linear interpolator is perfectly adequate.

As with analog switchers, chroma keying is also possible with composite digital inputs, but decoding must take place before it is possible to obtain the key signals. The video signals which are being keyed will, however, remain in the composite digital format.

In switcher/keyers, it is necessary to obtain a switching signal which ramps between two states from an input signal which can be any allowable video waveform. Manual controls are provided so that the operator can set thresholds and gains to obtain the desired effect. In the analog domain, these controls distort the transfer function of a video amplifier so that it is no longer linear. A digital keyer will perform the same functions using logic circuits.

Figure 4.11(a) shows the effect of a non-linear transfer function is to switch when the input signal passes through a particular level. The transfer function is implemented in a memory in digital systems. The incoming video sample value acts as the memory address, so that the selected memory location is proportional to the video level. At each memory location, the appropriate output level code is stored. If, for example, each memory location stored its own address, the output

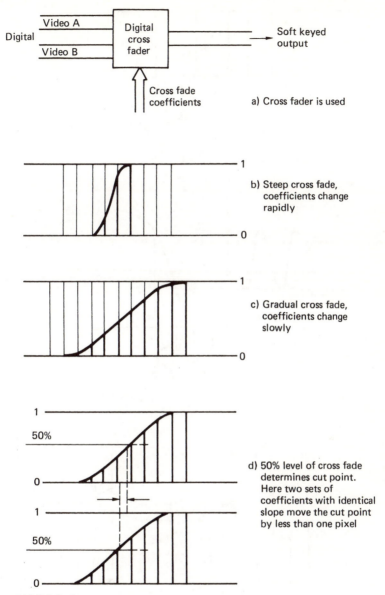

Figure 4.10 Soft keying.

would equal the input, and the device would be transparent. In practice, switching is obtained by distorting the transfer function to obtain more gain in one particular range of input levels at the expense of less gain at other input levels. With the transfer function shown in Figure 4.11(b), an input level change from a to b causes a smaller output change, whereas the same level change between c and d causes a considerable output change.

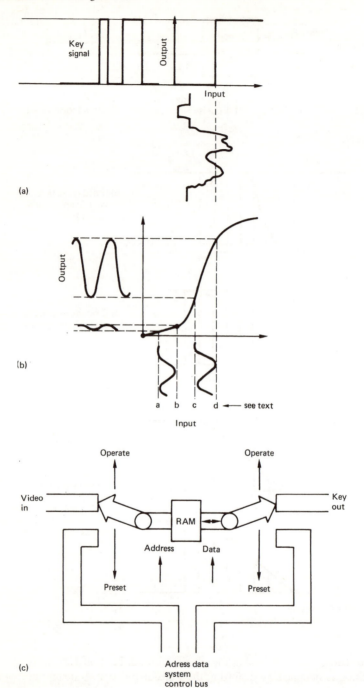

(a)

(b)

(c)

Figure 4.11 (a) A non-linear transfer function can be used to produce a keying signal. (b) The non-linear transfer function emphasizes contrast in part of the range but reduces it at other parts. (c) If a RAM is used as a flexible transfer function, it will be necessary to provide multiplexers so that the RAM can be preset with the desired values from the control system.

If the memory is RAM, different transfer functions can be loaded in by the control system, and this requires multiplexers in both data and address lines as shown in Figure 4.11(c). In practice such a RAM will be installed in Y, C_r and C_b channels, and the results will be combined to obtain the final switching coefficients.

4.3 Digital video effects

If a RAM of the type shown in Figure 4.11 is inserted in a digital luminance path, the result will be *solarizing*, which is a form of contrast enhancement. Figure 4.12 shows that a family of transfer functions can be implemented which control the degree of contrast enhancement. When the transfer function becomes so distorted that the slope reverses, the result is *luminance reversal*, where black and white are effectively interchanged. Solarizing can also be implemented in colour difference channels to obtain *chroma solarizing*. In effects machines, the degree of solarizing may need to change smoothly so that the effect can be gradually introduced. In this case the various transfer functions will be kept in different pages of a PROM, so that the degree of solarization can be selected immediately by changing the page address of the PROM. One page will have a straight transfer function, so the effect can be turned off by selecting that page.

In the digital domain it is easy to introduce various forms of quantizing distortion to obtain special effects. Figure 4.13 shows that eight-bit luminance allows 256 different brightnesses, which to the naked eye appears to be a continuous range. If some of the low-order bits of the samples are disabled, then a smaller number of brightness values describes the range from black to white. For example, if six bits are disabled, only two bits remain, and so only four possible brightness levels can be output. This gives an effect known as *contouring* since the visual effect somewhat resembles a relief map.

When the same process is performed with colour difference signals, the result is to limit the number of possible colours in the picture, which gives an effect known as *posterizing*, since the picture appears to have been coloured by paint from pots. Solarizing, contouring and posterizing cannot be performed in the composite digital domain, owing to the presence of the subcarrier in the sample values.

Figure 4.14 shows a latch in the luminance data which is being clocked at the sampling rate. It is transparent to the signal, but if the clock to the latch is divided down by some factor n, the result will be that the same sample value will be held on the output for n clock periods, giving the video waveform a staircase characteristic. This is the horizontal component of the effect known as *mosaicing*. The vertical component is obtained by feeding the output of the latch into a line memory, which stores one horizontally mosaiced line and then repeats that line m times. As n and m can be independently controlled, the mosaic tiles can be made to be of any size, and rectangular or square at will. Clearly, the mosaic circuitry must be implemented simultaneously in luminance and colour difference signal paths. It is not possible to perform mosaicing on a composite digital signal, since it will destroy the subcarrier. It is common to provide a bypass route which allows mosaiced and unmosaiced video to be simultaneously available. Dynamic switching between the two sources controlled by a separate key signal then allows mosaicing to be restricted to certain parts of the picture.

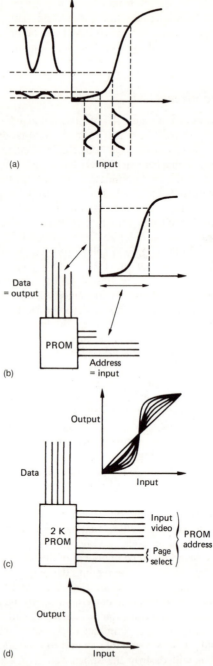

(a)

Input

(b)

Data
= output

PROM

Address
= input

Output

Input

(c)

Data

2 K
PROM

Input
video

Page
select

PROM
address

(d)

Output

Input

Figure 4.12 Solarization. (a) The non-linear transfer function emphasizes contrast in part of the range but reduces it at other parts. (b) The desired transfer function is implemented in a PROM. Each input sample value is used as the address to select a corresponding output value stored in the PROM. (c) A family of transfer functions can be accommodated in a larger PROM. Page select affects the high-order address bits. (d) Transfer function for luminance reversal.

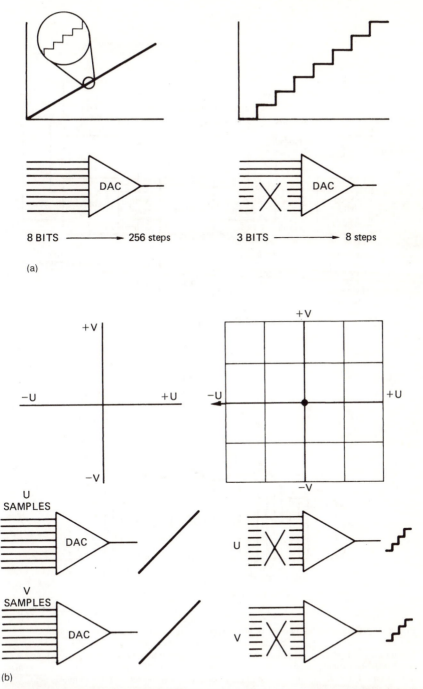

Figure 4.13 (a) In contouring, the least significant bits of the luminance samples are discarded, which reduces the number of possible output levels. (b) At left, the eight-bit colour difference signals allow 2^{16} different colours. At right, eliminating all but two bits of each colour difference signals allows only 2^4 different colours.

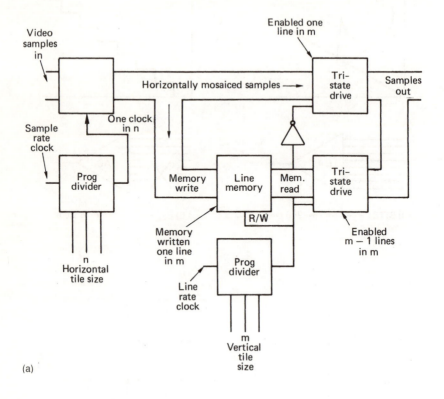

(a)

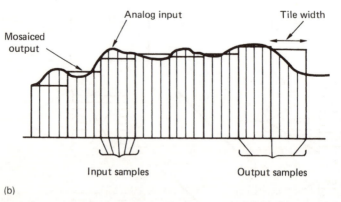

(b)

Figure 4.14 (a) Simplified diagram of mosaicing system. At the left-hand side, horizontal mosaicing is done by intercepting sample clocks. On one line in m, the horizontally mosaiced line becomes the output, and is simultaneously written into a one-line memory. On the remaining $(m-1)$ lines the memory is read to produce several identical successive lines to give the vertical dimensions of the tile. (b) In mosaicing, input samples are neglected, and the output is held constant by failing to clock a latch in the data stream for several sample periods. Heavy vertical lines here correspond to the clock signal occurring. Heavy horizontal line is resultant waveform.

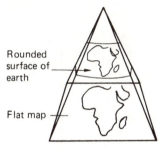

Rounded
surface of
earth

Flat map

Figure 4.15 Map projection is a close relative of video effects units which manipulate the shape of pictures.

The above simple effects did not change the shape or size of the picture, whereas the following class of manipulations will. Effects machines which manipulate video pictures are close relatives of the machines that produce computer-generated images.

The principle of all video manipulators is the same as the technique used by cartographers for centuries. Cartographers are faced with a continual problem in that the earth is round, and paper is flat. In order to produce flat maps, it is necessary to project the features of the round original onto a flat surface. Figure 4.15 shows an example of this. There are a number of different ways of projecting maps, and all of them must by definition produce distortion. The effect of this distortion is that distances measured near the extremities of the map appear further than they actually are. Another effect is that *great circle routes* (the shortest or longest path between two places on a planet) appear curved on a projected map. The type of projection used is usually printed somewhere on the map, a very common system being that due to Mercator. Clearly, the process of mapping involves some three-dimensional geometry in order to simulate the paths of light rays from the map so that they appear to have come from the curved surface. Video effects machines work in exactly the same way.

The distortion of maps means that things are not where they seem. In timesharing computers, every user appears to have his or her own identical address space in which the program resides, despite the fact that many different programs are simultaneously in the memory. In order to resolve this contradiction, memory management units are constructed which add a constant value to the address which the user thinks he has (the *virtual address*) in order to produce the *physical address*. As long as the unit gives each user a different constant, they can all program in the same virtual address space without one corrupting another's programs. Because the program is no longer where it seems to be, the term of 'mapping' was introduced. The address space of a computer is one-dimensional, but a video frame expressed as rows and columns of pixels can be considered to have a two-dimensional address as in Figure 4.16. Video manipulators work by mapping the pixel addresses in two dimensions.

All manipulators must begin with an array of pixels, in which the columns must be vertical. This can only be easily obtained if the sampling rate of the incoming video is a multiple of line rate, as is done in ITU-601. Composite video cannot be used directly, because the phase of the subcarrier would become meaningless after manipulation. If a composite video input is used, it must be

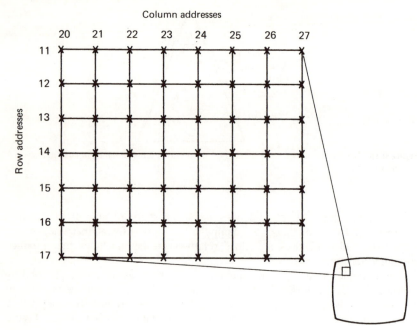

Figure 4.16 The entire TV picture can be broken down into uniquely addressable pixels.

decoded to baseband luminance and colour difference, so that in actuality there are three superimposed arrays of samples to be processed, one luminance, and two colour difference. Most production DVEs use ITU-601 sampling, so digital signals conforming to this standard could be used directly, from, for example, a DVTR or a hard-disk recorder.

DVEs work best with progressively scanned material and the use of interlace causes some additional problems. A DVE designed to work with interlaced inputs must de-interlace first so that all the vertical resolution information is available in every image to be mapped. The most accurate de-interlacing requires motion compensation which is considered in section 4.7.

The additional cost and complexity of de-interlacing is substantial, but this is essential for post-production work, where the utmost quality is demanded. Where low cost is the priority, a field-based DVE system may be acceptable.

The effect of de-interlacing is to produce an array of pixels which are the input data for the processor. In 13.5 MHz systems, the array will be about 720 pixels across and 600 down for 50 Hz systems, 500 down for 60 Hz systems. Most machines are set to strip out VITC or teletext data from the input, so that pixels representing genuine picture appear surrounded by blanking level.

Every pixel in the frame array has an address. The address is two-dimensional, because in order to uniquely specify one pixel, the column address and the row address must be supplied. It is possible to transform a picture by simultaneously addressing in rows and columns, but this is complicated, and very difficult to do in real time. It was discovered some time ago in connection with computer graphics that the two-dimensional problem can be converted with care into two one-dimensional problems.[1] Essentially if a horizontal transform affecting whole rows of pixels independently of other rows is performed on the array, followed by or preceded by a vertical transform which affects entire columns independently of other columns, the effect will be the same as if a two-dimensional transform had been performed. This is the principle of separability. From an academic standpoint, it does not matter which transform is performed first. In a world which is wedded to the horizontally scanned television set, there are practical matters to consider. In order to convert a horizontal raster input into a vertical column format, a memory is needed where the input signal is written in rows, but the output is read as columns of pixels.

The process of writing rows and reading columns in a memory is called transposition. Clearly, two stages of transposition are necessary to return to a horizontal raster output, as shown in Figure 4.17. The vertical transform must take place between the two transposes, but the horizontal transform could take place before the first transpose or after the second one. In practice, the horizontal transform cannot be placed before the first transpose, because it would interfere with the de-interlace and motion-sensing process. The horizontal transform is placed after the second transpose, and reads rows from the second transpose memory. As the output of the machine must be a horizontal raster, the horizontal transform can be made to work in synchronism with reference H-sync so that the digital output samples from the H-transform can be taken direct to a DAC to become analog video with a standard structure again. A further advantage is that the real-time horizontal output is one field at a time. The preceding vertical transform need only compute array values which lie in the next field to be needed at the output.

It is not possible to take a complete column from a memory until all the rows have entered. The presence of the two transposes in a DVE results in an unavoidable delay of one frame in the output image. Some DVEs can manage greater delay. The effect on lip-sync may have to be considered. It is not

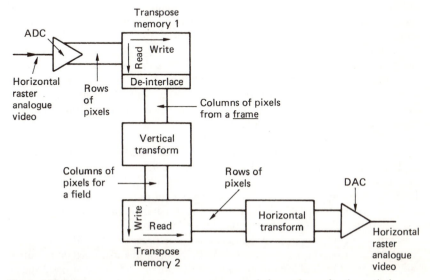

Figure 4.17 Two transposing memories are necessary, one before and one after the vertical transform.

advisable to cut from the input of a DVE to the output, since a frame will be lost, and on cutting back a frame will be repeated.

Address mapping is used to perform transforms. Now that rows and columns are processed individually, the mapping process becomes much easier to understand. Figure 4.18 shows a single row of pixels which are held in a buffer where each can be addressed individually and transferred to another. If a constant is added to the read address, the selected pixel will be to the right of the place where it will be put. This has the effect of moving the picture to the left. If the buffer represented a column of pixels, the picture would be moved vertically. As these two transforms can be controlled independently, the picture could be moved diagonally.

If the read address is multiplied by a constant, say 2, the effect is to bring samples from the input closer together on the output, so that the picture size is reduced. Again independent control of the horizontal and vertical transforms is possible, so that the aspect ratio of the picture can be modified. This is very useful for telecine work when CinemaScope films are to be broadcast. Clearly the secret of these manipulations is in the constants fed to the address generators. The added constant represents displacement, and the multiplied constant represents magnification. A multiplier constant of less than one will result in the picture getting larger. Figure 4.18 also shows, however, that there is a problem. If a constant of 0.5 is used, to make the picture twice as big, half of the addresses generated are not integers. A memory does not understand an address of two and a half!)

If an arbitrary magnification is used, nearly all the addresses generated are non-integer. A similar problem crops up if a constant of less than one is added to the address in an attempt to move the picture less than the pixel spacing. The solution to the problem is interpolation. Because the input image is spatially sampled, those samples contain enough information to represent the brightness and colour all over the screen. When the address generator comes up with an address of 2.5, it actually means that what is wanted is the value of the signal interpolated half-way between pixel two and pixel three. The output of the address generator will thus be split into two parts. The integer part will become the memory address, and the fractional part is the phase of the necessary interpolation. In order to interpolate pixel values a digital filter is necessary.

Figure 4.19 shows that the input and output of an effects machine must be at standard sampling rates to allow digital interchange with other equipment. When the size of a picture is changed, this causes the pixels in the picture to fail to register with output pixel spacing. The problem is exactly the same as sampling rate conversion, which produces a differently spaced set of samples that still represent the original waveform. One pixel value actually represents the peak brightness of a two-dimensional intensity function, which is the effect of the modulation transfer function of the system on an infinitely small point. As each dimension can be treated separately, the equivalent in one axis is that the pixel value represents the peak value of an infinitely short impulse which has been low-pass filtered to the system bandwidth and windowed.

In order to compute an interpolated value, it is necessary to add together the contribution from all relevant samples, at the point of interest. Each contribution can be obtained by looking up the value of the impulse response at the distance from the input pixel to the output pixel to obtain a coefficient, and multiplying the input pixel value by that coefficient. The process of taking several pixel

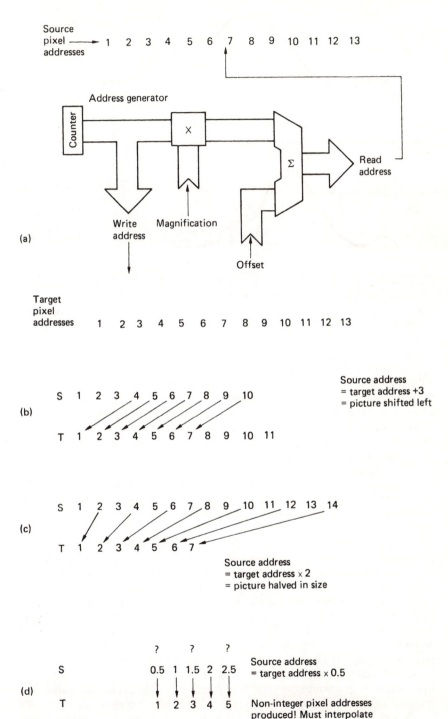

Figure 4.18 Address generation is the fundamental process behind transforms.

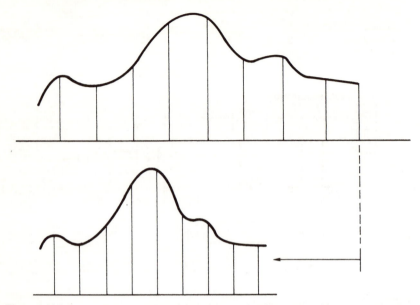

Figure 4.19 It is easy, almost trivial, to reduce the size of a picture by pushing the samples closer together, but this is not often of use, because it changes the sampling rate in proportion to the compression. Where a standard sampling-rate output is needed, interpolation must be used.

values, multiplying each by a different coefficient and summing the products can be performed by the FIR (finite impulse response) configuration described in Chapter 2. The impulse response of the filter necessary depends on the magnification. Where the picture is being enlarged, the impulse response can be the same as at normal size, but as the size is reduced, the impulse response has to become broader (corresponding to a reduced spatial frequency response) so that more input samples are averaged together to prevent aliasing. The coefficient store will need a two-dimensional structure, such that the magnification and the interpolation phase must both be supplied to obtain a set of coefficients. The magnification can easily be obtained by comparing successive outputs from the address generator.

The number of points in the filter is a compromise between cost and performance, eight being a typical number for high quality. As there are two transform processes in series, every output pixel will be the result of sixteen multiplications, so there will be 216 million multiplications per second taking place in the luminance channel alone for a 13.5 MHz sampling rate unit. The quality of the output video also depends on the number of different interpolation phases available between pixels. The address generator may compute fractional addresses to any accuracy, but these will be rounded off to the nearest available phase in the digital filter. The effect is that the output pixel value provided is actually the value a tiny distance away, and has the same result as sampling clock jitter, which is to produce program-modulated noise. The greater the number of phases provided, the larger will be the size of the coefficient store needed. As the coefficient store is two-dimensional, an increase in the number of filter points and phases causes an exponential growth in size and cost.

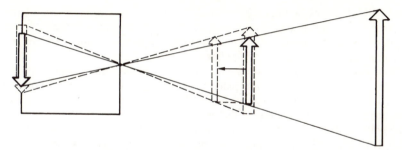

Figure 4.20 The image on the rear of the pinhole camera is identical for the two solid objects shown because the size of the object is proportional to distance, and the subtended angle remains the same. The image can be made larger (dotted) by making the object larger or moving it closer.

In order to follow the operation of a true perspective machine, some knowledge of perspective is necessary. Stated briefly, the phenomenon of perspective is due to the angle subtended to the eye by objects being a function not only of their size but also of their distance. Figure 4.20 shows that the size of an image on the rear wall of a pinhole camera can be increased either by making the object larger or bringing it closer. In the absence of stereoscopic vision, it is not possible to tell which has happened. The pinhole camera is very useful for study of perspective, and has indeed been used by artists for that purpose. The clinically precise perspective of Canaletto paintings was achieved through the use of the camera obscura (Latin: darkened room).[2]

It is sometimes claimed that the focal length of the lens used on a camera changes the perspective of a picture. This is not true, perspective is only a function of the relative positions of the camera and the subject. Fitting a wide-angle lens simply allows the camera to come near enough to keep dramatic perspective within the frame, whereas fitting a long-focus lens allows the camera to be far enough away to display a reasonably sized image with flat perspective.[3]

Since a single eye cannot tell distance unaided, all current effects machines work by simply producing the correct subtended angles which the brain perceives as a three-dimensional effect. Figure 4.21 shows that to a single eye, there is no

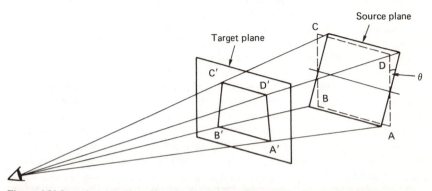

Figure 4.21 In a planar rotation effect the source plane ABCD is the rectangular input picture. If it is rotated through the angle θ, ray tracing to a single eye at left will produce a trapezoidal image A'B'C'D' on the target. Magnification will now vary with position on the picture.

difference between a three-dimensional scene and a two-dimensional image formed where rays traced from features to the eye intersect an imaginary plane. This is exactly the reverse of the map projection shown in Figure 4.15, and is the principle of all perspective manipulators.

The case of perspective rotation of a plane source will be discussed first. Figure 4.21 shows that when a plane input frame is rotated about a horizontal axis, the distance from the top of the picture to the eye is no longer the same as the distance from the bottom of the picture to the eye. The result is that the top and bottom edges of the picture subtend different angles to the eye, and where the rays cross the target plane, the image has become trapezoidal. There is now no such thing as the magnification of the picture. The magnification changes continuously from top to bottom of the picture, and if a uniform grid is input, after a perspective rotation it will appear non-linear, as the diagram shows.

The basic mechanism of the transform process has been described, but this is only half of the story, because these transforms have to be controlled. There is a lot of complex geometrical calculation necessary to perform even the simplest effect, and the operator cannot be expected to calculate directly the parameters required for the transforms. All effects machines require a computer of some kind, with which the operator communicates using keyboard entry or joystick/ trackball movements at high level. These high-level commands will specify such things as the position of the axis of rotation of the picture relative to the viewer, the position of the axis of rotation relative to the source picture, and the angle of rotation in the three axes.

An essential feature of this kind of effects machines is fluid movement of the source picture as the effect proceeds. If the source picture is to be made to move smoothly, then clearly the transform parameters will be different in each field. The operator cannot be expected to input the source position for every field, because this would be an enormous task. Additionally, storing the effect would require a lot of space. The solution is for the operator to specify the picture position at strategic points during the effect, and then digital filters are used to compute the intermediate positions so that every field will have different parameters.

The specified positions are referred to as knots, nodes or keyframes, the first being the computer graphics term. The operator is free to enter knots anywhere in the effect, and so they will not necessarily be evenly spaced in time, i.e. there may well be different numbers of fields between each knot. In this environment it is not possible to use conventional FIR-type digital filtering, because a fixed impulse response is inappropriate for irregularly spaced samples.

Interpolation of various orders is used ranging from zero-order hold for special jerky effects through linear interpolation to cubic interpolation for very smooth motion. The algorithms used to perform the interpolation are known as splines, a term which has come down from shipbuilding via computer graphics.[4] When a ship is designed, the draughtsman produces hull cross-sections at intervals along the keel, whereas the shipyard needs to re-create a continuous structure. The solution was a lead-filled bar, known as a spline, which could be formed to join up each cross-section in a smooth curve, and then used as a template to form the hull plating.

The filter which does not ring cannot be made, and so the use of spline algorithms for smooth motion sometimes results in unintentional overshoots of the picture position. This can be overcome by modifying the filtering algorithm.

Spline algorithms usually look ahead beyond the next knot in order to compute the degree of curvature in the graph of the parameter against time. If a break is put in that parameter at a given knot, the spline algorithm is prevented from looking ahead, and no overshoot will occur. In practice, the effect is created and run without breaks, and then breaks are added later where they are subjectively thought necessary.

It will be seen that there are several levels of control in an effects machine. At the highest level, the operator can create, store and edit knots, and specify the times which elapse between them. The next level is for the knots to be interpolated by spline algorithms to produce parameters for every field in the effect. The field frequency parameters are then used as the inputs to the geometrical computation of transform parameters which the lowest level of the machine will use as microinstructions to act upon the pixel data. Each of these layers will often have a separate processor, not just for speed, but also to allow software to be updated at certain levels without disturbing others.

4.4 Graphics

Although there is no easy definition of a video graphics system which distinguishes it from a graphic art system, for the purposes of discussion it can be said that graphics consists of generating alphanumerics on the screen, whereas graphic art is concerned with generating more general images. The simplest form of screen presentation of alphanumerics is the ubiquitous visual display unit (VDU) which is frequently necessary to control computer-based systems. The mechanism used for character generation in such devices is very simple, and thus makes a good introduction to the subject.

In VDUs, there is no grey scale, and the characters are formed by changing the video signal between two levels at the appropriate place in the line. Figure 4.22 shows how a character is built up in this way, and also illustrates how easy it is to obtain the reversed video used in some word processor displays to simulate dark characters on white paper. Also shown is the method of highlighting single characters or words by using localized reverse video.

Figure 4.23 is a representative character generator, as might be used in a VDU. The characters to be displayed are stored as ASCII symbols in a RAM, which has one location for each character position on each available text line on the screen. Each character must be used to generate a series of dots on the screen which will extend over several lines. Typically the characters are formed by an array five dots by nine. In order to convert from the ASCII code to a dot pattern, a ROM is programmed with a conversion. This will be addressed by the ASCII character, and the column and row addresses in the character array, and will output a high or low (bright or dark) output.

As the VDU screen is a raster-scanned device, the display scan will begin at the left-hand end of the top line. The first character in the ASCII RAM will be selected, and this and the first row and column addresses will be sent to the character generator, which ouputs the video level for the first pixel. The next column address will then be selected, and the next pixel will be output. As the scan proceeds, it will pass from the top line of the first character to the top line of the second character, so that the ASCII RAM address will need to be incremented. This process continues until the whole video line is completed. The next line on the screen is generated by repeating the selection of characters from

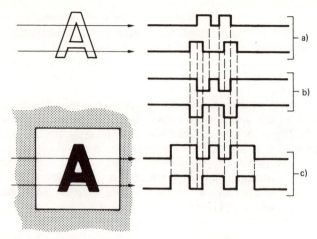

Figure 4.22 Elementary character generation. At (a), white on black waveform for two raster lines passing through letter A. At (b), black on white is simple inversion. At (c), reverse video highlight waveforms.

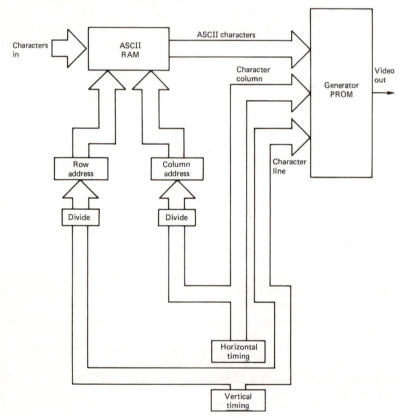

Figure 4.23 Simple character generator produces characters as rows and columns of pixels. See text for details.

the ASCII RAM, but using the second array line as the address to the character generator. This process will repeat until all the video lines needed to form one row of characters are complete. The next row of characters in the ASCII RAM can then be accessed to create the next line of text on the screen and so on.

The character quality of VDUs is adequate for the application, but is not satisfactory for high-quality broadcast graphics. The characters are monochrome, have a fixed, simple font, fixed size, no grey scale, and the sloping edges of the characters have the usual stepped edges due to the lack of grey scale. This stepping of diagonal edges is sometimes erroneously called aliasing. Since it is a result not of inadequate sampling rate, but of quantizing distortion, the use of the term is wholly inappropriate.

In a broadcast graphics unit, the characters will be needed in colour, and in varying sizes. Different fonts will be necessary, and additional features such as solid lines around characters and drop shadows are desirable. The complexity and cost of the necessary hardware is much greater than in the previous example.

In order to generate a character in a broadcast machine, a font and the character within that font are selected. The characters are actually stored as key signals, because the only difference between one character and another in the same font is the shape. A character is generated by specifying a constant background colour and luminance, and a constant character colour and luminance, and using the key signal to cut a hole in the background and insert the character colour. This is illustrated in Figure 4.24. The problem of stepped

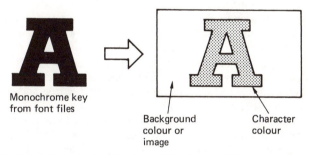

Monochrome key
from font files

Background
colour or
image

Character
colour

Figure 4.24 Font characters only store the shape of the character. This can be used to key any coloured character into a background.

diagonal edges is overcome by giving the key signal a grey scale which eliminates the quantizing distortion responsible for the stepped edges. The edge of the character now takes the form of a ramp, which has the desirable characteristic of limiting the bandwidth of the character generator output. Early character generators were notorious for producing out-of-band frequencies which drove equipment further down the line to distraction and in some cases would interfere with the sound channel on being broadcast. Figure 4.25 illustrates how in a system with grey scale and sloped edges, the edge of a character can be positioned to sub-pixel resolution, which completely removes the stepped effect on diagonals.

In a powerful system, the number of fonts available will be large, and all the necessary characters will be stored on disk drives. Some systems allow users to enter their own fonts using a rostrum camera. A frame grab is performed, but the

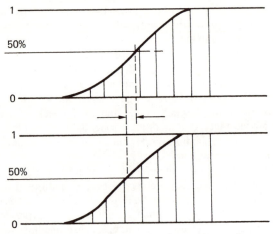

Figure 4.25 When a character has a ramped edge, the edge position can be moved in subpixel steps by changing the pixel values in the ramp.

system can be told to file the image as a font character key signal rather than as a still frame. This approach allows infinite flexibility if it is desired to work in Kanji or Cyrillic, and allows European graphics to be done with all necessary umlauts, tildes and cedillas.

In order to create a character string on the screen, it is necessary to produce a key signal which has been assembled from all the individual character keys. The keys are usually stored in a large format to give highest quality, and it will be necessary to reduce the size of the characters to fit the available screen area. The size reduction of a key signal in the digital domain is exactly the same as the zoom function of an effects machine, requiring FIR filtering and interpolation, but again, it is not necessary for it to be done in real time, and so less hardware can be used. The key source for the generation of the final video output is a RAM which has one location for every screen pixel. Position of the characters on the screen is controlled by changing the addresses in the key RAM into which the size reduced character keys are written.

The keying system necessary is shown in Figure 4.26. The character colour and the background colour are produced by latches on the control system bus, which output continuous digital parameters. The grey scale key signal obtained by scanning the key memory is used to provide coefficients for the digital crossfader which cuts between background and character colour to assemble the video signal in real time.

If characters with contrasting edges are required, an extra stage of keying can be used. The steps described above take place, but the background colour is replaced by the desired character edge colour. The size of each character key is then increased slightly, and the new key signal is used to cut the characters and a contrasting border into the final background.

Early character generators were based on a frame store which refreshes the dynamic output video. Recent devices abandon the framestore approach in favour of real-time synthesis. The symbols which make up a word can move on and turn with respect to the plane in which they reside as a function of time in

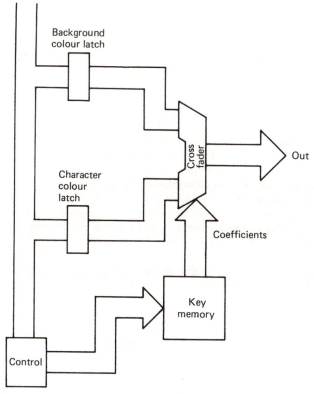

Figure 4.26 Simple character generator using keying. See text for details.

any way individually or together. Text can also be mapped onto an arbitrarily shaped line. The angle of the characters can follow a tangent to the line, or can remain at a fixed angle regardless of the line angle.

By controlling the size of planes, characters or words can appear to zoom into view from a distance and recede again. Rotation of the character planes off the plane of the screen allows the perspective effects to be seen. Rotating a plane back about a horizontal axis by 90° will reduce it to an edge-on line, but lowering the plane to the bottom of the screen allows the top surface to be seen, receding into the distance like a road. Characters or text strings can then roll off into the distance, getting smaller as they go. In fact the planes do not rotate, but a perspective transform is performed on them.

Since characters can be moved without restriction, it will be possible to make them overlap. Either one can be declared to be at the front, so that it cuts out the other, but if desired the overlapping area can be made a different colour from either of the characters concerned. If this is done with care, the overlapping colour will give the effect of transparency. In fact colour is attributed to characters flexibly so that a character may change colour with respect to time. If this is combined with a movement, the colour will appear to change with position. The background can also be allocated a colour in this way, or the background can be input video. Instead of filling characters with colour on a video background, the characters, or only certain characters, or only the overlapping areas of characters can be filled with video.

There will be several planes on which characters can move, and these are assigned a priority sequence so that the first one is essentially at the front of the screen and the last one is at the back. Where no character exists, a plane is transparent, and every character on a plane has an eight-bit transparency figure allocated to it. When characters on two different planes are moved until they overlap, the priority system and the transparency parameter decide what will be seen. An opaque front character will obscure any character behind it, whereas using a more transparent parameter will allow a proportion of a character to be seen through one in front of it. Since characters have multi-bit accuracy, all transparency effects are performed without degradation, and character edges always remain smooth, even at overlaps. Clearly, the character planes are not memories or frame stores, because perspective rotation of images held in frame stores in real time in this way would require staggering processing power. Instead, the contents of output fields are computed in real time. Every field is determined individually, so it is scarcely more difficult to animate by making successive fields different.

In a computer a symbol or character exists not as an array of pixels, but as an outline, rather as if a thin wire frame had been made to fit the edge of the symbol. The wire frame is described by a set of mathematical expressions. The operator positions the wire frame in space, and turns it to any angle. The computer then calculates what the wire frame will look like from the viewing position. Effectively the shape of the frame is projected onto a surface which will become the TV screen. The principle is shown in Figure 4.27, and will be seen to be

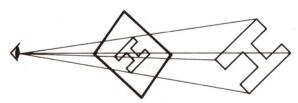

Figure 4.27 Perspective effects results from mapping or projecting the character outline on to a plane which represents the screen.

another example of mapping. If drop or plane shadows are being used, a further projection takes place which determines how the shadow of the wire frame would fall on a second plane. The only difference here between drop and plane shadows is the angle of the plane the shadow falls on. If it is parallel to the symbol frame, the result is a drop shadow. If it is at right angles, the result is a plane shadow.

Since the wire frame can be described by a minimum amount of data, the geometrical calculations needed to project onto the screen and shadow planes are quick, and are repeated for every symbol. This process also reveals where overlaps occur within a plane, since the projected frames will cross each other. This computation takes place for all planes. The positions of the symbol edges are then converted to a pixel array in a field-interlaced raster scan. Because the symbols are described by eight-bit pixels, edges can be positioned to sub-pixel

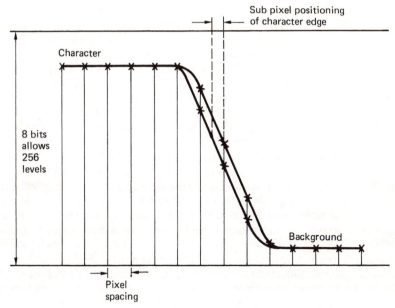

Figure 4.28 The eight-bit resolution of pixels allows the edge of characters to be placed accurately independent of the sampling points. This offers smooth character edges.

accuracy. An example is shown in Figure 4.28. The pixel sequence is now effectively a digital key signal, and the priority and transparency parameters are now used to reduce the amplitude of a given key when it is behind a symbol on a plane of higher priority which has less than 100 per cent transparency.

The key signals then pass to a device known as the filler which is a fast digital mixer that has as inputs the colour of each character, and the background, whether colour or video. On a pixel-by-pixel basis, the filler crossfades between different colours as the video line proceeds. The crossfading is controlled by the key signals. The output is three data streams, red, green and blue. The output board takes the data and converts them to the video standard needed, and the result can then be seen.

In graphic art systems, there is a requirement for disk storage of the generated images, and some art machines incorporate a still store unit, whereas others can be connected to a separate one by an interface. Disk-based stores are discussed in Chapter 7. The essence of an art system is that an artist can draw images which become a video signal directly with no intermediate paper and paint. Central to the operation of most art systems is a digitizing tablet, which is a flat surface over which the operator draws a stylus. The tablet can establish the position of the stylus in vertical and horizontal axes. One way in which this can be done is to launch ultrasonic pulses down the tablet, which are detected by a transducer in the stylus. The time taken to receive the pulse is proportional to the distance to the stylus. The coordinates of the stylus are converted to addresses in the frame store which correspond to the same physical position on the screen. In order to make a simple sketch, the operator specifies a background parameter, perhaps white, which would be loaded into every location in the frame store. A different parameter is then written into every location addressed by movement of the stylus, which results in a line drawing on the screen. The art world uses pens and brushes of different shapes and sizes to obtain a variety of effects, one common example being the rectangular pen nib where the width of the resulting line depends on the angle at which the pen is moved. This can be simulated on art systems, because the address derived from the tablet is processed to produce a range of addresses within a certain screen distance of the stylus. If all these locations are updated as the stylus moves, a broad stroke results.

If the address range is larger in the horizontal axis than in the vertical axis, for example, the width of the stroke will be a function of the direction of stylus travel. Some systems have a sprung tip on the stylus which connects to a force transducer, so that the system can measure the pressure the operator uses. By making the address range a function of pressure, broader strokes can be obtained simply by pressing harder. In order to simulate a set of colours available on a palette, the operator can select a mode where small areas of each colour are displayed in boxes on the monitor screen. The desired colour is selected by moving a screen cursor over the box using the tablet. The parameter to be written into selected locations in the frame RAM now reflects the chosen colour. In more advanced systems, simulation of airbrushing is possible. In this technique, the transparency of the stroke is great at the edge, where the background can be seen showing through, but transparency reduces to the centre of the stroke. A read–modify–write process is necessary in the frame memory, where background values are read, mixed with paint values with the appropriate transparency, and written back. The position of the stylus effectively determines the centre of a two-dimensional transparency contour, which is convolved with the memory contents as the stylus moves.

4.5 Applications of motion compensation

Section 2.10 introduced the concept of eye tracking and the optic flow axis. The optic flow axis is the locus of some point on a moving object which will be in a different place in successive pictures. Any device which computes with respect to the optic flow axis is said to be *motion compensated*. Until recently the amount of computation required in motion compensation was too expensive, but now this is no longer the case the technology has become very important in moving-image portrayal systems.

Figure 4.29(a) shows an example of a moving object which is in a different place in each of three pictures. The optic flow axis is shown. The object is not moving with respect to the optic flow axis and if this axis can be found some very useful results are obtained. The process of finding the optic flow axis is called *motion estimation*. Motion estimation is literally a process which analyses successive pictures and determines how objects move from one to the next. It is an important enabling technology because of the way it parallels the action of the human eye.

Figure 4.29(b) shows that if the object does not change its appearance as it moves, it can be portrayed in two of the pictures by using data from one picture only, simply by shifting part of the picture to a new location. This can be done using vectors as shown. Instead of transmitting a lot of pixel data, a few vectors are sent instead. This is the basis of motion-compensated compression which is used extensively in MPEG, as will be seen in Chapter 5.

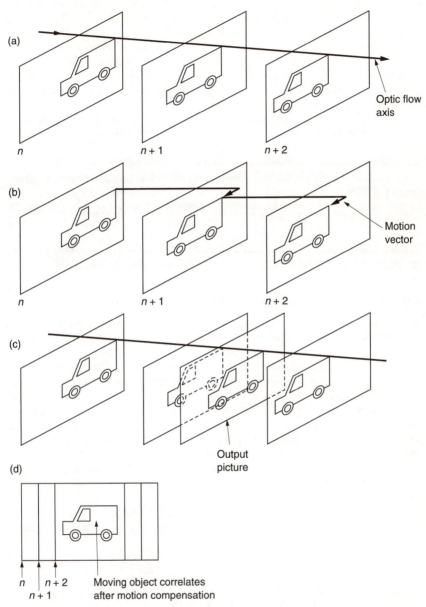

(a)

Optic flow
axis

n   *n* + 1   *n* + 2

(b)

Motion
vector

n   *n* + 1   *n* + 2

(c)

Output
picture

(d)

n   *n* + 2   Moving object correlates
   n + 1   after motion compensation

Figure 4.29 Motion compensation is an important technology. (a) The optic flow axis is found for a moving object. (b) The object in picture (*n* + 1) and (*n* + 2) can be re-created by shifting the object of picture *n* using motion vectors. MPEG uses this process for compression. (c) A standards convertor creates a picture on a new timebase by shifting object data along the optic flow axis. (d) With motion compensation a moving object can still correlate from one picture to the next so that noise reduction is possible.

Figure 4.29(c) shows that if a high-quality standards conversion is required between two different frame rates, the output frames can be synthesized by moving image data, not through time, but along the optic flow axis. This locates objects where they would have been if frames had been sensed at those times, and the result is a judder-free conversion. This process can be extended to drive image displays at a frame rate higher than the input rate so that flicker and background strobing are reduced. This technology is available in certain high-quality consumer television sets. This approach may also be used with 24 Hz film to eliminate judder in telecine machines.

Figure 4.29(d) shows that noise reduction relies on averaging two or more images so that the images add but the noise cancels. Conventional noise reducers fail in the presence of motion, but if the averaging process takes place along the optic flow axis, noise reduction can continue to operate.

The way in which eye tracking avoids aliasing is fundamental to the perceived quality of television pictures. Many processes need to manipulate moving images in the same way in order to avoid the obvious difficulty of processing with respect to a fixed frame of reference. Processes of this kind are referred to as *motion compensated* and rely on a quite separate process which has measured the motion. Motion compensation is also important where interlaced video needs to be processed as it allows the best possible de-interlacing performance.

4.6 Motion-compensated standards conversion

A conventional standards convertor is not transparent to motion portrayal, and the effect is judder and loss of resolution. Figure 4.30 shows what happens on the time axis in a conversion between 60 Hz and 50 Hz (in either direction). Fields in the two standards appear in different planes cutting through the spatio-temporal volume, and the job of the standards convertor is to interpolate along the time axis between input planes in one standard in order to estimate what an intermediate plane in the other standard would look like. With still images, this is easy, because planes can be slid up and down the time axis with no ill effect. If an object is moving, it will be in a different place in successive fields. Interpolating between several fields results in multiple images of the object. The position of the dominant image will not move smoothly, an effect which is perceived as judder. Motion compensation is designed to eliminate this undesirable judder.

A conventional standards convertor interpolates only along the time axis, whereas a motion-compensated standards convertor can swivel its interpolation

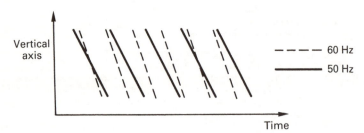

Figure 4.30 The different temporal distribution of input and output fields in a 50/60 Hz convertor.

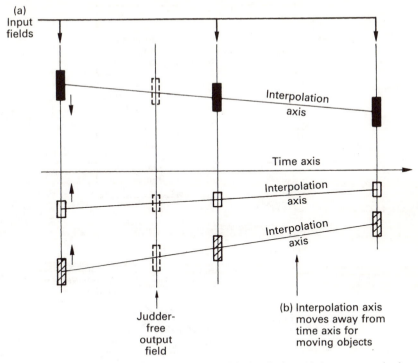

(a) Input fields

Interpolation axis

Time axis

Interpolation axis

Interpolation axis

Judder-free output field

(b) Interpolation axis moves away from time axis for moving objects

Figure 4.31 (a) Input fields with moving objects. (b) Moving the interpolation axes to make them parallel to the trajectory of each object.

axis off the time axis. Figure 4.31(a) shows the input fields in which three objects are moving in a different way. At (b) it will be seen that the interpolation axis is aligned with the optic flow axis of each moving object in turn.

Each object is no longer moving with respect to its own optic flow axis, and so on that axis it no longer generates temporal frequencies due to motion and temporal aliasing due to motion cannot occur.[5] Interpolation along the optic flow axes will then result in a sequence of output fields in which motion is properly portrayed. The process requires a standards convertor which contains filters that are modified to allow the interpolation axis to move dynamically within each output field. The signals which move the interpolation axis are known as motion vectors. It is the job of the motion-estimation system to provide these motion vectors. The overall performance of the convertor is determined primarily by the accuracy of the motion vectors. An incorrect vector will result in unrelated pixels from several fields being superimposed and the result is unsatisfactory.

Figure 4.32 shows the sequence of events in a motion-compensated standards convertor. The motion estimator measures movements between successive fields. These motions must then be attributed to objects by creating boundaries around sets of pixels having the same motion. The result of this process is a set of motion vectors, hence the term 'vector assignation'. The motion vectors are then input to a modified four-field standards convertor in order to deflect the inter-field interpolation axis.

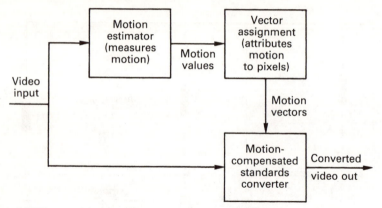

Figure 4.32 The essential stages of a motion-compensated standards convertor.

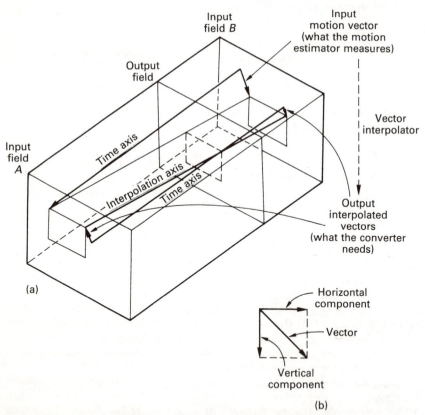

Figure 4.33 The motion vectors on the input field structure must be interpolated onto the output field structure as in (a). The field to be interpolated is positioned temporally between source fields and the motion vector between them is apportioned according to the location. Motion vectors are two dimensional, and can be transmitted as vertical and horizontal components shown at (b) which control the spatial shifting of input fields.

The vectors from the motion estimator actually measure the distance moved by an object from one input field to another. What the standards convertor requires is the value of motion vectors at an output field. A vector interpolation stage is required which computes where between the input fields A and B the current output field lies, and uses this to proportion the motion vector into two parts. Figure 4.33(a) shows that the first part is the motion between field A and the output field; the second is the motion between field B and the output field. Clearly, the difference between these two vectors is the motion between input fields. These processed vectors are used to displace parts of the input fields so that the axis of interpolation lies along the optic flow axis. The moving object is stationary with respect to this axis so interpolation between fields along it will not result in any judder.

Whilst a conventional convertor only needs to interpolate vertically and temporally, a motion-compensated convertor also needs to interpolate horizontally to account for lateral movement in images. Figure 4.33(b) shows that the motion vector from the motion estimator is resolved into two components, vertical and horizontal. The spatial impulse response of the interpolator is shifted in two dimensions by these components. This shift may be different in each of the fields which contribute to the output field.

When an object in the picture moves, it will obscure its background. The vector interpolator in the standards convertor handles this automatically provided the motion estimation has produced correct vectors. Figure 4.34 shows an example of background handling. The moving object produces a finite vector associated with each pixel, whereas the stationary background produces zero vectors except in the area O–X where the background is being obscured. Vectors

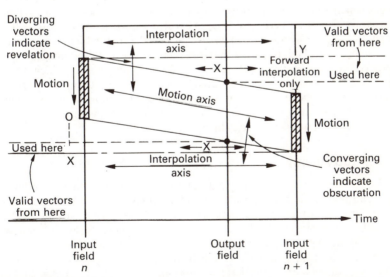

Figure 4.34 Background handling. When a vector for an output pixel near a moving object is not known, the vectors from adjacent background areas are assumed. Converging vectors imply obscuring is taking place which requires that interpolation can only use previous field data. Diverging vectors imply that the background is being revealed and interpolation can only use data from later fields.

converge in the area where the background is being obscured, and diverge where it is being revealed. Image correlation is poor in these areas so no valid vector is assigned.

An output field is located between input fields, and vectors are projected through it to locate the intermediate position of moving objects. These are interpolated along an axis which is parallel to the optic flow axis. This results in address mapping which locates the moving object in the input field RAMs. However, the background is not moving and so the optic flow axis is parallel to the time axis. The pixel immediately below the leading edge of the moving object does not have a valid vector because it is in the area O–X where forward image correlation failed.

The solution is for that pixel to assume the motion vector of the background below point X, but only to interpolate in a backwards direction, taking pixel data from previous fields. In a similar way, the pixel immediately behind the trailing edge takes the motion vector for the background above point Y and interpolates only in a forward direction, taking pixel data from future fields. The result is that the moving object is portrayed in the correct place on its trajectory, and the background around it is filled in only from fields which contain useful data.

The technology of the motion-compensated standards convertor can be used in other applications. When video recordings are played back in slow motion, the result is that the same picture is displayed several times, followed by a jump to the next picture. Figure 4.35 shows that a moving object would remain in the same place on the screen during picture repeats, but jump to a new position as a new picture was played. The eye attempts to track the moving object, but, as Figure 4.35 also shows, the location of the moving object wanders with respect to the trajectory of the eye, and this is visible as judder.

Motion-compensated slow-motion systems are capable of synthesizing new images which lie between the original images from a slow-motion source. Figure 4.36 shows that two successive images in the original recording (using DVE terminology, these are source fields) are fed into the unit, which then measures the distance travelled by all moving objects between those images. Using interpolation, intermediate fields (target fields) are computed in which moving objects are positioned so that they lie on the eye trajectory. Using the principles described above, background information is removed as moving objects conceal it, and replaced as the rear of an object reveals it. Judder is thus removed and motion with a fluid quality is obtained.

4.7 De-interlacing

Interlace is a compression technique which sends only half of the picture lines in each field. Whilst this works reasonably well for transmission, it causes difficulty in any process which requires image manipulation. This includes DVEs, standards convertors and display convertors. All these devices give better results when working with progressively scanned data and if the source material is interlaced, a de-interlacing process will be necessary.

Interlace distributes vertical detail information over two fields and for maximum resolution all that information is necessary. Unfortunately it is not possible to use the information from two different fields directly. Figure 4.37 shows a scene in which an object is moving. When the second field of the scene leaves the camera, the object will have assumed a different position from the one

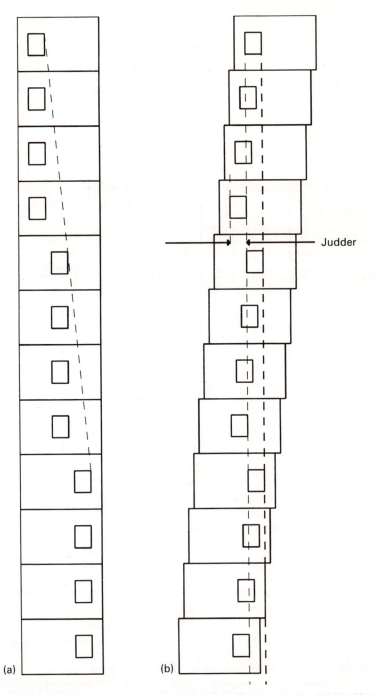

Figure 4.35 Conventional slow motion using field repeating with stationary eye shown at (a). With tracking eye at (b) the source of judder is seen.

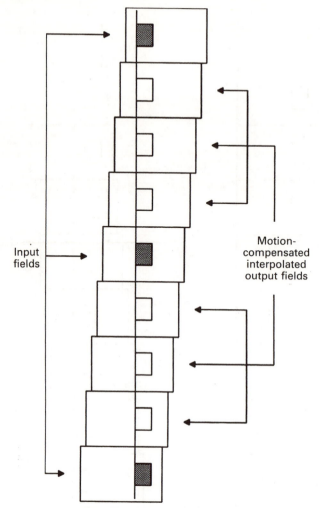

Input
fields

Motion-
compensated
interpolated
output fields

Figure 4.36 In motion-compensated slow motion, output fields are interpolated with moving objects displaying judder-free linear motion between input fields.

it had in the first field, and the result of combining the two fields to make a de-interlaced frame will be a double image. This effect can easily be demonstrated on any video recorder which offers a choice of still field or still frame. Stationary objects before a stationary camera, however, can be de-interlaced perfectly.

In simple de-interlacers, motion sensing is used so that de-interlacing can be disabled when movement occurs, and interpolation from a single field is used instead. Motion sensing implies comparison of one picture with the next. If interpolation is only to be used in areas where there is movement, it is necessary to test for motion over the entire frame. Motion can be simply detected by comparing the luminance value of a given pixel with the value of the same pixel two fields earlier. As two fields are to be combined, and motion can occur in either, then the comparison must be made between two odd fields and two even

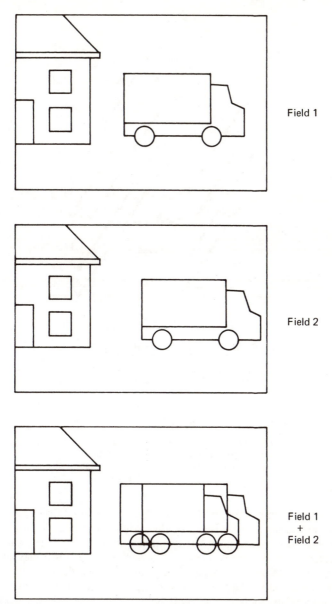

Field 1

Field 2

Field 1
+
Field 2

Figure 4.37 Moving object will be in a different place in two successive fields and will produce a double image.

fields. Thus four fields of memory are needed to correctly perform motion sensing. The luminance from four fields requires about a megabyte of storage.

At some point a decision must be made to abandon pixels from the previous field which are in the wrong place due to motion, and to interpolate them from adjacent lines in the current field. Switching suddenly in this way is visible, and

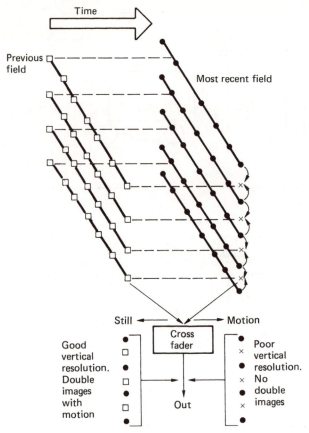

Figure 4.38 Pixels from the most recent field (●) are interpolated spatially to form low vertical resolution pixels (×) which will be used if there is excessive motion; pixels from the previous field (□) will be used to give maximum vertical resolution. The best possible de-interlaced frame results.

there is a more sophisticated mechanism which can be used. In Figure 4.38, two fields are shown, separated in time. Interlace can be seen by following lines from pixels in one field, which pass between pixels in the other field. If there is no movement, the fact that the two fields are separated in time is irrelevant, and the two can be superimposed to make a frame array. When there is motion, pixels from above and below the unknown pixels are added together and divided by two, to produce interpolated values. If both of these mechanisms work all the time, a better quality picture results if a crossfade is made between the two based on the amount of motion. A suitable digital crossfader was shown in section 4.1. At some motion value, or some magnitude of pixel difference, the loss of resolution due to a double image is equal to the loss of resolution due to interpolation. That amount of motion should result in the crossfader arriving at a 50/50 setting. Any less motion will result in a fade towards both fields, any more motion in a fade towards the interpolated values.

The most efficient way of de-interlacing is to use motion compensation. Figure 4.39 shows that when an object moves in an interlaced system, the interlace

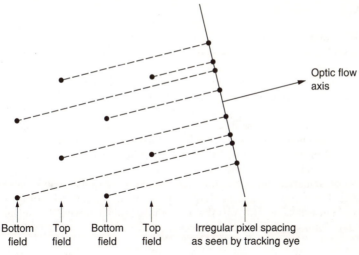

Figure 4.39 In the presence of vertical motion or motion having a vertical component, interlace breaks down and the pixel spacing with respect to the tracking eye becomes irregular.

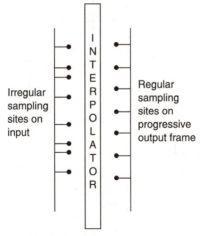

Figure 4.40 A de-interlacer needs an interpolator which can operate with input samples which are positioned arbitrarily rather than regularly.

breaks down with respect to the optic flow axis as was seen in section 2.11. If the motion is known, two or more fields can be shifted so that a moving object is in the same place in both. Pixels from both field can then be used to describe the object with better resolution than would be possible from one field alone. It will be seen from Figure 4.40 that the combination of two fields in this way will result in pixels having a highly irregular spacing and a special type of filter is needed to convert this back to a progressive frame with regular pixel spacing. At some critical vertical speeds there will be alignment between pixels in adjacent fields and no improvement is possible, but at other speeds the process will always give better results.

4.8 Noise reduction

The basic principle of all video noise reducers is that there is a certain amount of correlation between the video content of successive frames, whereas there is no correlation between the noise content.

A basic recursive device is shown in Figure 4.41. There is a frame store which acts as a delay, and the output of the delay can be fed back to the input through an attenuator, which in the digital domain will be a multiplier. In the case of a still picture, successive frames will be identical, and the recursion will be large. This means that the output video will actually be the average of many frames. If there is movement of the image, it will be necessary to reduce the amount of recursion to prevent the generation of trails or smears. Probably the most famous examples of recursion smear are the television pictures sent back of astronauts walking on the moon. The received pictures were very noisy and needed a lot of averaging to make them viewable. This was fine until the astronaut moved. The technology of the day did not permit motion sensing.

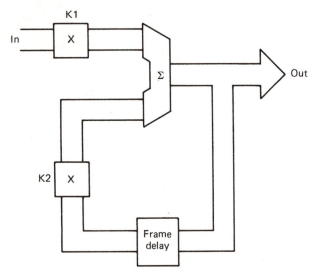

Figure 4.41 A basic recursive device feeds back the output to the input via a frame store which acts as a delay. The characteristics of the device are controlled totally by the values of the two coefficients K1 and K2 which control the multipliers.

The noise reduction increases with the number of frames over which the noise is integrated, but image motion prevents simple combining of frames. If motion estimation is available, the image of a moving object in a particular frame can be integrated from the images in several frames which have been superimposed on the same part of the screen by displacements derived from the motion measurement. The result is that greater reduction of noise becomes possible.[6] In fact a motion-compensated standards convertor performs such a noise reduction process automatically and can be used as a noise reducer, albeit an expensive one, by setting both input and output to the same standard.

References

1. Newman, W.M. and Sproull, R.F., *Principles of Interactive Computer graphics*, Tokyo: McGraw-Hill (1979)
2. Gernsheim, H., *A Concise History of Photography*, London: Thames and Hudson, 9–15 (1971)
3. Hedgecoe, J., *The Photographer's Handbook*, London: Ebury Press, 104–105 (1977)
4. de Boor, C., *A Practical Guide to Splines*, Berlin: Springer (1978)
5. Lau, H. and Lyon, D. Motion compensated processing for enhanced slow motion and standards conversion. *IEE Conf. Publ. No. 358*, 62–66 (1992)
6. Weiss, P. and Christensson, J., Real time implementation of sub-pixel motion estimation for broadcast applications. *IEE Digest*, 1990/128

Chapter 5

Video compression and MPEG

5.1 Introduction to compression

Compression allows the same (or nearly the same) information to be represented by a smaller quantity or rate of data. There are several reasons why compression techniques are popular:

1 Compression extends the playing time of a given storage device.
2 Compression allows miniaturization. With less data to store, the same playing time is obtained with smaller hardware. This is useful in ENG (electronic news gathering) and consumer devices.
3 Tolerances can be relaxed. With fewer data to record, storage density can be reduced making equipment which is more resistant to adverse environments and which requires less maintenance.
4 In transmission systems, compression allows a reduction in bandwidth which will generally result in a reduction in cost.
5 If a given bandwidth is available to an uncompressed signal, compression allows faster than real-time transmission in the same bandwidth.
6 If a given bandwidth is available, compression allows a better quality signal in the same bandwidth.

Compression is summarized in Figure 5.1. It will be seen in (a) that the data rate is reduced at source by the *compressor*. The compressed data are then passed through a communication channel and returned to the original rate by the *expander*. The ratio between the source data rate and the channel data rate is called the *compression factor*. The term *coding gain* is also used. Sometimes a compressor and expander in series are referred to as a *compander*. The compressor may equally well be referred to as a *coder* and the expander a *decoder* in which case the tandem pair may be called a *codec*.

Where the encoder is more complex than the decoder the system is said to be asymmetrical as in Figure 5.1(b). The encoder needs to be algorithmic or adaptive whereas the decoder is 'dumb' and carries out fixed actions. This is advantageous in applications such as broadcasting where the number of expensive complex encoders is small but the number of simple inexpensive decoders is large. In point-to-point applications the advantage of asymmetrical coding is not so great.

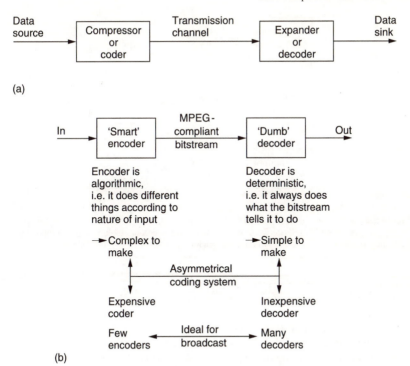

Data source → Compressor or coder → Transmission channel → Expander or decoder → Data sink

(a)

In → 'Smart' encoder → MPEG-compliant bitstream → 'Dumb' decoder → Out

Encoder is algorithmic, i.e. it does different things according to nature of input

Decoder is deterministic, i.e. it always does what the bitstream tells it to do

→ Complex to make

→ Simple to make

Asymmetrical coding system

Expensive coder

Inexpensive decoder

Few encoders ← Ideal for broadcast → Many decoders

(b)

Figure 5.1 In (a) a compression system consists of compressor or coder, a transmission channel and a matching expander or decoder. The combination of coder and decoder is known as a codec. (b) MPEG is asymmetrical since the encoder is much more complex than the decoder.

Although there are many different coding techniques, all of them fall into one or other of these categories. In *lossless* coding, the data from the expander are identical bit-for-bit with the original source data. The so-called 'stacker' programs which increase the apparent capacity of disk drives in personal computers use lossless codecs. Clearly, with computer programs the corruption of a single bit can be catastrophic. Lossless coding is generally restricted to compression factors of around 2:1.

In *lossy* coding data from the expander are not identical bit-for-bit with the source data and as a result comparing the input with the output is bound to reveal differences. Lossy codecs are not suitable for computer data, but are used in MPEG as they allow greater compression factors than lossless codecs. Successful lossy codecs are those in which the errors are arranged so that a human viewer or listener finds them subjectively difficult to detect. Thus lossy codecs must be based on an understanding of psychoacoustic and psychovisual perception and are often called *perceptive* codes.

In perceptive coding, the greater the compression factor required, the more accurately must the human senses be modelled. Perceptive coders can be forced to operate at a fixed compression factor. This is convenient for practical transmission applications where a fixed data rate is easier to handle than a variable rate. The result of a fixed compression factor is that the subjective quality can vary with the 'difficulty' of the input material. Perceptive codecs

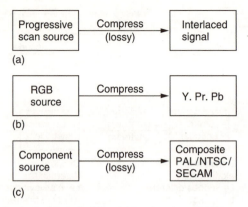

Figure 5.2 Compression is as old as television. (a) Interlace is a primitive way of having the bandwidth. (b) Colour difference working invisibly reduces colour resolution. (c) Composite video transmits colour in the same bandwidth as monochrome.

should not be concatenated indiscriminately especially if they use different algorithms.

Although the adoption of digital techniques is recent, compression itself is as old as television. Figure 5.2 shows some of the compression techniques used in traditional television systems. One of the oldest techniques is interlace which has been used in analog television from the very beginning as a primitive way of reducing bandwidth. As Chapter 2 showed, interlace is not without its problems, particularly in motion rendering. MPEG-2 supports interlace simply because legacy interlaced signals exist and there is a requirement to compress them.

The generation of colour difference signals from *RGB* in video represents an application of perceptive coding. The human visual system (HVS) sees no change in quality although the bandwidth of the colour difference signals is reduced. This is because human perception of detail in colour changes is much less than in brightness changes. This approach is sensibly retained in MPEG.

Composite video systems such as PAL, NTSC and SECAM are all analog compression schemes which embed a subcarrier in the luminance signal so that colour pictures are available in the same bandwidth as monochrome. In comparison with a progressive scan *RGB* picture, interlaced composite video has a compression factor of 6:1.

In many respects MPEG-2 is a modern digital equivalent of analog composite video as it has most of the same attributes. For example, the eight-field sequence of a PAL subcarrier which makes editing diffficult has its equivalent in the GOP (group of pictures) of MPEG.[1]

In a PCM digital system the bit rate is the product of the sampling rate and the number of bits in each sample and this is generally constant. Nevertheless the *information* rate of a real signal varies. In all real signals, part of the signal is obvious from what has gone before or what may come later and a suitable receiver can predict that part so that only the true information actually has to be sent. If the characteristics of a predicting receiver are known, the transmitter can omit parts of the message in the knowledge that the receiver has the ability to re-create it. Thus all encoders must contain a model of the decoder.

The difference between the information rate and the overall bit rate is known as the redundancy. Compression systems are designed to eliminate as much of that redundancy as practicable or perhaps affordable. One way in which this can be done is to exploit statistical predictability in signals. The information content or *entropy* of a sample is a function of how different it is from the predicted value. Most signals have some degree of predictability.

At the opposite extreme a signal such as noise is completely unpredictable and as a result all codecs find noise *difficult*. There are two consequences of this characteristic. First, a codec which is designed using the statistics of real material should not be tested with random noise because it is not a representative test. Second, a codec which performs well with clean source material may perform badly with source material containing superimposed noise. Most practical compression units require some form of preprocessing before the compression stage proper and appropriate noise reduction should be incorporated into the preprocessing if noisy signals are anticipated. It will also be necessary to restrict the degree of compression applied to noisy signals.

All real signals fall part-way between the extremes of total predictability and total unpredictability or noisiness. If the bandwidth (set by the sampling rate) and the dynamic range (set by the wordlength) of the transmission system are used to delineate an area, this sets a limit on the information capacity of the system. Figure 5.3(a) shows that most real signals occupy only part of that area. The signal may not contain all frequencies, or it may not have full dynamics at certain frequencies.

Entropy can be thought of as a measure of the actual area occupied by the signal. This is the area that *must* be transmitted if there are to be no subjective differences or *artifacts* in the received signal. The remaining area is called the *redundancy* because it adds nothing to the information conveyed. Thus an ideal coder could be imagined which miraculously sorts out the entropy from the redundancy and sends only the former. An ideal decoder would then re-create the original impression of the information quite perfectly.

As the ideal is approached, the coder complexity and the latency or delay both rise. Figure 5.3(b) shows how complexity increases with compression factor. Figure 5.3(c) shows how increasing the codec latency can improve the compression factor. Obviously we would have to provide a channel which could accept whatever entropy the coder extracts in order to have transparent quality. As a result, moderate coding gains which only remove redundancy need not cause artifacts and result in systems which are described as *subjectively lossless*.

If the channel capacity is not sufficient for that, then the coder will have to discard some of the entropy and with it useful information. Larger coding gains which remove some of the entropy must result in artifacts. It will also be seen from Figure 5.3 that an imperfect coder will fail to separate the redundancy and may discard entropy instead, resulting in artifacts at a suboptimal compression factor.

A single variable-rate transmission or recording channel is inconvenient and unpopular with channel providers because it is difficult to police. The requirement can be overcome by combining several compressed channels into one constant rate transmission in a way which flexibly allocates data rate between the channels. Provided the material is unrelated, the probability of all channels reaching peak entropy at once is very small and so those channels which

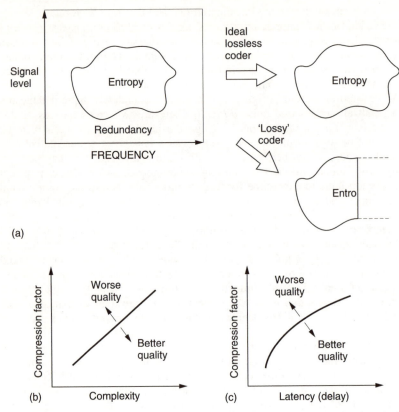

Figure 5.3 (a) A perfect coder removes only the redundancy from the input signal and results in subjectively lossless coding. If the remaining entropy is beyond the capacity of the channel some of it must be lost and the codec will then be lossy. An imperfect coder will also be lossy as it fails to keep all entropy. (b) As the compression factor rises, the complexity must also rise to maintain quality. (c) High compression factors also tend to increase latency or delay through the system.

are at one instant passing easy material will free up transmission capacity for those channels which are handling difficult material. This is the principle of statistical multiplexing.

Where the same type of source material is used consistently, e.g. English text, then it is possible to perform a statistical analysis on the frequency with which particular letters are used. Variable-length coding is used in which frequently used letters are allocated short codes and letters which occur infrequently are allocated long codes. This results in a lossless code. The well-known Morse code used for telegraphy is an example of this approach. The letter e is the most frequent in English and is sent with a single dot. An infrequent letter such as z is allocated a long complex pattern. It should be clear that codes of this kind which rely on a prior knowledge of the statistics of the signal are only effective with signals actually having those statistics. If Morse code is used with another language, the transmission becomes significantly less efficient because the statistics are quite different; the letter z, for example, is quite common in Czech.

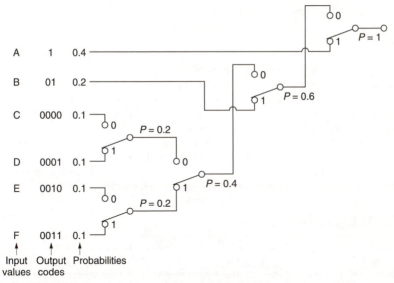

Figure 5.4 The Huffman code achieves compression by allocating short codes to frequent values. To aid deserializing the short codes are not prefixes of longer codes.

The Huffman code[2] is one which is designed for use with a data source having known statistics and shares the same principles with the Morse code. The probability of the different code values to be transmitted is studied, and the most frequent codes are arranged to be transmitted with short wordlength symbols. As the probability of a code value falls, it will be allocated a longer wordlength. The Huffman code is used in conjunction with a number of compression techniques and is shown in Figure 5.4.

The input or *source* codes are assembled in order of descending probability. The two lowest probabilities are distinguished by a single code bit and their probabilities are combined. The process of combining probabilities is continued until unity is reached and at each stage a bit is used to distinguish the path. The bit will be a zero for the most probable path and one for the least. The compressed output is obtained by reading the bits which describe which path to take going from right to left.

In the case of computer data, there is no control over the data statistics. Data to be recorded could be instructions, images, tables, text files and so on; each having their own code value distributions. In this case a coder relying on fixed source statistics will be completely inadequate. Instead a system is used which can learn the statistics as it goes along. The Lempel–Ziv–Welch (LZW) lossless codes are in this category. These codes build up a conversion table between frequent long source data strings and short transmitted data codes at both coder and decoder and initially their compression factor is below unity as the contents of the conversion tables are transmitted along with the data. However, once the tables are established, the coding gain more than compensates for the initial loss. In some applications, a continuous analysis of the frequency of code selection is made and if a data string in the table is no longer being used with sufficient frequency it can be deselected and a more common string substituted.

Lossless codes are less common for audio and video coding where perceptive codes are permissible. The perceptive codes often obtain a coding gain by shortening the wordlength of the data representing the signal waveform. This must increase the noise level and the trick is to ensure that the resultant noise is placed at frequencies where human senses are least able to perceive it. As a result although the received signal is measurably different from the source data, it can *appear* the same to the human listener or viewer at moderate compressions factors. As these codes rely on the characteristics of human sight and hearing, they can only be fully tested subjectively.

The compression factor of such codes can be set at will by choosing the wordlength of the compressed data. Whilst mild compression will be undetectable, with greater compression factors, artifacts become noticeable. Figure 5.3 shows that this is inevitable from entropy considerations.

5.2 What is MPEG?

MPEG is actually an acronym for the Moving Pictures Experts Group which was formed by the ISO (International Standards Organization) to set standards for audio and video compression and transmission. The first compression standard for audio and video was MPEG-1[3,4] but this was of limited application and the subsequent MPEG-2 standard was considerably broader in scope and of wider appeal. For example, MPEG-2 supports interlace whereas MPEG-1 did not.

The approach of the ISO to standardization in MPEG is novel because it is not the encoder which is standardized. Figure 5.5(a) shows that instead the way in which a decoder will interpret the bitstream is defined. A decoder which can successfully interpret the bitstream is said to be *compliant*. Figure 5.5(b) shows that the advantage of standardizing the decoder is that over time, encoding algorithms can improve yet compliant decoders will continue to function with them.

Manufacturers can supply encoders using algorithms which are proprietary and their details do not need to be published. A useful result is that there can be competition between different encoder designs which means that better designs will evolve. The user will have greater choice because different levels of cost and complexity can exist in a range of coders yet a compliant decoder will operate with them all.

MPEG is, however, much more than a compression scheme as it also standardizes the protocol and syntax under which it is possible to combine or multiplex audio data with video data to produce a digital equivalent of a television program. Many such programs can be combined in a single multiplex and MPEG defines the way in which such multiplexes can be created and transported. The definitions include the metadata which decoders require to demultiplex correctly and which users will need to locate programs of interest.

As with all video systems there is a requirement for synchronizing or genlocking and this is particularly complex when a multiplex is assembled from many signals which are not necessarily synchronized to one another.

The applications of audio and video compression are limitless and the ISO has done well to provide standards which are appropriate to the wide range of possible compression products.

MPEG-2 embraces video pictures from the tiny screen of a videophone to the high-definition images needed for electronic cinema. Audio coding stretches from speech-grade mono to multichannel surround sound.

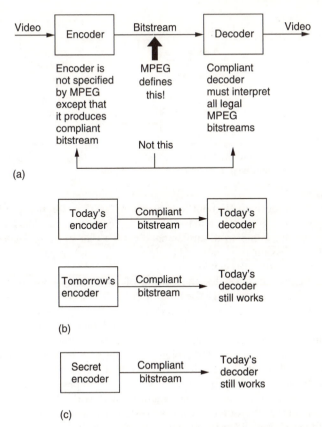

Figure 5.5 (a) MPEG defines the protocol of the bitstream between encoder and decoder. The decoder is defined by implication, the encoder is left very much to the designer. (b) This approach allows future encoders of better performance to remain compatible with existing decoders. (c) This approach also allows an encoder to produce a standard bitstream while its technical operation remains a commercial secret.

Figure 5.6 shows the use of a codec with a recorder. The playing time of the medium is extended in proportion to the compression factor. In the case of tapes, the access time is improved because the length of tape needed for a given recording is reduced and so it can be rewound more quickly.

In the case of DVD (digital video disk, aka digital versatile disk) the challenge was to store an entire movie on one 12 cm disk. The storage density available with today's optical disk technology is such that recording of conventional uncompressed video would be out of the question.

In communications, the cost of data links is often roughly proportional to the data rate and so there is simple economic pressure to use a high compression factor. However, it should be borne in mind that implementing the codec also has a cost which rises with compression factor and so a degree of compromise will be inevitable.

In the case of video-on-demand, technology exists to convey full bandwidth video to the home, but to do so for a single individual at the moment would be

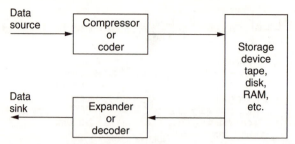

Figure 5.6 Compression can be used around a recording medium. The storage capacity may be increased or the access time reduced according to the application.

prohibitively expensive. Without compression, HDTV (high-definition tele-vision) requires too much bandwidth. With compression, HDTV can be transmitted to the home in a similar bandwidth to an existing analog SDTV channel. Compression does not make video-on-demand or HDTV possible, it makes them economically viable.

In workstations designed for the editing of audio and/or video, the source material is stored on hard disks for rapid access. Whilst top-grade systems may function without compression, many systems use compression to offset the high cost of disk storage. When a workstation is used for *off-line* editing, a high compression factor can be used and artifacts will be visible in the picture.

This is of no consequence as the picture is only seen by the editor who uses it to make an EDL (edit decision list) which is no more than a list of actions and the timecodes at which they occur. The original uncompressed material is then *conformed* to the EDL to obtain a high-quality edited work. When *on-line* editing is being performed, the output of the workstation is the finished product and clearly a lower compression factor will have to be used.

Perhaps it is in broadcasting where the use of compression will have its greatest impact. There is only one electromagnetic spectrum and pressure from other services such as cellular telephones makes efficient use of bandwidth mandatory. Analog television broadcasting is an old technology and makes very inefficient use of bandwidth. Its replacement by a compressed digital transmis-sion will be inevitable for the practical reason that the bandwidth is needed elsewhere.

Fortunately in broadcasting there is a mass market for decoders and these can be implemented as low-cost integrated circuits. Fewer encoders are needed and so it is less important if these are expensive. Whilst the cost of digital storage goes down year on year, the cost of electromagnetic spectrum goes up. Consequently in the future the pressure to use compression in recording will ease or even cease whereas the pressure to use it in radio communications will increase.

5.3 Spatial and temporal redundancy in MPEG

Video signals exist in four dimensions: these are the attributes of the sample, the horizontal and vertical spatial axes and the time axis. Compression can be applied in any or all of those four dimensions. MPEG-2 assumes an eight-bit colour difference signal as the input, requiring rounding if the source is ten-bit.

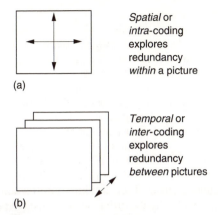

Spatial or
intra-coding
explores
redundancy
within a picture

(a)

Temporal or
inter-coding
explores
redundancy
between pictures

(b)

Figure 5.7 (a) Spatial or intra-coding works on individual images. (b) Temporal or inter-coding works on successive images.

The sampling rate of the colour signals is less than that of the luminance. This is done by downsampling the colour samples horizontally and generally vertically as well. Essentially an MPEG-2 system has three parallel simultaneous channels, one for luminance and two for colour difference, which after coding are multiplexed into a single bitstream.

Figure 5.7(a) shows that when individual pictures are compressed without reference to any other pictures, the time axis does not enter the process which is therefore described as *intra-coded* (intra = within) compression. The term *spatial coding* will also be found. It is an advantage of intra-coded video that there is no restriction to the editing which can be carried out on the picture sequence. As a result compressed VTRs such as Digital Betacam, DVC and D-9 use spatial coding. Cut editing may take place on the compressed data directly if necessary. As spatial coding treats each picture independently, it can employ certain techniques developed for the compression of still pictures.

The ISO JPEG (Joint Photographic Experts Group) compression standards[5,6] are in this category. Where a succession of JPEG coded images are used for television, the term 'Motion JPEG' will be found.

Greater compression factors can be obtained by taking account of the redundancy from one picture to the next. This involves the time axis, as Figure 5.7(b) shows, and the process is known as *inter-coded* (inter = between) or *temporal* compression.

Temporal coding allows a higher compression factor, but has the disadvantage that an individual picture may exist only in terms of the differences from a previous picture. Clearly, editing must be undertaken with caution and arbitrary cuts simply cannot be performed on the MPEG bitstream. If a previous picture is removed by an edit, the difference data will then be insufficient to re-create the current picture.

Intra-coding works in three dimensions on the horizontal and vertical spatial axes and on the sample values. Analysis of typical television pictures reveals that whilst there is a high spatial frequency content due to detailed areas of the picture, there is a relatively small amount of energy at such frequencies. Often pictures contain sizable areas in which the same or similar pixel values exist. This gives rise to low spatial frequencies. The average brightness of the picture results in a substantial zero frequency component. Simply omitting the high-frequency components is unacceptable as this causes an obvious softening of the picture.

A coding gain can be obtained by taking advantage of the fact that the amplitude of the spatial components falls with frequency. It is further possible to take advantage of the eye's reduced sensitivity to noise in high spatial frequencies. If the spatial frequency spectrum is divided into frequency bands the high-frequency bands can be described by fewer bits not only because their amplitudes are smaller but also because more noise can be tolerated. The discrete cosine transform introduced in Chapter 2 is used in MPEG to allows two-dimensional pictures to be described in the frequency domain.

Inter-coding takes further advantage of the similarities between successive pictures in real material. Instead of sending information for each picture separately, inter-coders will send the difference between the previous picture and the current picture in a form of differential coding. Figure 5.8 shows the principle. A picture store is required at the coder to allow comparison to be made between successive pictures and a similar store is required at the decoder to make the previous picture available. The difference data may be treated as a picture itself and subjected to some form of transform-based spatial compression.

The simple system of Figure 5.8(a) is of limited use as in the case of a transmission error, every subsequent picture would be affected. Channel switching in a television set would also be impossible. In practical systems a modification is required. The approach used in MPEG is that periodically some absolute picture data are transmitted in place of difference data.

Figure 5.8(b) shows that absolute picture data, known as *I* or *intra pictures* are interleaved with pictures which are created using difference data, known as *P* or *predicted* pictures. The *I* pictures require a large amount of data, whereas the *P* pictures require less data. As a result the instantaneous data rate varies dramatically and buffering has to be used to allow a constant transmission rate.

The *I* picture and all the *P* pictures prior to the next *I* picture are called a group of pictures (GOP). For a high compression factor, a large number of *P* pictures should be present between *I* pictures, making a long GOP. However, a long GOP delays recovery from a transmission error. The compressed bitstream can only be edited at *I* pictures as shown.

In the case of moving objects, although their appearance may not change greatly from picture to picture, the data representing them on a fixed sampling grid will change and so large differences will be generated between successive pictures. It is a great advantage if the effect of motion can be removed from difference data so that they only reflect the changes in appearance of a moving object since a much greater coding gain can then be obtained. This is the objective of motion compensation introduced in section 4.5.

It will be clear that the data values representing a moving object change with respect to the time axis. However, looking along the optic flow axis the appearance of an object only changes if it deforms, moves into shadow or rotates. For simple translational motions the data representing an object are highly redundant with respect to the optic flow axis. Thus if the optic flow axis can be located, coding gain can be obtained in the presence of motion.

A motion-compensated coder works as follows. An *I* picture is sent, but is also locally stored so that it can be compared with the next input picture to find

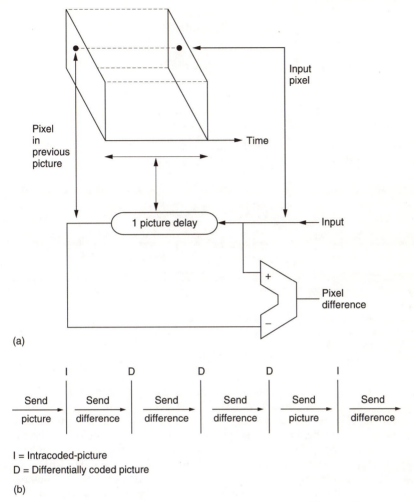

(a)

I = Intracoded-picture
D = Differentially coded picture

(b)

Figure 5.8 An inter-coded system (a) uses a delay to calculate the pixel differences between successive pictures. To prevent error propagation, intra-coded pictures (b) may be used periodically.

motion vectors for various areas of the picture. The *I* picture is then shifted according to these vectors to cancel inter-picture motion. The resultant *predicted* picture is compared with the actual picture to produce a *prediction error* also called a *residual*. The prediction error is transmitted with the motion vectors. At the receiver the original *I* picture is also held in a memory. It is shifted according to the transmitted motion vectors to create the predicted picture and then the prediction error is added to it to re-create the original. When a picture is encoded in this way MPEG calls it a *P* picture.

Figure 5.9(a) shows that spatial redundancy is redundancy within a single image, for example repeated pixel values in a large area of blue sky. Temporal redundancy (b) exists between successive images.

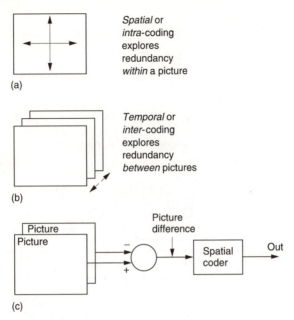

Figure 5.9 (a) Spatial or intra-coding works on individual images. (b) Temporal or inter-coding works on successive images. (c) In MPEG inter-coding is used to create difference images. These are then compressed spatially.

Where temporal compression is used, the current picture is not sent in its entirety; instead the difference between the current picture and the previous picture is sent. The decoder already has the previous picture, and so it can add the difference to make the current picture. A difference picture is created by subtracting every pixel in one picture from the corresponding pixel in another pixel. This is trivially easy in a progressively scanned system, but MPEG-2 has had to develop greater complexity so that this can also be done with interlaced pictures. The handling of interlace in MPEG will be detailed later.

A difference picture is an image of a kind, although not a viewable one, and so should contain some kind of spatial redundancy. Figure 5.9(c) shows that MPEG-2 takes advantage of both forms of redundancy. Picture differences are spatially compressed prior to transmission. At the decoder the spatial compression is decoded to re-create the difference picture, then this difference picture is added to the previous picture to complete the decoding process.

Whenever objects move they will be in a different place in successive pictures. This will result in large amounts of difference data. MPEG-2 overcomes the problem using motion compensation. The encoder contains a motion estimator which measures the direction and distance of motion between pictures and outputs these as vectors which are sent to the decoder. When the decoder receives the vectors it uses them to shift data in a previous picture to more closely resemble the current picture. Effectively the vectors are describing the optic flow axis of some moving screen area, along which axis the image is highly redundant. Vectors are bipolar codes which determine the amount of horizontal and vertical shift required.

In real images, moving objects do not necessarily maintain their appearance as they move. For example, objects may turn, move into shade or light, or move behind other objects. Consequently motion compensation can never be ideal and it is still necessary to send a picture difference to make up for any shortcomings in the motion compensation.

Figure 5.10 shows how this works. In addition to the motion encoding system, the coder also contains a motion decoder. When the encoder outputs motion vectors, it also uses them locally in the same way that a real decoder will, and is able to produce a *predicted picture* based solely on the previous picture shifted by motion vectors. This is then subtracted from the *actual* current picture to produce a *prediction error* or *residual* which is an image of a kind that can be spatially compressed.

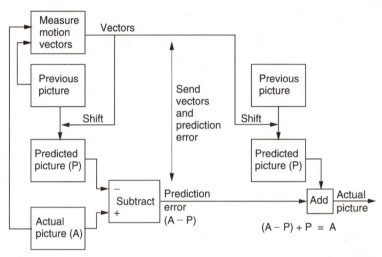

Figure 5.10 A motion-compensated compression system. The coder calculates motion vectors which are transmitted as well as being used locally to create a predicted picture. The difference between the predicted picture and the actual picture is transmitted as a prediction error.

The decoder takes the previous picture, shifts it with the vectors to re-create the predicted picture and then decodes and adds the prediction error to produce the actual picture. Picture data sent as vectors plus prediction error are said to be *P* coded.

The concept of sending a prediction error is a useful approach because it allows both the motion estimation and compensation to be imperfect.

A good motion-compensation system will send just the right amount of vector data. With insufficient vector data, the prediction error will be large, but transmission of excess vector data will also cause the the bit rate to rise. There will be an optimum balance which minimizes the sum of the prediction error data and the vector data.

In MPEG-2 the balance is obtained by dividing the screen into areas called *macroblocks* which are 16 luminance pixels square. Each macroblock is steered by a vector. The location of the boundaries of a macroblock are fixed and so the vector does not move the macroblock. Instead the vector tells the decoder where

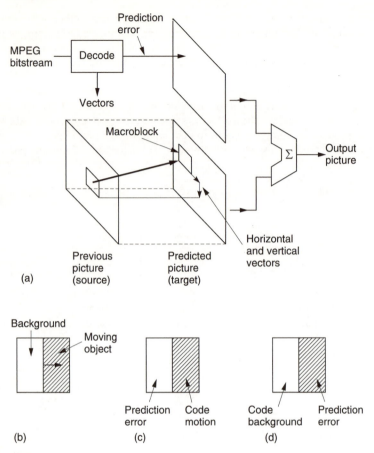

Figure 5.11 (a) In motion compensation, pixel data are brought to a fixed macroblock in the target picture from a variety of places in another picture. (b) Where only part of a macroblock is moving, motion compensation is non-ideal. The motion can be coded (c), causing a prediction error in the background, or the background can be coded (d) causing a prediction error in the moving object.

to look in another frame to find pixel data to *fetch* to the macroblock. Figure 5.11(a) shows this concept. The shifting process is generally done by modifying the read address of a RAM using the vector. This can shift by one-pixel steps. MPEG-2 vectors have half-pixel resolution so it is necessary to interpolate between pixels from RAM to obtain half-pixel shifted values.

Real moving objects will not coincide with macroblocks and so the motion compensation will not be ideal but the prediction error makes up for any shortcomings. Figure 5.11(b) shows the case where the boundary of a moving object bisects a macroblock. If the system measures the moving part of the macroblock and sends a vector, the decoder will shift the entire block making the stationary part wrong. If no vector is sent, the moving part will be wrong. Both approaches are legal in MPEG-2 because the prediction error sorts out the incorrect values. An intelligent coder might try both approaches to see which required the least prediction error data.

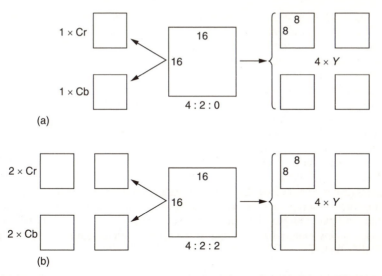

Figure 5.12 The structure of a macroblock. (A macroblock is the screen area steered by *one* vector.) (a) In 4:2:0, there are two chroma DCT blocks per macroblock whereas in 4:2:2 (b) there are four. 4:2:2 needs 33% more data than 4:2:0.

The prediction error concept also allows the use of simple but inaccurate motion estimators in low-cost systems. The greater prediction error data is handled using a higher bit rate. On the other hand, if a precision motion estimator is available, a very high compression factor may be achieved because the prediction error data are minimized. MPEG-2 does not specify how motion is to be measured; it simply defines how a decoder will interpret the vectors. Encoder designers are free to use any motion-estimation system provided that the right vector protocol is created.

Figure 5.12(a) shows that a macroblock contains both luminance and colour difference data at different resolutions. Most of the MPEG-2 Profiles use a 4:2:0 structure which means that the colour is downsampled by a factor of two in both axes. Thus in a 16×16 pixel block, there are only 8×8 colour difference sampling sites. MPEG-2 is based upon the 8×8 DCT (see section 2.24) and so the 16×16 block is the screen area which contains an 8×8 colour difference sampling block. Thus in 4:2:0 in each macroblock there are four luminance DCT blocks, one $R-Y$ DCT block and one $B-Y$ DCT block, all steered by the same vector.

In the 4:2:2 Profile of MPEG-2, shown in Figure 5.12(b), the chroma is not downsampled vertically, and so there is twice as much chroma data in each macroblock which is otherwise substantially the same.

5.4 *I* and *P* coding

Predictive (*P*) coding cannot be used indefinitely, as it is prone to error propagation. A further problem is that it becomes impossible to decode the transmission if reception begins part-way through. In real video signals, cuts or edits can be present across which there is little redundancy and which make motion estimators throw up their hands.

In the absence of redundancy over a cut, there is nothing to be done but to send the new picture information in absolute form. This is called *I* coding where *I* is an abbreviation of *intra* coding. As *I* coding needs no previous picture for decoding, then decoding can begin at *I* coded information.

MPEG-2 is effectively a toolkit and there is no compulsion to use all the tools available. Thus an encoder may choose whether to use *I* or *P* coding, either once and for all or dynamically on a macroblock-by-macroblock basis.

For practical reasons, an entire frame may be encoded as *I* macroblocks periodically. This creates a *re-entry point* where the bitstream might be edited or where decoding could begin.

Figure 5.13 shows a typical application of the Simple Profile of MPEG-2. Periodically an *I* picture is created. Between *I* pictures are *P* pictures which are based on the picture before. These *P* pictures predominantly contain macroblocks having vectors and prediction errors. However, it is perfectly legal for *P* pictures to contain *I* macroblocks. This might be useful where, for example, a camera pan introduces new material at the edge of the screen which cannot be created from an earlier picture.

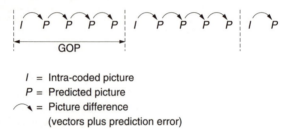

I   = Intra-coded picture
P   = Predicted picture
⤷   = Picture difference
      (vectors plus prediction error)

Figure 5.13 A Simple Profile MPEG-2 signal may contain periodic *I* pictures with a number of *P* pictures between.

Note that although what is sent is called a *P* picture, it is not a picture at all. It is a set of instructions to convert the previous picture into the current picture. If the previous picture is lost, decoding is impossible. An *I* picture together with all the pictures before the next *I* picture form a *group of pictures* (GOP).

5.5 Bidirectional coding

Motion-compensated predictive coding is a useful compression technique, but it does have the drawback that it can only take data from a previous picture. Where moving objects reveal a background this is completely unknown in previous pictures and forward prediction fails. However, more of the background is visible in later pictures. Figure 5.14 shows the concept. In the centre of the diagram, a moving object has revealed some background. The previous picture can contribute nothing, whereas the next picture contains all that is required.

Bidirectional coding is shown in Figure 5.15. A bidirectional or *B* macroblock can be created using a combination of motion compensation and the addition of a prediction error. This can be done by forward prediction from a previous picture or backward prediction from a subsequent picture. It is also possible to use an average of both forward and backward prediction. On noisy material this may

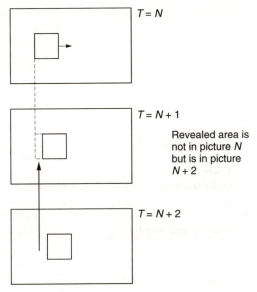

$T = N$

$T = N + 1$

Revealed area is
not in picture N
but is in picture
$N + 2$

$T = N + 2$

Figure 5.14 In bidirectional coding the revealed background can be efficiently coded by bringing data back from a future picture.

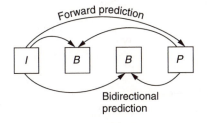

Forward prediction

I B B P

Bidirectional
prediction

I = Intra- or spatially coded
    'anchor' picture

P = Forward predicted. Coder sends
    difference between I and P decoder.
    Adds difference to create P

B = Bidirectionally coded picture can be
    coded from a previous
    I or P picture or a later I or P picture.
    B pictures are not coded from each other

Figure 5.15 In bidirectional coding, a number of B pictures can be inserted between periodic forward predicted pictures. See text.

result in some reduction in bit rate. The technique is also a useful way of portraying a dissolve.

The averaging process in MPEG-2 is a simple linear interpolation which works well when only one B picture exists between the reference pictures before and after. A larger number of B pictures would require weighted interpolation but MPEG-2 does not support this.

Typically two *B* pictures are inserted between *P* pictures or between *I* and *P* pictures. As can be seen, *B* pictures are never predicted from one another, only from *I* or *P* pictures. A typical GOP for broadcasting purposes might have the structure *IBBPBBPBBPBB*. Note that the last *B* pictures in the GOP require the *I* picture in the next GOP for decoding and so the GOPs are not truly independent. Independence can be obtained by creating a *closed GOP* which may contain *B* pictures but which ends with a *P* picture. It is also legal to have a *B* picture in which every macroblock is forward predicted, needing no future picture for decoding.

Bidirectional coding is very powerful. Figure 5.16 is a constant quality curve showing how the bit rate changes with the type of coding. On the left, only *I* or spatial coding is used, whereas on the right an *IBBP* structure is used. This means that there are two bidirectionally coded pictures in between a spatially coded picture (*I*) and a forward predicted picture (*P*). Note how for the same quality the system which only uses spatial coding needs two and a half times the bit rate that the bidirectionally coded system needs.

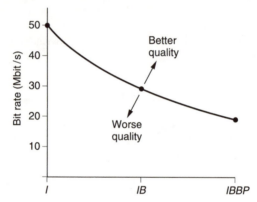

Figure 5.16 Bidirectional coding is very powerful as it allows the same quality with only 40 per cent of the bit rate of intra-coding. However, the encoding and decoding delays must increase. Coding over a longer time span is more efficient but editing is more difficult.

Clearly information in the future has yet to be transmitted and so is not normally available to the decoder. MPEG-2 gets around the problem by sending pictures in the wrong order. Picture reordering requires delay in the encoder and a delay in the decoder to put the order right again. Thus the overall codec delay must rise when bidirectional coding is used. This is quite consistent with Figure 5.3 which showed that as the compression factor rises the latency must also rise.

Figure 5.17 shows that although the original picture sequence is *IBBPBBPB-BIBB* ..., this is transmitted as *IPBBPBBIBB* ... so that the future picture is already in the decoder before bidirectional decoding begins. Note that the *I* picture of the next GOP is actually sent before the last *B* pictures of the current GOP.

Figure 5.17 also shows that the amount of data required by each picture is dramatically different. *I* pictures have only spatial redundancy and so need a lot

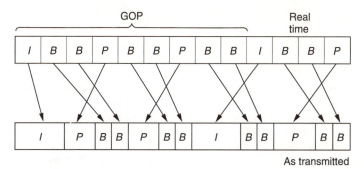

Figure 5.17 Comparison of pictures before and after compression showing sequence change and varying amount of data needed by each picture type. *I, P, B* pictures use unequal amounts of data.

of data to describe them. *P* pictures need fewer data because they are created by shifting the *I* picture with vectors and then adding a prediction error picture. *B* pictures need the least data of all because they can be created from *I* or *P*.

With pictures requiring a variable length of time to transmit, arriving in the wrong order, the decoder needs some help. This takes the form of picture-type flags and time stamps.

Figure 5.18 shows a variety of GOP structures. The simplest is the *III* . . . sequence in which every picture is intra-coded. Pictures can be fully decoded without reference to any other pictures and so editing is straightforward. However, this approach requires about two and one half times the bit rate of a full bidirectional system. Bidirectional coding is most useful for final delivery of post-produced material either by broadcast or on prerecorded media as there is then no editing requirement. As a compromise the *IBIB* structure can be used which has some of the bit rate advantage of bidirectional coding but without too much latency. It is possible to edit an *IBIB* stream by performing some processing. If it is required to remove the video following a *B* picture, that *B* picture could not be decoded because it needs *I* pictures either side of it for bidirectional decoding. The solution is to decode the *B* picture first, and then re-encode it with forward prediction only from the previous *I* picture. The subsequent *I* picture can then be replaced by an edit process. Some quality loss

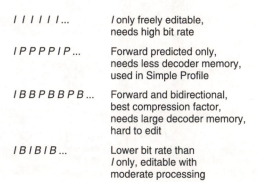

Figure 5.18 Various possible GOP structures used with MPEG. See text for details.

is inevitable in this process but this is acceptable in applications such as ENG and industrial video.

5.6 Spatial compression

Spatial compression in MPEG-2 is used in *I* pictures on actual picture data and in *P* and *B* pictures on prediction error data. MPEG-2 uses the discrete cosine transform described in section 2.24. The DCT works on blocks and in MPEG-2 these are 8 × 8 pixels. The macroblocks of the motion-compensation structure are designed so they can be broken down into 8 × 8 DCT blocks. In a 4:2:0 macroblock there will be six DCT blocks whereas in a 4:2:2 macroblock there will be eight.

Figure 5.19 shows the table of basis functions or *wave table* for an 8 × 8 DCT. Adding these two-dimensional waveforms together in different proportions will give any original 8 × 8 pixel block. The coefficients of the DCT simply control the proportion of each wave which is added in the inverse transform. The top-left wave has no modulation at all because it conveys the DC component of the block. This coefficient will be a unipolar (positive only) value in the case of luminance and will typically be the largest value in the block as the spectrum of typical video signals is dominated by the DC component.

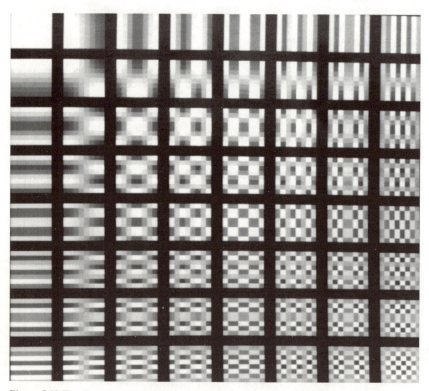

Figure 5.19 The discrete cosine transform breaks up an image area into discrete frequencies in two dimensions. The lowest frequency can be seen here at the top-left corner. Horizontal frequency increases to the right and vertical frequency increases downwards.

Increasing the DC coefficient adds a constant amount to every pixel. Moving to the right the coefficients represent increasing horizontal spatial frequencies and moving downwards they represent increasing vertical spatial frequencies. The bottom-right coefficient represents the highest diagonal frequencies in the block. All these coefficients are bipolar, where the polarity indicates whether the original spatial waveform at that frequency was inverted.

Figure 5.20 shows a one-dimensional example of an inverse transform. The DC coefficient produces a constant level throughout the pixel block. The remaining waves in the table are AC coefficients. A zero coefficient would result in no modulation, leaving the DC level unchanged. The wave next to the DC component represents the lowest frequency in the transform which is half a cycle per block. A positive coefficient would make the left side of the block brighter and the right side darker whereas a negative coefficient would do the opposite. The magnitude of the coefficient determines the amplitude of the wave which is added. Figure 5.20 also shows that the next wave has a frequency of one cycle per block. i.e. the block is made brighter at both sides and darker in the middle.

Consequently an inverse DCT is no more than a process of mixing various pixel patterns from the wave table where the relative amplitudes and polarity of these patterns are controlled by the coefficients. The original transform is simply

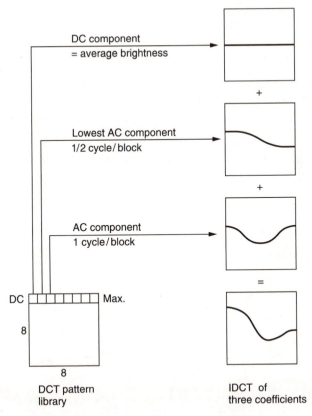

Figure 5.20 A one-dimensional inverse transform. See text for details.

a mechanism which finds the coefficient amplitudes from the original pixel block.

The DCT itself achieves no compression at all. Sixty-four pixels are converted to sixty-four coefficients. However, in typical pictures, not all coefficients will have significant values; there will often be a few dominant coefficients. The coefficients representing the higher two-dimensional spatial frequencies will often be zero or of small value in large areas, due to blurring or simply plain undetailed areas before the camera.

Statistically, the further from the top-left corner of the wave table the coefficient is, the smaller will be its magnitude. Coding gain (the technical term for reduction in the number of bits needed) is achieved by transmitting the low-valued coefficients with shorter wordlengths. The zero-valued coefficients need not be transmitted at all. Thus it is not the DCT which compresses the data, it is the subsequent processing. The DCT simply expresses the data in a form which makes the subsequent processing easier.

Higher compression factors require the coefficient wordlength to be further reduced using requantizing. Coefficients are divided by some factor which increases the size of the quantizing step. The smaller number of steps which results permits coding with fewer bits, but, of course, with an increased quantizing error. The coefficients will be multiplied by a reciprocal factor in the decoder to return to the correct magnitude.

Inverse transforming a requantized coefficient means that the frequency it represents is reproduced in the output with the wrong amplitude. The difference between original and reconstructed amplitude is regarded as a noise added to the wanted data. Figure 5.21 shows that the visibility of such noise is far from uniform. The maximum sensitivity is found at DC and falls thereafter. As a result, the top-left coefficient is often treated as a special case and left unchanged. It may warrant more error protection than other coefficients.

MPEG-2 takes advantage of the falling sensitivity to noise. Prior to requantizing, each coefficient is divided by a different weighting constant as a function of its frequency. Figure 5.22 shows a typical weighting process. Naturally the decoder must have a corresponding inverse weighting. This weighting process has the effect of reducing the magnitude of high-frequency

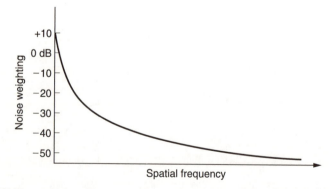

Figure 5.21 The sensitivity of the eye to noise is greatest at low frequencies and drops rapidly with increasing frequency. This can be used to mask quantizing noise caused by the compression process.

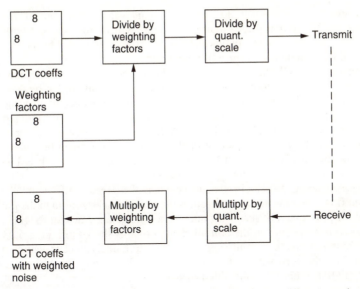

Figure 5.22 Weighting is used to make the noise caused by requantizing different at each frequency.

coefficients disproportionately. Clearly, different weighting will be needed for colour difference data as colour is perceived differently.

P and *B* pictures are decoded by adding a prediction error image to a reference image. That reference image will contain weighted noise. One purpose of the prediction error is to cancel that noise to prevent tolerance build-up. If the prediction error were also to contain weighted noise this result would not be obtained. Consequently prediction error coefficients are flat weighted.

When forward prediction fails, such as in the case of new material introduced in a *P* picture by a pan, *P* coding would set the vectors to zero and encode the new data entirely as an unweighted prediction error. In this case it is better to encode that material as an *I* macroblock because then weighting can be used and this will require fewer bits.

Requantizing increases the step size of the coefficients, but the inverse weighting in the decoder results in step sizes which increase with frequency. The larger step size increases the quantizing noise at high frequencies where it is less visible. Effectively the noise floor is shaped to match the sensitivity of the eye. The quantizing table in use at the encoder can be transmitted to the decoder periodically in the bitstream.

Study of the signal statistics gained from extensive analysis of real material is used to measure the probability of a given coefficient having a given value. This probability turns out to be highly non-uniform, suggesting the possibility of a variable-length encoding for the coefficient values. On average, the higher the spatial frequency, the lower the value of a coefficient will be. This means that the value of a coefficient falls as a function of its radius from the DC coefficient.

Typical material often has many coefficients which are zero valued, especially after requantizing. The distribution of these also follows a pattern. The non-zero values tend to be found in the top-left-hand corner of the DCT block, but as the radius increases, not only do the coefficient values fall, but it becomes increasingly likely that these small coefficients will be interspersed with zero-valued coefficients. As the radius increases further it is probable that a region where all coefficients are zero will be entered.

MPEG-2 uses all these attributes of DCT coefficients when encoding a coefficient block. By sending the coefficients in an optimum order, by describing their values with Huffman coding and by using run-length encoding for the zero-valued coefficients it is possible to achieve a significant reduction in coefficient data which remains entirely lossless. Despite the complexity of this process, it does contribute to improved picture quality because for a given bit rate lossless coding of the coefficients must be better than requantizing, which is lossy. Of course, for lower bit rates both will be required.

It is an advantage to scan in a sequence where the largest coefficient values are scanned first. Then the next coefficient is more likely to be zero than the previous one. With progressively scanned material, a regular zig-zag scan begins in the top-left corner and ends in the bottom-right corner as shown in Figure 5.23. Zig-zag scanning means that significant values are more likely to be transmitted first, followed by the zero values. Instead of coding these zeros, an unique 'end of block' (EOB) symbol is transmitted instead.

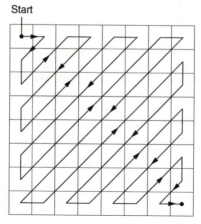

Figure 5.23 The zig-zag scan for a progressively scanned image.

As the zig-zag scan approaches the last finite coefficient it is increasingly likely that some zero-value coefficients will be scanned. Instead of transmitting the coefficients as zeros, the *zero-run-length*, i.e. the number of zero valued coefficients in the scan sequence, is encoded into the next non-zero coefficient which is itself variable-length coded. This combination of run-length and variable-length coding is known as RLC/VLC in MPEG-2.

The DC coefficient is handled separately because it is differentially coded and this discussion relates to the AC coefficients. Three items need to be handled for each coefficient: the zero-run-length prior to this coefficient, the wordlength and the coefficient value itself. The wordlength needs to be known by the decoder so that it can correctly parse the bitstream. The wordlength of the coefficient is expressed directly as an integer called the *size*.

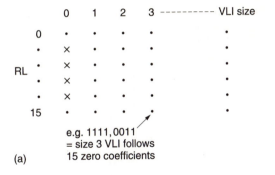

e.g. 1111,0011
= size 3 VLI follows
15 zero coefficients

(a)

Run \ Size	0	1	2	3 → etc.
0	1010 (EOB)	00	01	100
1	–	1100	11011	
2	–	11100	11111001	
3	–	111010	111110111	
4	–	111011	1111111000	
5	–	1111010		

etc.

(b)

Figure 5.24 Run-length and variable-length coding simultaneously compresses runs of zero-valued coefficients and describes the wordlength of a non-zero coefficient.

Figure 5.24(a) shows that a two-dimensional run/size table is created. One dimension expresses the zero-run-length; the other the size. A run length of zero is obtained when adjacent coefficients are non-zero, but a code of 0/0 has no meaningful run/size interpretation and so this bit pattern is used for the EOB symbol.

In the case where the zero-run-length exceeds 14, a code of 15/0 is used, signifying that there are fifteen zero-valued coefficients. This is then followed by another run/size parameter whose run-length value is added to the previous fifteen.

The run/size parameters contain redundancy because some combinations are more common than others. Figure 5.24(b) shows that each run/size value is converted to a variable-length Huffman codeword for transmission. The Huffman codes are designed so that short codes are never a prefix of long codes so that the decoder can deduce the parsing by testing an increasing number of bits until a match with the look-up table is found. Having parsed and decoded the Huffman run/size code, the decoder then knows what the coefficient wordlength will be and can correctly parse that.

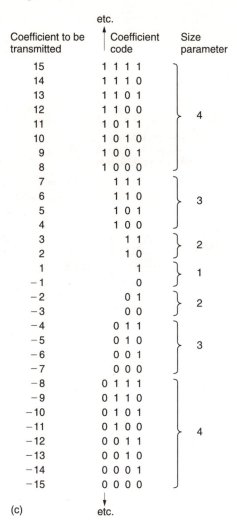

(c)

Figure 5.24 (*Continued*) (c)

The variable-length coefficient code has to describe a bipolar coefficient, i.e one which can be positive or negative. Figure 5.24(c) shows that for a particular size, the coding scale has a certain gap in it. For example, all values from − 7 to + 7 can be sent by a size 3 code, so a size 4 code only has to send the values of − 15 to − 8 and + 8 to + 15. The coefficient code is sent as a pure binary number whose value ranges from all zeros to all ones where the maximum value is a function of the size. The number range is divided into two, the lower half of the codes specifying negative values and the upper half specifying positive.

In the case of positive numbers, the transmitted binary value is the actual coefficient value, whereas in the case of negative numbers a constant must be subtracted which is a function of the size. In the case of a size 4 code, the constant is 15_{10}. Thus a size 4 parameter of 0111_2 (7_{10}) would be interpreted as

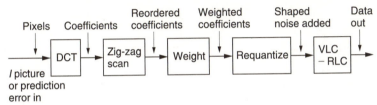

Figure 5.25 A complete spatial coding system which can compress an *I* picture or the prediction error in *P* and *B* pictures. See text for details.

$7 - 15 = - 8$. A size of 5 has a constant of 31 so a transmitted coded of $01010_2(10_2)$ would be interpreted as $10 - 31 = -21$.

This technique saves a bit because, for example, 63 values from $- 31$ to $+ 31$ are coded with only 5 bits having only 32 combinations. This is possible because that extra bit is effectively encoded into the run/size parameter.

Figure 5.25 shows the whole spatial coding subsystem. Macroblocks are subdivided into DCT blocks and the DCT is calculated. The resulting coefficients are multiplied by the weighting matrix and then requantized. The coefficients are then reordered by the zig-zag scan so that full advantage can be taken of run-length and variable-length coding. The last non-zero coefficient in the scan is followed by the EOB symbol.

In predictive coding, sometimes the motion-compensated prediction is nearly exact and so the prediction error will be almost zero. This can also happen on still parts of the scene. MPEG-2 takes advantage of this by sending a code to tell the decoder there is no prediction error data for the macroblock concerned.

The success of temporal coding depends on the accuracy of the vectors. Trying to reduce the bit rate by reducing the accuracy of the vectors is false economy as this simply increases the prediction error. Consequently for a given GOP structure it is only in the spatial coding that the overall bit rate is determined. The RLC/VLC coding is lossless and so its contribution to the compression cannot be varied. If the bit rate is too high, the only option is to increase the size of the coefficient-requantizing steps. This has the effect of shortening the wordlength of large coefficients, and rounding small coefficients to zero, so that the bit rate goes down. Clearly, if taken too far the picture quality will also suffer because at some point the noise floor will become visible as some form of artifact.

5.7 A bidirectional coder

MPEG-2 does not specify how an encoder is to be built or what coding decisions it should make. Instead it specifies the protocol of the bitstream at the output. As a result the coder shown in Figure 5.26 is only an example.

Figure 5.26(a) shows the component parts of the coder. At the input is a chain of picture stores which can be bypassed for reordering purposes. This allows a picture to be encoded ahead of its normal timing when bidirectional coding is employed.

At the centre is a dual-motion estimator which can simultaneously measure motion between the input picture an earlier picture and a later picture. These reference pictures are held in frame stores. The vectors from the motion estimator are used locally to shift a picture in a frame store to form a predicted picture. This

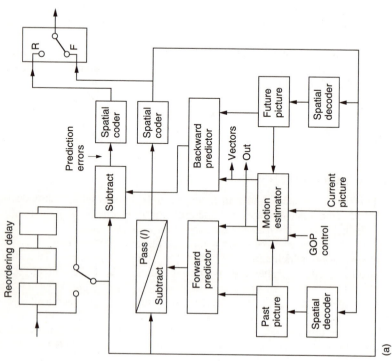

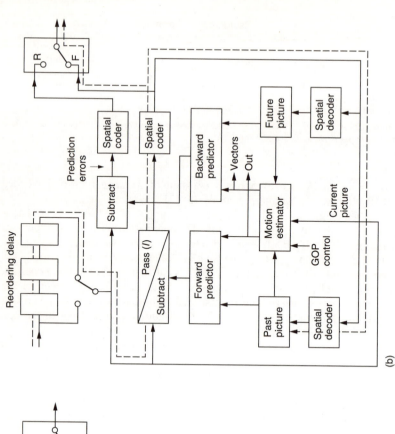

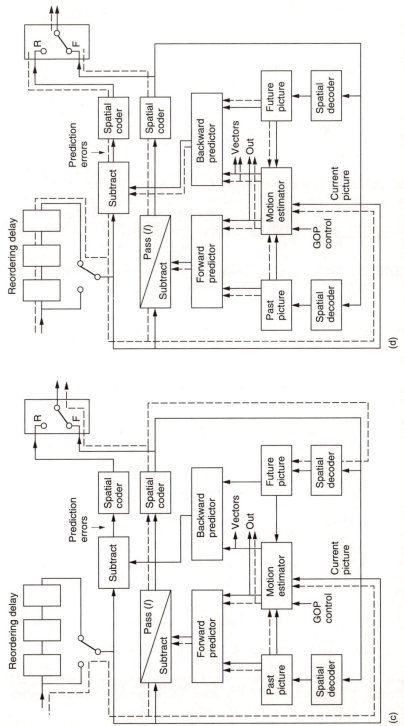

Figure 5.26 A bidirectional coder. (a) The essential components. (b) Signal flow when coding an *I* picture. (c) Signal flow when coding a *P* picture. (d) Signal flow when bidirectional coding.

is subtracted from the input picture to produce a prediction error picture which is then spatially coded.

The bidirectional encoding process will now be described. A GOP begins with an *I* picture which is intra coded. In Figure 5.26(b) the *I* picture emerges from the reordering delay. No prediction is possible on an *I* picture so the motion estimator is inactive. There is no predicted picture and so the prediction error subtractor is set simply to pass the input. The only processing which is active is the forward spatial coder which describes the picture with DCT coefficients. The output of the forward spatial coder is locally decoded and stored in the past picture frame store.

The reason for the spatial encode/decode is that the past picture frame store now contains exactly what the decoder frame store will contain, including the effects of any requantizing errors. When the same picture is used as a reference at both ends of a differential coding system, the errors will cancel out.

Having encoded the *I* picture, attention turns to the *P* picture. The input sequence is *IBBP*, but the transmitted sequence must be *IPBB*. Figure 5.26(c) shows that the reordering delay is bypassed to select the *P* picture. This passes to the motion estimator which compares it with the *I* picture and outputs a vector for each macroblock. The forward predictor uses these vectors to shift the *I* picture so that it more closely resembles the *P* picture. The predicted picture is then subtracted from the actual picture to produce a forward prediction error. This is then spatially coded. Thus the *P* picture is transmitted as a set of vectors and a prediction error image.

The *P* picture is locally decoded in the right-hand decoder. This takes the forward predicted picture and adds the decoded prediction error to obtain exactly what the decoder will obtain.

Figure 5.26(d) shows that the encoder now contains a *I* picture in the left store and a *P* picture in the right store. The reordering delay is reselected so that the first *B* picture can be input. This passes to the motion estimator where it is compared with both the *I* and *P* pictures to produce forward and backward vectors. The forward vectors go to the forward predictor to make a *B* prediction from the *I* picture. The backward vectors go to the backward predictor to make a *B* prediction from the *P* picture. These predictions are simultaneously subtracted from the actual *B* picture to produce a forward prediction error and a backward prediction error. These are then spatially encoded. The encoder can then decide which direction of coding resulted in the best prediction; i.e. the smallest prediction error.

Not shown in the interests of clarity is a third signal path which creates a predicted *B* picture from the average of forward and backward predictions. This is subtracted from the input picture to produce a third prediction error. In some circumstances this prediction error may use fewer data than either forward of backward prediction alone.

As *B* pictures are never used to create other pictures, the decoder does not locally decode the *B* picture. After decoding and displaying the *B* picture the decoder will discard it. At the encoder the *I* and *P* pictures remain in their frame stores and the second *B* picture is input from the reordering delay.

Following the encoding of the second *B* picture, the encoder must reorder again to encode the second *P* picture in the GOP. This will be locally decoded and will replace the *I* picture in the left store. The stores and predictors switch designation because the left store is now a future *P* picture and the right store is now a past *P* picture. *B* pictures between them are encoded as before.

There is still some redundancy in the output of a bidirectional coder and MPEG-2 is remarkably diligent in finding it. In *I* pictures, the DC coefficient describes the average brightness of an entire DCT block. In real video the DC component of adjacent blocks will be similar much of the time. A saving in bit rate can be obtained by differentially coding the DC coefficient.

In *P* and *B* pictures this is not done because these are prediction errors not actual images and the statistics are different. However, *P* and *B* pictures send vectors and instead the redundancy in these is explored. In a large moving object, many macroblocks will be moving at the same velocity and their vectors will be the same. Thus differential vector coding will be advantageous.

As has been seen above, differential coding cannot be used indiscriminately as it is prone to error propagation. Periodically absolute DC coefficients and vectors must be sent and the *slice* is the logical structure which supports this mechanism. In *I* pictures, the first DC coefficient in a slice is sent in absolute form, whereas the subsequent coefficients are sent differentially. In *P* or *B* pictures, the first vector in a slice is sent in absolute form, but the subsequent vectors are differential.

Slices are horizontal picture strips which are one macroblock (16 pixels) high and which proceed from left to right across the screen. The sides of the picture must coincide with the beginning or the end of a slice in MPEG-2, but otherwise the encoder is free to decide how big slices should be and where they begin.

In the case of a central dark building silhouetted against the bright sky, there would be two large changes in the DC coefficients, one at each edge of the building. It may be advantageous to the encoder to break the width of the picture into three slices, one each for the left and right areas of sky and one for the building. In the case of a large moving object, different slices may be used for the object and the background.

Each slice contains its own synchronizing pattern, so following a transmission error, correct decoding can resume at the next slice. Slice size can also be matched to the characteristics of the transmission channel. For example, in an error-free transmission system the use of a large number of slices in a packet simply wastes data capacity on surplus synchronizing patterns. However, in a non-ideal system it might be advantageous to have frequent resynchronizing.

5.8 Handling interlaced pictures

Spatial coding, predictive coding and motion compensation can still be performed using interlaced source material at the cost of considerable complexity. Despite that complexity, MPEG-2 cannot be expected to perform as well with interlaced material.

Figure 5.27 shows that in an incoming interlaced frame there are two fields each of which contain half of the lines in the frame. In MPEG-2 these are known as the *top field* and the *bottom field*. In video from a camera, these fields represent the state of the image at two different times. Where there is little image motion, this is unimportant and the fields can be combined obtaining more effective compression. However, in the presence of motion the fields become increasingly decorrelated because of the displacement of moving objects from one field to the next.

This characteristic determines that MPEG-2 must be able to handle fields independently or together. This dual approach permeates all aspects of MPEG-2

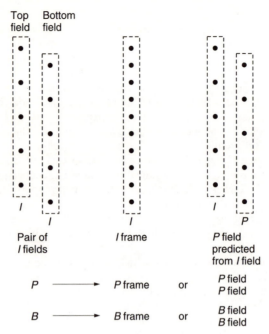

Figure 5.27 An interlaced frame consists of top and bottom fields. MPEG-2 can code a frame in the ways shown here.

and affects the definition of pictures, macroblocks, DCT blocks and zig-zag scanning.

Figure 5.27 also shows how MPEG-2 designates interlaced fields. In picture types *I*, *P* and *B*, the two fields can be superimposed to make a *frame-picture* or the two fields can be coded independently as two *field-pictures*. As a third possibility, in *I* pictures only, the bottom field-picture can be predictively coded from the top field-picture to make an *IP* frame-picture.

A frame-picture is one in which the macroblocks contain lines from both field types over a picture area sixteen scan lines high. Each luminance macroblock contains the usual four DCT blocks but there are two ways in which these can be assembled. Figure 5.28(a) shows how a frame is divided into *frame DCT* blocks. This is identical to the progressive scan approach in that each DCT block contains eight contiguous picture lines. In 4:2:0, the colour difference signals have been downsampled by a factor of two and shifted. Figure 5.28(a) also shows how one 4:2:0 DCT block contains the chroma data from sixteen lines in two fields.

Even small amounts of motion in any direction can destroy the correlation between odd and even lines and a frame DCT will result in an excessive number of coefficients. Figure 5.28(b) shows that instead the luminance component of a frame can also be divided into *field DCT* blocks. In this case one DCT block contains odd lines and the other contains even lines. In this mode the chroma still produces one DCT block from both fields as in Figure 5.28(a).

When an input frame is designated as two field-pictures, the macroblocks come from a screen area which is thirty two lines high. Figure 5.28(c) shows that

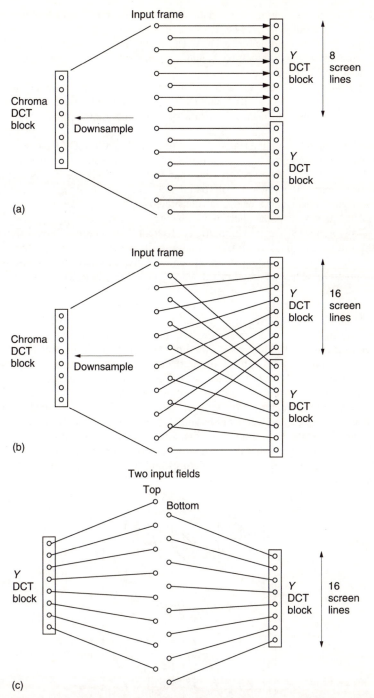

Figure 5.28 (a) In Frame-DCT, a picture is effectively de-interlaced. (b) In Field-DCT, each DCT block only contains lines from one field, but over twice the screen area. (c) The same DCT content results when field-pictures are assembled into blocks.

the DCT blocks contain the same data as if the input frame had been designated a *frame-picture* but with *field DCT*. Consequently it is only frame-pictures which have the option of field or frame DCT. These may be selected by the encoder on a macroblock-by-macroblock basis and, of course, the resultant bitstream must specify what has been done.

In a frame which contains a small moving area, it may be advantageous to encode as a frame-picture with frame DCT except in the moving area where field DCT is used. This approach may result in fewer bits than coding as two field-pictures.

In a field-picture and in a frame-picture using field DCT, a DCT block contains lines from one field type only and this must have come from a screen area sixteen scan lines high, whereas in progressive scan and frame DCT the area is only eight scan lines high. A given vertical spatial frequency in the image is sampled at points twice as far apart which is interpreted by the field DCT as a doubled spatial frequency, whereas there is no change in the horizontal spectrum.

Following the DCT calculation, the coefficient distribution will be different in field-pictures and field DCT frame-pictures. In these cases, the probability of coefficients is not a constant function of radius from the DC coefficient as it is in progressive scan, but is elliptical where the ellipse is twice as high as it is wide.

Using the standard 45° zig-zag scan with this different coefficient distribution would not have the required effect of putting all the significant coefficients at the beginning of the scan. To achieve this requires a different zig-zag scan, which is shown in Figure 5.29. This scan, sometimes known as the Yeltsin walk, attempts to match the elliptical probability of interlaced coefficients with a scan slanted at 67.5° to the vertical. This is clearly suboptimal, and is one of the reasons why MPEG-2 does not work so well with interlaced video.

Motion estimation is more difficult in an interlaced system. Vertical detail can result in differences between fields and this reduces the quality of the match. Fields are vertically subsampled without filtering and so contain alias products. This aliasing will mean that the vertical waveform representing a moving object will not be the same in successive pictures and this will also reduce the quality of the match.

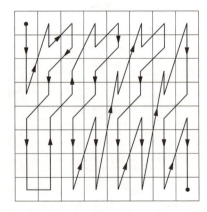

Figure 5.29 The zig-zag scan for an interlaced image has to favour vertical frequencies twice as much as horizontal.

Even when the correct vector has been found, the match may be poor so the estimator fails to recognize it. If it is recognized, a poor match means that the quality of the prediction in P and B pictures will be poor and so a large prediction error or residual has to be transmitted. In an attempt to reduce the residual, MPEG-2 allows field-pictures to use motion-compensated prediction from either the adjacent field or from the same field type in another frame. In this case the encoder will use the better match. This technique can also be used in areas of frame-pictures which use field DCT.

The motion compensation of MPEG-2 has half-pixel resolution and this is inherently compatible with an interlace because an interpolator must be present to handle the half-pixel shifts. Figure 5.30(a) shows that in an interlaced system, each field contains half of the frame lines and so interpolating half-way between lines of one field type will actually create values lying on the sampling structure of the other field type. Thus it is equally possible for a predictive system to decode a given field type based on pixel data from the other field type or of the same type.

If when using predictive coding from the other field type the vertical motion vector contains a half-pixel component, then no interpolation is needed because the act of transferring pixels from one field to another results in such a shift.

Figure 5.30(b) shows that a macroblock in a given P field-picture can be encoded using a vector which shifts data from the previous field or from the field before that, irrespective of which frames these fields occupy. As noted above, field-

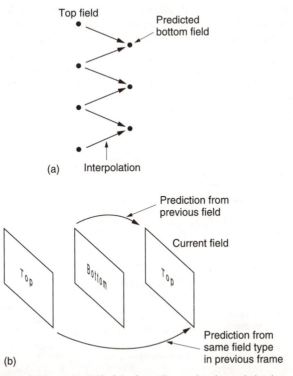

Figure 5.30 (a) Each field contains half of the frame lines and so interpolation is needed to create values lying on the sampling structure of the other field type. (b) Prediction can use data from the previous field or the one before that.

picture macroblocks come from an area of screen thirty-two lines high and this means that the vector density is halved resulting in larger prediction errors at the boundaries of moving objects.

As an option, field-pictures can restore the vector density by using 16×8 motion compensation where separate vectors are used for the top and bottom halves of the macroblock. Frame-pictures can also use 16×8 motion compensation in conjunction with field DCT. Whilst the 2×2 DCT block luminance structure of a macroblock can easily be divided vertically in two, in 4:2:0 the same screen area is represented by only one chroma macroblock of each component type. As it cannot be divided in half, this chroma is deemed to belong to the luminance DCT blocks of the upper field. In 4:2:2 no such difficulty arises.

MPEG-2 supports interlace simply because interlaced video exists in legacy systems and there is a requirement to compress it. However, where the opportunity arises to define a new system, interlace should be avoided. Legacy interlaced source material should be handled using a motion-compensated de-interlacer prior to compression in the progressive domain.

5.9 An MPEG-2 coder

Figure 5.31 shows the complete coder. The bidirectional coder outputs coefficients and vectors, and the quantizing table in use. The vectors of P and B pictures and the DC coefficients of I pictures are differentially encoded in slices and the remaining coefficients are RLC/VLC coded. The multiplexer assembles all these data into a single bitstream called an elementary stream. The output of the encoder is a buffer which absorbs the variations in bit rate between different picture types. The buffer output has a constant bit rate determined by the demand clock. This comes from the transmission channel or storage device. If the bit rate is low, the buffer will tend to fill up, whereas if it is high the buffer will tend to empty. The buffer content is used to control the severity of the requantizing in the spatial coders. The more the buffer fills, the bigger the requantizing steps get.

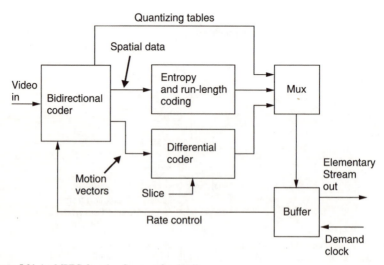

Figure 5.31 An MPEG 2 coder. See text for details.

The buffer in the decoder has a finite capacity and the encoder must model the decoder's buffer occupancy so that it neither overflows nor underflows. An overflow might occur if an *I* picture is transmitted when the buffer content is already high. The buffer occupancy of the decoder depends somewhat on the memory access strategy of the decoder. Instead of defining a specific buffer size, MPEG-2 defines the size of a particular mathematical model of a hypothetical buffer. The decoder designer can use any strategy which implements the model, and the encoder can use any strategy which doesn't overflow or underflow the model. The elementary stream has a parameter called the video buffer verifier (VBV) which defines the minimum buffering assumptions of the encoder.

Buffering is one way of ensuring constant quality when picture entropy varies. An intelligent coder may run down the buffer contents in anticipation of a difficult picture sequence so that a large amounts of data can be sent.

MPEG-2 does not define what a decoder should do if a buffer underflow or overflow occurs, but since both irrecoverably lose data it is obvious that there will be more or less of an interruption to the decoding. Even a small loss of data may cause loss of synchronization and in the case of long GOP the lost data may make the rest of the GOP undecodable. A decoder may chose to repeat the last properly decoded picture until it can begin to operate correctly again.

Buffer problems occur if the VBV model is violated. If this happens then more than one underflow or overflow can result from a single violation. Switching an MPEG bitstream can cause a violation because the two encoders concerned may have radically different buffer occupancy at the switch.

5.10 The elementary stream

Figure 5.32 shows the structure of the elementary stream from an MPEG-2 encoder. The structure begins with a set of coefficients representing a DCT block. Six or eight DCT blocks form the luminance and chroma content of one

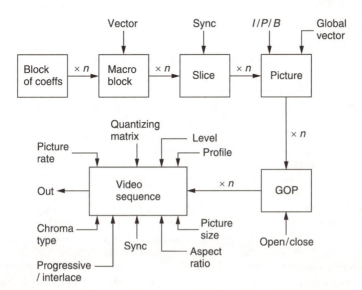

Figure 5.32 The structure of an elementary stream. MPEG defines the syntax precisely.

macroblock. In *P* and *B* pictures a macroblock will be associated with a vector for motion compensation. Macroblocks are associated into slices in which DC coefficients of *I* pictures and vectors in *P* and *B* pictures are differentially coded. An arbitrary number of slices forms a picture and this needs *I/P/B* flags describing the type of picture it is. The picture may also have a global vector which efficiently deals with pans.

Several pictures form a GOP. The GOP begins with an *I* picture and may or may not include *P* and *B* pictures in a structure which may vary dynamically.

Several GOPs form a Sequence which begins with a Sequence header containing important data to help the decoder. It is possible to repeat the header within a sequence, and this helps lock-up in random access applications. The Sequence header describes the MPEG-2 profile and level, whether the video is progressive or interlaced, whether the chroma is 4:2:0 or 4:2:2, the size of the picture and the aspect ratio of the pixels. The quantizing matrix used in the spatial coder can also be sent. The sequence begins with a standardized bit pattern which is detected by a decoder to synchronize the deserialization.

5.11 An MPEG-2 decoder

The decoder is only defined by implication from the definitions of syntax and any decoder which can correctly interpret all combinations of syntax at a particular profile will be deemed compliant however it works.

The first problem a decoder has is that the input is an endless bitstream which contains a huge range of parameters many of which have variable length. Unique synchronizing patterns must be placed periodically throughout the bitstream so that the decoder can identify a known starting point. The pictures which can be sent under MPEG-2 are so flexible that the decoder must first find a Sequence header so that it can establish the size of the picture, the frame rate, the colour coding used, etc.

The decoder must also be supplied with a 27 MHz system clock. In a DVD player, this would come from a crystal, but in a transmission system this would be provided by a numerically locked loop running from a clock reference parameter in the bitstream (see Chapter 9). Until this loop has achieved lock the decoder cannot function properly.

Figure 5.33 shows a bidirectional decoder. The decoder can only begin decoding with a *I* picture and as this only uses intra-coding there will be no vectors. An *I* picture is transmitted as a series of slices. These slices begin with subsidiary synchronizing patterns. The first macroblock in the slice contains an absolute DC coefficient, but the remaining macroblocks code the DC coefficient differentially so the decoder must subtract the differential values from the previous value to obtain the absolute value.

The AC coefficients are sent as Huffman coded run/size parameters followed by coefficient value codes. The variable-length Huffman codes are decoded by using a look-up table and extending the number of bits considered until a match is obtained. This allows the zero-run-length and the coefficient size to be established. The right number of bits is taken from the bitstream corresponding to the coefficient code and this is decoded to the actual coefficient using the size parameter.

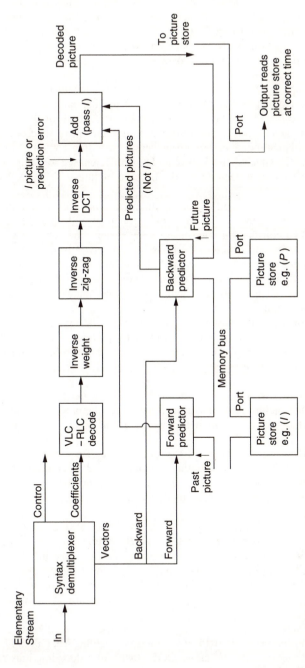

Figure 5.33 A bidirectional MPEG-2 decoder. See text for details.

If the correct number of bits has been taken from the stream, the next bit must be the beginning of the next run/size code and so on until the EOB (end of block) symbol is reached. The decoder uses the coefficient values and the zero-run-lengths to populate a DCT coefficient block following the appropriate zig-zag scanning sequence. Following EOB, the bitstream then continues with the next DCT block. Clearly this Huffman decoding will work perfectly or not at all. A single bit slippage in synchronism or a single corrupted data bit can cause a spectacular failure.

Once a complete DCT coefficient block has been received, the coefficients need to be inverse quantized and inverse weighted. Then an inverse DCT can be performed and this will result in an 8×8 pixel block. A series of DCT blocks will allow the luminance and colour information for an entire macroblock to be decoded and this can be placed in a framestore. Decoding continues in this way until the end of the slice when an absolute DC coefficient will once again be sent. Once all the slices have been decoded, an entire picture will be resident in the framestore.

The amount of data needed to decode the picture is variable and the decoder just keeps going until the last macroblock is found. It will obtain data from the input buffer. In a constant bit rate transmission system, the decoder will remove more data to decode an I picture than has been received in one picture period, leaving the buffer emptier than it began. Subsequent P and B pictures need much less data and allow the buffer to fill again. The picture will be output when the time stamp (see Chapter 9) sent with the picture matches the state of the decoder's time count.

Following the I picture may be another I picture or a P picture. Assuming a P picture, this will be predictively coded from the I picture. The P picture will be divided into slices as before. The first vector in a slice is absolute, but subsequent vectors are sent differentially. However, the DC coefficients are not differential.

Each macroblock may contain a forward vector. The decoder uses this to shift pixels from the I picture into the correct position for the predicted P picture. The vectors have half-pixel resolution and where a half-pixel shift is required, an interpolator will be used.

The DCT data are sent much as for an I picture. They will require inverse quantizing, but not inverse weighting because P and B coefficients are flat-weighted. When decoded this represents an error-cancelling picture which is added pixel-by-pixel to the motion predicted picture. This results in the output picture.

If bidirectional coding is being used, the P picture may be stored until one or more B pictures have been decoded. The B pictures are sent essentially as a P picture might be, except that the vectors can be forward, backward or bidirectional. The decoder must take pixels from the I picture, the P picture, or both, and shift them according to the vectors to make a predicted picture. The DCT data decode to produce an error-cancelling image as before.

In an interlaced system, the prediction mechanism may alternatively obtain pixel data from the previous field or the field before that. Vectors may relate to macroblocks or to 16×8 pixel areas. DCT blocks after decoding may represent frame lines or field lines. This adds up to a lot of different possibilities for a decoder handling an interlaced input.

5.12 Coding artifacts

This section describes the visible results of imperfect coding. Imperfect coding may be where the coding algorithm is sub-optimal, where the coder latency is too short or where the compression factor in use is simply too great for the material.

In motion-compensated systems such as MPEG, the use of periodic intra-fields means that the coding noise varies from picture to picture and this may be visible as noise pumping. Noise pumping may also be visible where the amount of motion changes. If a pan is observed, as the pan speed increases the motion vectors may become less accurate and reduce the quality of the prediction processes. The prediction errors will get larger and will have to be more coarsely quantized. Thus the picture gets noisier as the pan accelerates and the noise reduces as the pan slows down. The same result may be apparent at the edges of a picture during zooming. The problem is worse if the picture contains fine detail. Panning on grass or trees waving in the wind taxes most coders severely. Camera shake from a hand-held camera also increases the motion vector data and results in more noise as does film weave.

Input video noise or film grain degrades inter-coding as there is less redundancy between pictures and the difference data become larger, requiring coarse quantizing and adding to the existing noise.

Where a codec is really fighting the quantizing may become very coarse and as a result the video level at the edge of one DCT block may not match that of its neighbour. As a result, the DCT block structure becomes visible as a mosaicing or tiling effect. Coarse quantizing also causes some coefficients to be rounded up and appear larger than they should be. High-frequency coefficients may be eliminated by heavy quantizing and this forces the DCT to act as a steep-cut low-pass filter. This causes fringeing or ringing around sharp edges and extra shadowy edges which were not in the original. This is most noticeable on text.

Excess compression may also result in colour bleed where fringeing has taken place in the chroma or where high-frequency chroma coefficients have been discarded. Graduated colour areas may reveal banding or posterizing as the colour range is restricted by requantizing. These artifacts are almost impossible to measure with conventional test gear.

Neither noise pumping nor blocking are visible on analog video recorders and so it is nonsense to liken the performance of a codec to the quality of a VCR. In fact noise pumping is extremely objectionable because, unlike steady noise, it attracts attention in peripheral vision and may result in viewing fatigue.

In addition to highly detailed pictures with complex motion, certain types of video signal are difficult for MPEG-2 to handle and will usually result in a higher level of artifacts than usual. Noise has already been mentioned as a source of problems. Timebase error from, for example, VCRs is undesirable because this puts succesive lines in different horizontal positions. A straight vertical line becomes jagged and this results in high spatial frequencies in the DCT process. Spurious coefficients are created which need to be coded.

Much archive video is in composite form and MPEG-2 can only handle this after it has been decoded to components. Unfortunately many general-purpose composite decoders have a high level of residual subcarrier in the outputs. This is normally not a problem because the subcarrier is designed to be invisible to the

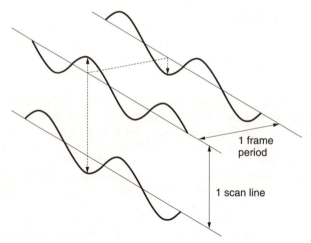

1 frame period

1 scan line

Figure 5.34 In composite video the subcarrier frequency is arranged so that inversions occur between adjacent lines and pictures to help reduce the visibility of the chroma.

naked eye. Figure 5.34 shows that in PAL and NTSC the subcarrier frequency is selected so that a phase reversal is achieved between successive lines and frames.

Whilst this makes the subcarrier invisible to the eye, it is not invisible to an MPEG decoder. The subcarrier waveform is interpreted as a horizontal frequency, the vertical phase reversals are interpreted as a vertical spatial frequency and the picture-to-picture reversals increase the magnitude of the prediction errors. The subcarrier level may be low but it can be present over the whole screen and require an excess of coefficients to describe it.

Composite video should not in general be used as a source for MPEG-2 encoding, but where this is inevitable the standard of the decoder must be much higher than average, especially in the residual subcarrier specification. Some MPEG preprocessors support high-grade composite decoding options.

Judder from conventional linear standards convertors degrades the performance of MPEG-2. The optic flow axis is corrupted and linear filtering causes multiple images which confuse motion estimators and result in larger prediction errors. If standards conversion is necessary, the MPEG-2 system must be used to encode the signal in its original format and the standards convertor should be installed after the decoder. If a standards convertor has to be used before the encoder, then it must be a type which has effective motion compensation.

Film weave causes movement of one picture with respect to the next and this results in more vector activity and larger prediction errors. Movement of the centre of the film frame along the optical axis causes magnification changes which also result in excess prediction error data. Film grain has the same effect as noise: it is random and so cannot be compressed.

Perhaps because it is relatively uncommon, MPEG-2 cannot handle image rotation well because the motion-compensation system is only designed for translational motion. Where a rotating object is highly detailed, such as in certain fairground rides, the motion-compensation failure requires a significant amount of prediction error data and if a suitable bit rate is not available the level of artifacts will rise.

Flash guns used by still photographers are a serious hazard to MPEG-2 especially when long GOPs are used. At a press conference where a series of flashes may occur, the resultant video contains intermittent white frames which defeat prediction. A huge prediction error is required to return to the original picture from a white picture. The output buffer fills and heavy requantizing is employed. After a few flashes the picture has generally gone to tiles.

5.13 Processing MPEG-2 and concatenation

Concatenation loss occurs when the losses introduced by one codec are compounded by a second codec. All practical compressers, MPEG-2 included, are lossy because what comes out of the decoder is not bit-identical to what went into the encoder. The bit differences are controlled so that they have minimum visibility to a human viewer.

MPEG-2 is a toolbox which allows a variety of manipulations to be performed in both the spatial and the temporal domains. There is a limit to the compression which can be used on a single frame, and if higher compression factors are needed, temporal coding will have to be used. The longer the run of pictures considered, the lower the bit rate needed, but the harder it becomes to edit.

The most editable form of MPEG-2 is to use *I* pictures only. As there is no temporal coding, pure cut edits can be made between pictures. The next best thing is to use a repeating *IB* structure which is locked to the odd/even field structure. Cut edits cannot be made as the *B* pictures are bidirectionally coded and need data from both adjacent *I* pictures for decoding. The *B* picture has to be decoded prior to the edit and re-encoded after the edit. This will cause a small concatenation loss.

Beyond the *IB* structure processing gets harder. If a long GOP is used for the best compression factor, an *IBBPBBP* . . . structure results. Editing this is very difficult because the pictures are sent out of order so that bidirectional decoding can be used. MPEG allows closed GOPs where the last *B* picture is coded wholly from the previous pictures and does not need the *I* picture in the next GOP. The bitstream can be switched at this point but only if the GOP structures in the two source video signals are synchronized (makes colour framing seem easy). Consequently in practice a long GOP bitstream will need to be decoded prior to any production step. Afterwards it will need to be re-encoded.

This is known as *naive* concatenation and an enormous pitfall awaits. Unless the GOP structure of the output is identical to and synchronized with the input the results will be disappointing. The worst case is where an *I* picture is encoded from a picture which was formerly a *B* picture. It is easy enough to lock the GOP structure of a coder to a single input, but if an edit is made between two inputs, the GOP timings could well be different.

As there are so many structures allowed in MPEG, there will be a need to convert between them. If this has to be done, it should only be in the direction which increases the GOP length and reduces the bit rate. Going the other way is inadvisable. The ideal way of converting from, say, the *IB* structure of a news system to the *IBBP* structure of an emission system is to use a recompressor. This is a kind of standards convertor which will give better results than a decode followed by an encode.

The DCT part of MPEG-2 itself is lossless. If all the coefficients are preserved intact an inverse transform yields the same pixel data. Unfortunately this does not yield enough compression for many applications. In practice the coefficients are made less accurate by removing bits starting at the least significant end and working upwards. This process is weighted, or made progressively more aggressive as spatial frequency increases.

Small-value coefficients may be truncated to zero and large-value coefficients are most coarsely truncated at high spatial frequencies where the effect is least visible.

Figure 5.35(a) shows what happens in the ideal case where two *identical* coders are put in tandem and synchronized. The first coder quantizes the coefficients to finite accuracy and causes a loss on decoding. However, when the second coder performs the DCT calculation, the coefficients obtained will be identical to the quantized coefficients in the first coder and so if the second weighting and requantizing step is identical the same truncated coefficient data will result and there will be no further loss of quality.[7]

In practice this ideal situation is elusive. If the two DCTs become non-identical for any reason, the second requantizing step will introduce further error in the coefficients and the artifact level goes up. Figure 5.35(b) shows that non-identical concatenation can result from a large number of real-world effects.

An intermediate processing step such as a fade will change the pixel values and thereby the coefficients. A DVE resize or shift will move pixels from one DCT block to another. Even if there is no processing step, this effect will also occur if the two codecs disagree on where the MPEG picture boundaries are

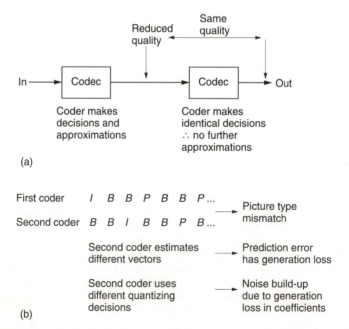

(a)

(b)

Figure 5.35 (a) Two identical coders in tandem which are synchronized make similar coding decisions and cause little loss. (b) There are various ways in which concatenated coders can produce non-ideal performance.

within the picture. If the boundaries are correct there will still be concatenation loss if the two codecs use different weightings.

One problem with MPEG is that the compressor design is unspecified. Whilst this has advantages, it does mean that the chances of finding identical coders is minute because each manufacturer will have their own views on the best compression algorithm. In a large system it may be worth obtaining the coders from a single supplier.

It is now increasingly accepted that concatenation of compression techniques is potentially damaging, and results are worse if the codecs are different. Clearly, feeding a digital coder such as MPEG-2 with a signal which has been subject to analog compression comes into the category of worse. Using interlaced video as a source for MPEG coding is sub-optimal and using decoded composite video is even worse.

One way of avoiding concatenation is to stay in the compressed data domain. If the goal is just to move pictures from one place to another, decoding to traditional video so an existing router can be used is not ideal, although substantially better than going through the analog domain.

Figure 5.36 shows some possibilities for picture transport. Clearly, if the pictures exist as a compressed file on a server, a file transfer is the right way to do it as there is no possibility of loss because there has been no concatenation. File transfer is also quite indifferent to the picture format. It doesn't care whether the pictures are interlaced or not, whether the colour is 4:2:0 or 4:2:2.

Decoding to SDI (serial digital interface) standard is sometimes done so that existing serial digital routing can be used. This is concatenation and has to be done carefully. The compressed video can only use interlace with non-square

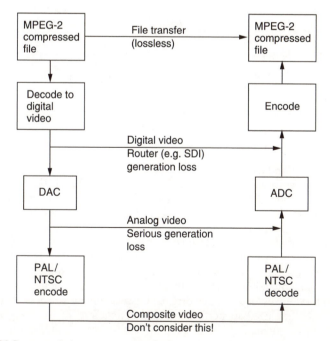

Figure 5.36 Compressed picture transport mechanisms contrasted.

pixels and the colour coding has to be 4:2:2 because SDI only allows that. If a compressed file has 4:2:0 the chroma has to be interpolated up to 4:2:2 for SDI transfer and then subsampled back to 4:2:0 at the second coder and this will cause generation loss. An SDI transfer also can only be performed in real time, thus negating one of the advantages of compression. In short, traditional SDI is not really at home with compression.

As 4:2:0 progressive scan gains popularity and video production moves steadily towards non-format-specific hardware using computers and data networks, use of the serial digital interface will eventually decline. In the short term, if an existing SDI router has to be used, one solution is to produce a bitstream which is sufficiently similar to SDI that a router will pass it. In other words, the signal level, frequency and impedance is pure SDI, but the data protocol is different so that a bit-accurate file transfer can be performed. This has two advantages over SDI. First, the compressed data format can be anything appropriate and non-interlaced and/or 4:2:0 can be handled in any picture size, aspect ratio or frame rate. Second, a faster than real-time transfer can be used depending on the compression factor of the file. Equipment which allows this is becoming available and its use can mean that the full economic life of a SDI routing installation can be obtained.

An improved way of reducing concatenation loss has emerged from the ATLANTIC research project.[8] Figure 5.37 shows that the second encoder in a concatenated scheme does not make its own decisions from the incoming video, but is instead steered by information from the first bitstream. As the second encoder has less intelligence, it is known as a *dim* encoder.

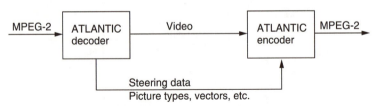

Figure 5.37 In an ATLANTIC system, the second encoder is steered by information from the decoder.

The information bus carries all the structure of the original MPEG-2 bitstream which would be lost in a conventional decoder. The ATLANTIC decoder does more than decode the pictures. It also places on the information bus all parameters needed to make the dim encoder re-enact what the initial MPEG-2 encode did as closely as possible.

The GOP structure is passed on so that pictures are re-encoded as the same type. Positions of macroblock boundaries become identical so that DCT blocks contain the same pixels and motion vectors relate to the same screen data. The weighting and quantizing tables are passed so that coefficient truncation is identical. Motion vectors from the original bitsream are passed on so that the dim encoder does not need to perform motion estimation. In this way predicted pictures will be identical to the original prediction and the prediction error data will be the same.

One application of this approach is in recompression, where an MPEG-2 bitstream has to have its bit rate reduced. This has to be done by heavier requantizing of coefficients, but if as many other parameters as possible can be kept the same, such as motion vectors, the degradation will be minimized. In a simple recompressor just requantizing the coefficients means that the predictive coding will be impaired. In a proper encode, the quantizing error due to coding, say, an *I* picture is removed from the *P* picture by the prediction process. The prediction error of *P* is obtained by subtracting the decoded *I* picture rather than the original *I* picture.

In simple recompression this does not happen and there may be a tolerance build-up known as drift.[9] A more sophisticated recompressor will need to repeat the prediction process using the decoded output pictures as the prediction reference.

MPEG-2 bitstreams will often be decoded for the purpose of switching. Local insertion of commercial breaks into a centrally originated bitstream is one obvious requirement. If the decoded video signal is switched, the information bus must also be switched. At the switch point identical re-encoding becomes

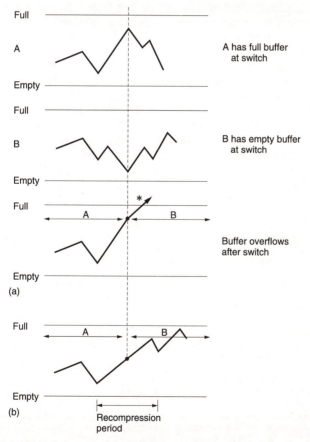

Figure 5.38 (a) A bitstream switch at a different level of buffer occupancy can cause a decoder overflow. (b) Recompression after a switch to return to correct buffer occupancy.

impossible because prior pictures required for predictive coding will have disappeared. At this point the dim encoder has to become bright again because it has to create an MPEG-2 bitstream without assistance.

It is possible to encode the information bus into a form which allows it to be invisibly carried in the serial digital interface. Where a production process such as a vision mixer or DVE performs no manipulation, i.e. becomes bit transparent, the subsequent encoder can extract the information bus and operate in 'dim' mode. Where a manipulation is performed, the information bus signal will be corrupted and the encoder has to work in 'bright' mode. The encoded information signal is known as a 'mole'[10] because it burrows through the processing equipment!

There will be a generation loss at the switch point because the re-encode will be making different decisions in bright mode. This may be difficult to detect because the human visual system is slow to react to a vision cut and defects in the first few pictures after a cut are masked.

In addition to the video computation required to perform a cut, the process has to consider the buffer occupancy of the decoder. A downstream decoder has finite buffer memory, and individual encoders model the decoder buffer occupancy to ensure that it neither overflows nor underflows. At any instant the decoder buffer can be nearly full or nearly empty without a problem provided there is a subsequent correction. An encoder which is approaching a complex I picture may run down the buffer so it can send a lot of data to describe that picture. Figure 5.38(a) shows that if a decoder with a nearly full buffer is suddenly switched to an encoder which has been running down its buffer occupancy, the decoder buffer will overflow when the second encoder sends a lot of data.

An MPEG-2 switcher will need to monitor the buffer occupancy of its own output to avoid overflow of downstream decoders. Where this is a possibility the second encoder will have to recompress to reduce the output bit rate temporarily. In practice there will be a recovery period where the buffer occupancy of the newly selected signal is matched to that of the previous signal. This is shown in Figure 5.38(b).

References

1. MPEG Video Standard: ISO/IEC 13818–2: Information technology – generic coding of moving pictures and associated audio information: Video (1996) (aka ITU-T Rec. H–262 (1996)
2. Huffman, D.A., A method for the construction of minimum redundancy codes. *Proc. IRE*, **40** 1098–1101 (1952)
3. LeGall, D., MPEG: a video compression standard for multimedia applications. *Communications of the ACM*, **34**, No.4, 46–58 (1991)
4. ISO/IEC JTC1/SC29/WG11 MPEG, International standard ISO 11172 'Coding of moving pictures and associated audio for digital storage media up to 1.5 Mbits/s' (1992)
5. ISO Joint Photographic Experts Group standard JPEG–8-R8
6. Wallace, G.K., Overview of the JPEG (ISO/CCITT) still image compression standard. ISO/ JTC1/SC2/WG8 N932 (1989)
7. Stone, J. and Wilkinson, J., Concatenation of video compression systems. Presented at 137th SMPTE Tech. Conf. New Orleans (1995)
8. Wells, N.D., The ATLANTIC project: Models for programme production and distribution. *Proc. Euro. Conf. Multimedia Applications Services and Techniques (ECMAST)*, 243–253 (1996)
9. Werner, O., Drift analysis and drift reduction for multiresolution hybrid video coding. *Image Communication*, **8**, 387–409 (1996)
10. Knee, M.J. and Wells, N.D., Seamless concatenation – a 21st century dream. Presented at Int. Television. Symp. Montreux (1997)

Chapter 6

Digital coding principles

Recording and communication are quite different tasks, but they have a great deal in common. Digital transmission consists of converting data into a waveform suitable for the path along which they are to be sent. Digital recording is basically the process of recording a digital transmission waveform on a suitable medium. In this chapter the fundamentals of digital recording and transmission are introduced along with descriptions of the coding and error-correction techniques used in practical applications.

6.1 Introduction

Data can be recorded on many different media and conveyed using many forms of transmission. The generic term for the path down which the information is sent is the *channel*. In a transmission application, the channel may be a point-to-point cable, a network stage or a radio link. In a recording application the channel will include the record head, the medium and the replay head.

In digital circuitry there is a great deal of noise immunity because the signal has only two states, which are widely separated compared with the amplitude of noise. In both digital recording and transmission this is not always the case. In magnetic recording, noise immunity is a function of track width and reduction of the working SNR of a digital track allows the same information to be carried in a smaller area of the medium, improving economy of operation. In broadcasting, the noise immunity is a function of the transmitter power and reduction of working SNR allows lower power to be used with consequent economy. These reductions also increase the random error rate, but, as was seen in Chapter 1, an error-correction system may already be necessary in a practical system and it is simply made to work harder.

In real channels, the signal may *originate* with discrete states which change at discrete times, but the channel will treat it as an analog waveform and so it will not be *received* in the same form. Various frequency-dependent loss mechanisms will reduce the amplitude of the signal. Noise will be picked up as a result of stray electric fields or magnetic induction and in radio receivers the circuitry will contribute some noise. As a result, the received voltage will have an infinitely varying state along with a degree of uncertainty due to the noise. Different frequencies can propagate at different speeds in the channel; this is the phenomenon of group delay. An alternative way of considering group delay is that there will be frequency-dependent phase shifts in the signal and these will result in uncertainty in the timing of pulses.

In digital circuitry, the signals are generally accompanied by a separate clock signal which reclocks the data to remove jitter as was shown in Chapter 1. In contrast, it is generally not feasible to provide a separate clock in recording and transmission applications. In the transmission case, a separate clock line would not only raise cost, but is impractical because at high frequency it is virtually impossible to ensure that the clock cable propagates signals at the same speed as the data cable except over short distances. In the recording case, provision of a separate clock track is impractical at high density because mechanical tolerances cause phase errors between the tracks. The result is the same; timing differences between parallel channels which are known as skew.

The solution is to use a self-clocking waveform and the generation of this is a further essential function of the coding process. Clearly, if data bits are simply clocked serially from a shift register in so-called direct recording or transmission this characteristic will not be obtained. If all the data bits are the same, for example all zeros, there is no clock when they are serialized.

It is not the channel which is digital; instead the term describes the way in which the received signals are *interpreted*. When the receiver makes discrete decisions from the input waveform it attempts to reject the uncertainties in voltage and time. The technique of channel coding is one where transmitted waveforms are restricted to those which still allow the receiver to make discrete decisions despite the degradations caused by the analog nature of the channel.

6.2 Types of transmission channel

Transmission can be by electrical conductors, radio or optical fibre. Although these appear to be completely different, they are in fact just different examples of electromagnetic energy travelling from one place to another. If the energy is made time-variant, information can be carried.

Electromagnetic energy propagates in a manner which is a function of frequency, and our partial understanding requires it to be considered as electrons, waves or photons so that we can predict its behaviour in given circumstances.

At DC and at the low frequencies used for power distribution, electromagnetic energy is called electricity and needs to be transported completely inside conductors. It has to have a complete circuit to flow in, and the resistance to current flow is determined by the cross-sectional area of the conductor. The insulation around the conductor and the spacing between the conductors has no effect on the ability of the conductor to pass current. At DC an inductor appears to be a short circuit, and a capacitor appears to be an open circuit.

As frequency rises, resistance is exchanged for impedance. Inductors display increasing impedance with frequency, capacitors show falling impedance. Electromagnetic energy increasingly tends to leave the conductor. The first symptom is the skin effect: the current flows only in the outside layer of the conductor effectively causing the resistance to rise.

As the energy is starting to leave the conductors, the characteristics of the space between them become important. This determines the impedance. A change of impedance causes reflections in the energy flow and some of it heads back towards the source. Constant impedance cables with fixed conductor spacing are necessary, and these must be suitably terminated to prevent reflections. The most important characteristic of the insulation is its thickness as this determines the spacing between the conductors.

As frequency rises still further, the energy travels less in the conductors and more in the insulation between them, and their composition becomes important and they begin to be called dielectrics. A poor dielectric like PVC absorbs high-frequency energy and attenuates the signal. So-called low-loss dielectrics such as PTFE are used, and one way of achieving low loss is to incorporate as much air into the dielectric as possible by making it in the form of a foam or extruding it with voids.

High-frequency signals can also be propagated without a medium, and are called radio. As frequency rises further the electromagnetic energy is termed 'light' which can also travel without a medium, but can be also be guided through a suitable medium. Figure 6.1(a) shows an early type of optical fibre in which total internal reflection is used to guide the light. It will be seen that the length of the optical path is a function of the angle at which the light is launched. Thus at the end of a long fibre sharp transitions would be smeared by this effect. Later optical fibres are made with a radially varying refractive index such that light diverging from the axis is automatically refracted back into the fibre. Figure 6.1(b) shows that in single-mode fibre light can only travel down one path and so the smearing of transitions is avoided.

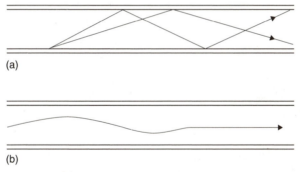

(a)

(b)

Figure 6.1 (a) Early optical fibres operated on internal reflection, and signals could take a variety of paths along the fibre, hence multi-mode. (b) Later fibres used graduated refractive index which meant that light was guided to the centre of the fibre and only one mode was possible.

6.3 Transmission lines

Frequency-dependent behaviour is the most important factor in deciding how best to harness electromagnetic energy flow for information transmission. It is obvious that the higher the frequency, the greater the possible information rate, but in general, losses increase with frequency, and flat frequency response is elusive. The best that can be managed is that over a narrow band of frequencies, the response can be made reasonably constant with the help of equalization. Unfortunately raw data when serialized have an unconstrained spectrum. Runs of identical bits can produce frequencies much lower than the bit rate would suggest. One of the essential steps in a transmission system is to modify the spectrum of the data into something more suitable.

At moderate bit rates, say a few megabits per second, and with moderate cable lengths, say a few metres, the dominant effect will be the capacitance of the cable due to the geometry of the space between the conductors and the dielectric between. The capacitance behaves under these conditions as if it were a single capacitor connected across the signal. The effect of the series source resistance and the parallel capacitance is that signal edges or transitions are turned into exponential curves as the capacitance is effectively being charged and discharged through the source impedance.

As cable length increases, the capacitance can no longer be lumped as if it were a single unit; it has to be regarded as being distributed along the cable. With rising frequency, the cable inductance also becomes significant, and it too is distributed.

The cable is now a transmission line and pulses travel down it as current loops which roll along as shown in Figure 6.2. If the pulse is positive, as it is launched along the line, it will charge the dielectric locally as at (a). As the pulse moves along, it will continue to charge the local dielectric as at (b). When the driver finishes the pulse, the trailing edge of the pulse follows the leading edge along the line. The voltage of the dielectric charged by the leading edge of the pulse is now higher than the voltage on the line, and so the dielectric discharges into the line as at (c). The current flows forward as it is in fact the same current which is

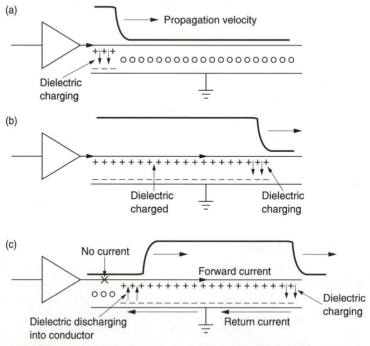

Figure 6.2 A transmission line conveys energy packets which appear with respect to the dielectric. In (a) the driver launches a pulse which charges the dielectric at the beginning of the line. As it propagates the dielectric is charged further along as in (b). When the driver ends the pulse, the charged dielectric discharges into the line. A current loop is formed where the current in the return loop flows in the opposite direction to the current in the 'hot' wire.

flowing into the dielectric at the leading edge. There is thus a loop of current rolling down the line flowing forward in the 'hot' wire and backwards in the return.

The constant to-ing and fro-ing of charge in the dielectric results in dielectric loss of signal energy. Dielectric loss increases with frequency and so a long transmission line acts as a filter. Thus the term 'low-loss' cable refers primarily to the kind of dielectric used.

Transmission lines which transport energy in this way have a characteristic impedance caused by the interplay of the inductance along the conductors with the parallel capacitance. One consequence of that transmission mode is that correct termination or matching is required between the line and both the driver

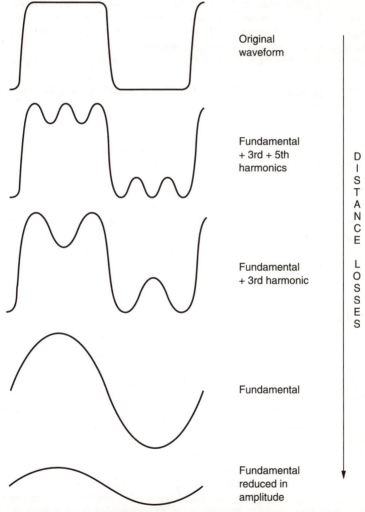

Original waveform

Fundamental + 3rd + 5th harmonics

Fundamental + 3rd harmonic

Fundamental

Fundamental reduced in amplitude

DISTANCE LOSSES

Figure 6.3 A signal may be square at the transmitter, but losses increase with frequency, and as the signal propagates, more of the harmonics are lost until only the fundamental remains. The amplitude of the fundamental then falls with further distance.

and the receiver. When a line is correctly matched, the rolling energy rolls straight out of the line into the load and the maximum energy is available. If the impedance presented by the load is incorrect, there will be reflections from the mismatch. An open circuit will reflect all the energy back in the same polarity as the original, whereas a short circuit will reflect all the energy back in the opposite polarity. Thus impedances above or below the correct value will have a tendency towards reflections whose magnitude depends upon the degree of mismatch and whose polarity depends upon whether the load is too high or too low. In practice it is the need to avoid reflections which is the most important reason to terminate correctly.

A perfectly square pulse contains an indefinite series of harmonics, but the higher ones suffer progressively more loss. A square pulse at the driver becomes less and less square with distance as Figure 6.3 shows. The harmonics are progressively lost until in the extreme case all that is left is the fundamental. A transmitted square wave is received as a sine wave. Fortunately data can still be recovered from the fundamental signal component.

Once all the harmonics have been lost, further losses cause the amplitude of the fundamental to fall. The effect worsens with distance and it is necessary to ensure that data recovery is still possible from a signal of unpredictable level.

6.4 Types of recording medium

Digital media do not need to have linear transfer functions, nor do they need to be noise-free or continuous. All they need to do is to allow the player to be able to distinguish the presence or absence of replay events, such as the generation of pulses, with reasonable (rather than perfect) reliability. In a magnetic medium, the event will be a flux change from one direction of magnetization to another. In an optical medium, the event must cause the pickup to perceive a change in the intensity of the light falling on the sensor. This may be obtained through selective absorption of light by dyes, or by phase contrast (see Chapter 7). In magneto-optical disks the recording itself is magnetic, but it is made and read using light.

6.5 Magnetic recording

Magnetic recording relies on the hysteresis of certain magnetic materials. After an applied magnetic field is removed, the material remains magnetized in the same direction. By definition, the process is non-linear.

Figure 6.4 shows the construction of a typical digital record head. A magnetic circuit carries a coil through which the record current passes and generates flux. A non-magnetic gap forces the flux to leave the magnetic circuit of the head and penetrate the medium. The current through the head must be set to suit the coercivity of the tape, and is arranged to almost saturate the track. The amplitude of the current is constant, and recording is performed by reversing the direction of the current with respect to time. As the track passes the head, this is converted to the reversal of the magnetic field left on the tape with respect to distance. The magnetic recording is therefore bipolar. Figure 6.5 shows that the recording is actually made just after the trailing pole of the record head where the flux strength from the gap is falling. As in analog recorders, the width of the gap is generally made quite large to ensure that the full thickness of the magnetic

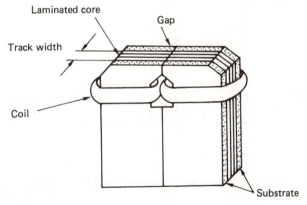

Figure 6.4 A digital record head is similar in principle to an analog head but uses much narrower tracks.

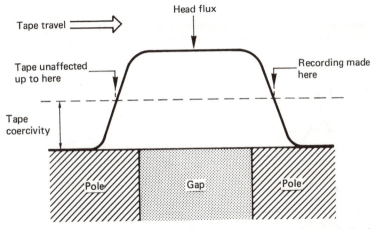

Figure 6.5 The recording is actually made near the trailing pole of the head where the head flux falls below the coercivity of the tape.

coating is recorded, although this cannot be done if the same head is intended to replay.

Figure 6.6 shows what happens when a conventional inductive head, i.e. one having a normal winding, is used to replay the bipolar track made by reversing the record current. The head output is proportional to the rate of change of flux and so only occurs at flux reversals. In other words, the replay head differentiates the flux on the track. The polarity of the resultant pulses alternates as the flux changes and changes back. A circuit is necessary which locates the peaks of the pulses and outputs a signal corresponding to the original record current waveform. There are two ways in which this can be done.

The amplitude of the replay signal is of no consequence and often an AGC system is used to keep the replay signal constant in amplitude. What matters is the time at which the write current, and hence the flux stored on the medium, reverses. This can be determined by locating the peaks of the replay impulses,

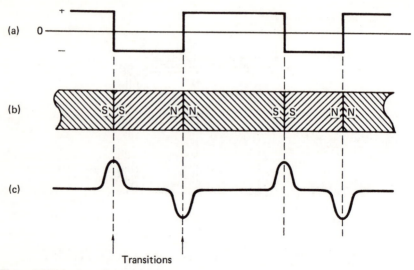

Figure 6.6 Basic digital recording. At (a) the write current in the head is reversed from time to time, leaving a binary magnetization pattern shown at (b). When replayed, the waveform at (c) results because an output is only produced when flux in the head changes. Changes are referred to as transitions.

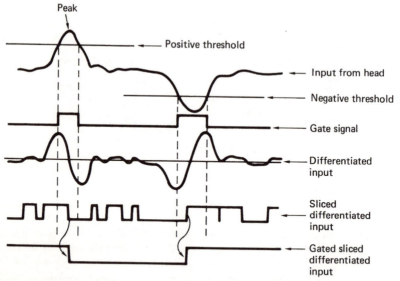

Figure 6.7 Gated peak detection rejects noise by disabling the differentiated output between transitions.

which can conveniently be done by differentiating the signal and looking for zero crossings. Figure 6.7 shows that this results in noise between the peaks. This problem is overcome by the gated peak detector, where only zero crossings from a pulse which exceeds the threshold will be counted. The AGC system allows the thresholds to be fixed. As an alternative, the record waveform can also be

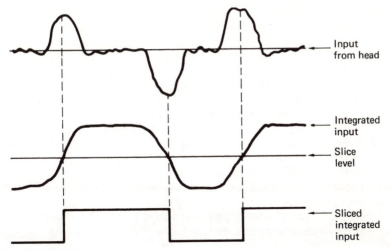

Figure 6.8 Integration method for re-creating write-current waveform.

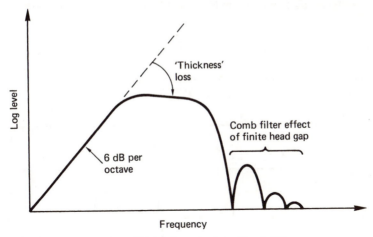

Figure 6.9 The major mechanisms defining magnetic channel bandwidth.

restored by integration, which opposes the differentiation of the head as in Figure 6.8.[1]

The head shown in Figure 6.4 has a frequency response shown in Figure 6.9. At DC there is no change of flux and no output. As a result, inductive heads are at a disadvantage at very low speeds. The output rises with frequency until the rise is halted by the onset of thickness loss. As the frequency rises, the recorded wavelength falls and flux from the shorter magnetic patterns cannot be picked up so far away. At some point, the wavelength becomes so short that flux from the back of the tape coating cannot reach the head and a decreasing thickness of tape contributes to the replay signal.[2] In digital recorders using short wavelengths to obtain high density, there is no point in using thick coatings. As wavelength further reduces, the familiar gap loss occurs, where the head gap is too big to resolve detail on the track. The construction of the head results in the same action as that of a two-point transversal filter, as the two poles of the head see the tape with a small delay interposed due to the finite gap. As expected, the head response is like a comb filter with the well-known nulls where flux cancellation takes place across the gap. Clearly, the smaller the gap, the shorter the wavelength of the first null. This contradicts the requirement of the record head to have a large gap. In quality analog audio recorders, it is the norm to have different record and replay heads for this reason, and the same will be true in digital machines which have separate record and playback heads. Clearly, where the same heads are used for record and play, the head gap size will be determined by the playback requirement.

As can be seen, the frequency response is far from ideal, and steps must be taken to ensure that recorded data waveforms do not contain frequencies which suffer excessive losses.

A more recent development is the magneto-resistive (M-R) head. This is a head which measures the flux on the tape rather than using it to generate a signal directly. Flux-measurement works down to DC and so offers advantages at low tape speeds. Unfortunately flux-measuring heads are not polarity-conscious but sense the modulus of the flux and if used directly they respond to positive and negative flux equally, as shown in Figure 6.10. This is overcome by using a small extra winding in the head carrying a constant current. This creates a steady bias field which adds to the flux from the tape. The flux seen by the head is now unipolar and changes between two levels and a more useful output waveform results.

Recorders which have low head-to-medium speed use M-R heads, whereas recorders with high bit rates, such as DVTRs tend to use inductive heads.

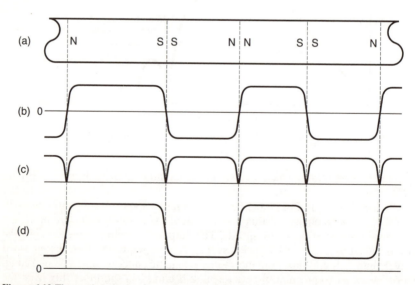

Figure 6.10 The sensing element in a magneto-resistive head is not sensitive to the polarity of the flux, only the magnitude. At (a) the track magnetization is shown and this causes a bidirectional flux variation in the head as at (b), resulting in the magnitude output at (c). However, if the flux in the head due to the track is biased by an additional field, it can be made unipolar as at (d) and the correct waveform is obtained.

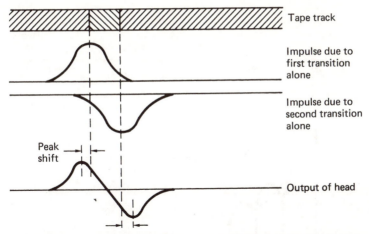

Figure 6.11 Readout pulses from two closely recorded transitions are summed in the head and the effect is that the peaks of the waveform are moved outwards. This is known as peak-shift distortion and equalization is necessary to reduce the effect.

Heads designed for use with tape work in actual contact with the magnetic coating. The tape is tensioned to pull it against the head. There will be a wear mechanism and need for periodic cleaning. In the hard disk, the rotational speed is high in order to reduce access time, and the drive must be capable of staying on-line for extended periods. In this case the heads do not contact the disk surface, but are supported on a boundary layer of air. The presence of the air film causes spacing loss, which restricts the wavelengths at which the head can replay. This is the penalty of rapid access.

Digital media must operate at high density for economic reasons. This implies that the shortest possible wavelengths will be used. Figure 6.11 shows that when two flux changes, or transitions, are recorded close together, they affect each other on replay. The amplitude of the composite signal is reduced, and the position of the peaks is pushed outwards. This is known as inter-symbol interference, or peak-shift distortion and it occurs in all magnetic media.

The effect is primarily due to high-frequency loss and it can be reduced by equalization on replay, as is done in most tapes, or by precompensation on record as is done in hard disks.

6.6 Azimuth recording and rotary heads

Figure 6.12(a) shows that in azimuth recording, the transitions are laid down at an angle to the track by using a head which is tilted. Machines using azimuth recording must always have an even number of heads, so that adjacent tracks can be recorded with opposite azimuth angle. The two track types are usually referred to as A and B. Figure 6.12(b) shows the effect of playing a track with the wrong type of head. The playback process suffers from an enormous azimuth error. The effect of azimuth error can be understood by imagining the tape track to be made from many identical parallel strips. In the presence of azimuth error, the strips at one edge of the track are played back with a phase shift relative to strips at the other side. At some wavelengths, the phase shift will be 180°, and there will be

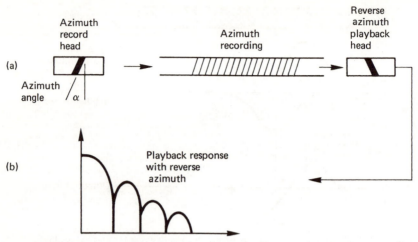

Figure 6.12 In azimuth recording (a), the head gap is tilted. If the track is played with the same head, playback is normal, but the response of the reverse azimuth head is attenuated (b).

no output; at other wavelengths, especially long wavelengths, some output will reappear. The effect is rather like that of a comb filter, and serves to attenuate crosstalk due to adjacent tracks so that no guard bands are required. Since no tape is wasted between the tracks, more efficient use is made of the tape. The term 'guard-bandless recording' is often used instead of, or in addition to, 'azimuth recording'. The failure of the azimuth effect at long wavelengths is a characteristic of azimuth recording, and it is necessary to ensure that the spectrum of the signal to be recorded has a small low-frequency content. The signal will need to pass through a rotary transformer to reach the heads, and cannot therefore contain a DC component.

In some rotary head recorders there is no separate erase process, and erasure is achieved by overwriting with a new waveform. Overwriting is only successful when there are no long wavelengths in the earlier recording, since these penetrate deeper into the tape, and the short wavelengths in a new recording will not be able to erase them. In this case the ratio between the shortest and longest wavelengths recorded on tape should be limited.

Restricting the spectrum of the code to allow erasure by overwrite also eases the design of the rotary transformer.

6.7 Optical and magneto-optical disks

Optical recorders have the advantage that light can be focused at a distance whereas magnetism cannot. This means that there need be no physical contact between the pickup and the medium and no wear mechanism.

In the same way that the recorded wavelength of a magnetic recording is limited by the gap in the replay head, the density of optical recording is limited by the size of light spot which can be focused on the medium. This is controlled by the wavelength of the light used and by the aperture of the lens. When the light spot is as small as these limits allow, it is said to be diffraction limited.

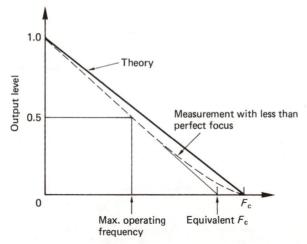

Figure 6.13 Frequency response of laser pickup. Maximum operating frequency is about half of cut-off frequency F_c.

The frequency response of an optical disk is shown in Figure 6.13. The response is best at DC and falls steadily to the optical cut-off frequency. Although the optics work down to DC, this cannot be used for the data recording. DC and low frequencies in the data would interfere with the focus and tracking servos and, as will be seen, difficulties arise when attempting to demodulate a unipolar signal. In practice the signal from the pickup is split by a filter. Low frequencies go to the servos, and higher frequencies go to the data circuitry. As a result, the optical disk channel has the same inability to handle DC as does a magnetic recorder, and the same techniques are needed to overcome it.

When a magnetic material is heated above its Curie temperature, it becomes demagnetized, and on cooling will assume the magnetization of an applied field which would be too weak to influence it normally. This is the principle of magneto-optical recording. The heat is supplied by a finely focused laser, the field is supplied by a coil which is much larger.

Figure 6.14 shows that the medium is initially magnetized in one direction only. In order to record, the coil is energized with a current in the opposite direction. This is too weak to influence the medium in its normal state, but when it is heated by the recording laser beam the heated area will take on the magnetism from the coil when it cools. Thus a magnetic recording with very small dimensions can be made even though the magnetic circuit involved is quite large in comparison.

Readout is obtained using the Kerr effect or the Faraday effect, which are phenomena whereby the plane of polarization of light can be rotated by a magnetic field. The angle of rotation is very small and needs a sensitive pickup. The pickup contains a polarizing filter before the sensor. Changes in polarization change the ability of the light to get through the polarizing filter and results in an intensity change which once more produces a unipolar output.

The magneto-optic recording can be erased by reversing the current in the coil and operating the laser continuously as it passes along the track. A new recording can then be made on the erased track.

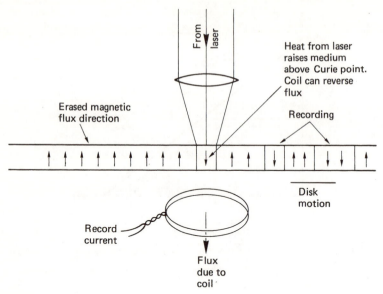

Figure 6.14 The thermomagneto-optical disk uses the heat from a laser to allow magnetic field to record on the disk.

A disadvantage of magneto-optical recording is that all materials having a Curie point low enough to be useful are highly corrodible by air and need to be kept under an effectively sealed protective layer. The magneto-optical channel has the same frequency response as that shown in Figure 2.69.

6.8 Equalization and data separation

The characteristics of most channels are that signal loss occurs which increases with frequency. This has the effect of slowing down rise times and thereby sloping off edges. If a signal with sloping edges is sliced, the time at which the waveform crosses the slicing level will be changed, and this causes jitter. Figure 6.15 shows that slicing a sloping waveform in the presence of baseline wander causes more jitter.

On a long cable, high-frequency roll-off can cause sufficient jitter to move a transition into an adjacent bit period. This is called inter-symbol interference and the effect becomes worse in signals which have greater asymmetry, i.e. short pulses alternating with long ones. The effect can be reduced by the application of

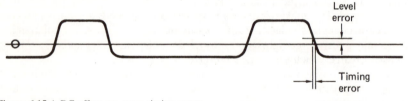

Figure 6.15 A DC offset can cause timing errors.

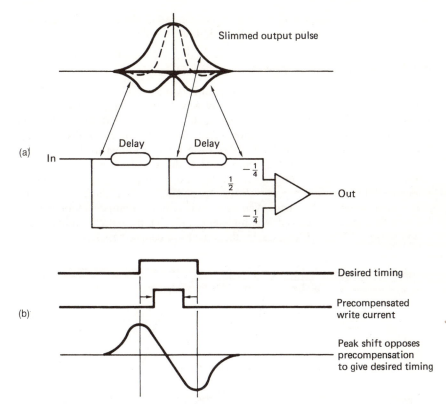

Figure 6.16 Peak-shift distortion is due to the finite width of replay pulses. The effect can be reduced by the pulse slimmer shown in (a) which is basically a transversal filter. The use of a linear operational amplifier emphasizes the analog nature of channels. Instead of replay pulse slimming, transitions can be written with a displacement equal and opposite to the anticipated peak shift as shown in (b).

equalization, which is typically a high-frequency boost, and by choosing a channel code which has restricted asymmetry.

Compensation for peak shift distortion in recording requires equalization of the channel,[3] and this can be done by a network after the replay head, termed an equalizer or pulse sharpener,[4] as in Figure 6.16(a). This technique uses transversal filtering to oppose the inherent transversal effect of the head. As an alternative, precompensation in the record stage can be used as shown in Figure 6.16(b). Transitions are written in such a way that the anticipated peak shift will move the readout peaks to the desired timing.

The important step of information recovery at the receiver or replay circuit is known as data separation. The data separator is rather like an analog-to-digital convertor because the two processes of sampling and quantizing are both present. In the time domain, the sampling clock is derived from the clock content of the channel waveform. In the voltage domain, the process of *slicing* converts the analog waveform from the channel back into a binary representation. The slicer is thus a form of quantizer which has only one-bit resolution. The slicing process makes a discrete decision about the voltage of the incoming signal in order to reject noise. The sampler makes discrete decisions along the time axis in order to reject jitter. These two processes will be described in detail.

6.9 Slicing and jitter rejection

The slicer is implemented with a comparator which has analog inputs but a binary output. In a cable receiver, the input waveform can be sliced directly. In an inductive magnetic replay system, the replay waveform is differentiated and must first pass through a peak detector (Figure 6.7) or an integrator (Figure 6.8). The signal voltage is compared with the midway voltage, known as the threshold, baseline or slicing level by the comparator. If the signal voltage is above the threshold, the comparator outputs a high level, if below, a low level results.

Figure 6.17 shows some waveforms associated with a slicer. At (a) the transmitted waveform has an uneven duty cycle. The DC component, or average level, of the signal is received with high amplitude, but the pulse amplitude falls as the pulse gets shorter. Eventually the waveform cannot be sliced.

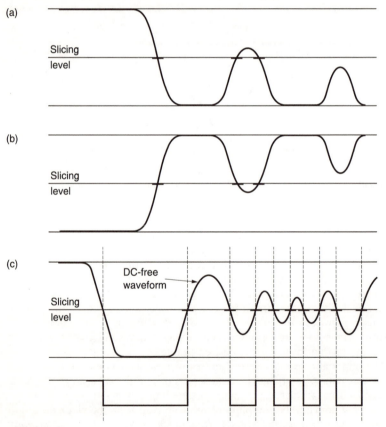

Figure 6.17 Slicing a signal which has suffered losses works well if the duty cycle is even. If the duty cycle is uneven, as at (a), timing errors will become worse until slicing fails. With the opposite duty cycle, the slicing fails in the opposite direction as at (b). If, however, the signal is DC free, correct slicing can continue even in the presence of serious losses, as (c) shows.

At (b) the opposite duty cycle is shown. The signal level drifts to the opposite polarity and once more slicing is impossible. The phenomenon is called baseline wander and will be observed with any signal whose average voltage is not the same as the slicing level.

At (c) it will be seen that if the transmitted waveform has a relatively constant average voltage, slicing remains possible up to high frequencies even in the presence of serious amplitude loss, because the received waveform remains symmetrical about the baseline.

It is clearly not possible simply to serialize data in a shift register for so-called direct transmission, because successful slicing can only be obtained if the number of ones is equal to the number of zeros; there is little chance of this happening consistently with real data. Instead, a modulation code or channel code is necessary. This converts the data into a waveform which is DC-free or nearly so for the purpose of transmission.

The slicing threshold level is naturally zero in a bipolar system such as magnetic inductive replay or a cable. When the amplitude falls it does so symmetrically and slicing continues. The same is not true of M-R heads and optical pickups, which both respond to intensity and therefore produce a unipolar output. If the replay signal is sliced directly, the threshold cannot be zero, but must be some level approximately half the amplitude of the signal as shown in Figure 6.18(a). Unfortunately when the signal level falls it falls towards zero and not towards the slicing level. The threshold will no longer be appropriate for the signal as can be seen at (b). This can be overcome by using a DC-free coded waveform. If a series capacitor is connected to the unipolar signal from an optical pickup, the waveform is rendered bipolar because the capacitor blocks any DC component in the signal. The DC-free channel waveform passes through unaltered. If an amplitude loss is suffered, Figure 6.18(c) shows that the resultant bipolar signal now reduces in amplitude about the slicing level and slicing can continue.

The binary waveform at the output of the slicer will be a replica of the transmitted waveform, except for the addition of jitter or time uncertainty in the position of the edges due to noise, baseline wander, intersymbol interference and imperfect equalization.

Binary circuits reject noise by using discrete voltage levels which are spaced further apart than the uncertainty due to noise. In a similar manner, digital coding combats time uncertainty by making the time axis discrete using events, known as transitions, spaced apart at integer multiples of some basic time period, called a detent, which is larger than the typical time uncertainty. Figure 6.19 shows how this jitter-rejection mechanism works. All that matters is to identify the detent in which the transition occurred. Exactly where it occurred within the detent is of no consequence.

As ideal transitions occur at multiples of a basic period, an oscilloscope, which is repeatedly triggered on a channel-coded signal carrying random data, will show an eye pattern if connected to the output of the equalizer. Study of the eye pattern reveals how well the coding used suits the channel. In the case of transmission, with a short cable, the losses will be small, and the eye opening will be virtually square except for some edge-sloping due to cable capacitance. As cable length increases, the harmonics are lost and the remaining fundamental gives the eyes a diamond shape. The same eye pattern will be obtained with a recording channel where it is uneconomic to provide bandwidth much beyond the fundamental.

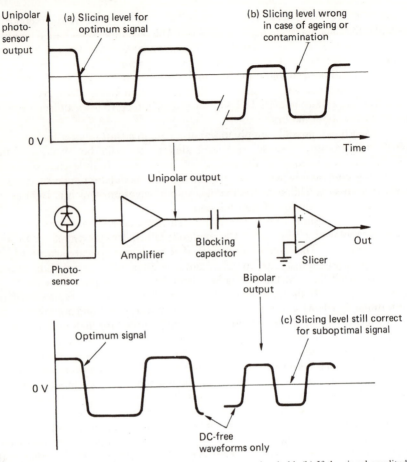

Figure 6.18 (a) Slicing a unipolar signal requires a non-zero threshold. (b) If the signal amplitude changes, the threshold will then be incorrect. (c) If a DC-free code is used, a unipolar waveform can be converted to a bipolar waveform using a series capacitor. A zero threshold can be used and slicing continues with amplitude variations.

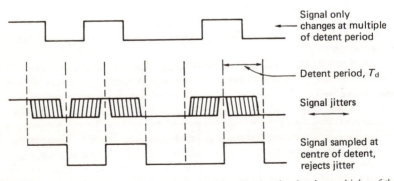

Figure 6.19 A certain amount of jitter can be rejected by changing the signal at multiples of the basic detent period T_d.

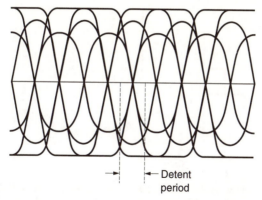

←| |←— Detent
period

Figure 6.20 A transmitted waveform will appear like this on an oscilloscope as successive parts of the waveform are superimposed on the tube. When the waveform is rounded off by losses, diamond-shaped eyes are left in the centre, spaced apart by the detent period.

Noise closes the eyes in a vertical direction, and jitter closes the eyes in a horizontal direction, as in Figure 6.20. If the eyes remain sensibly open, data separation will be possible. Clearly, more jitter can be tolerated if there is less noise, and vice versa. If the equalizer is adjustable, the optimum setting will be where the greatest eye opening is obtained.

In the centre of the eyes, the receiver must make binary decisions at the channel bit rate about the state of the signal, high or low, using the slicer output. As stated, the receiver is sampling the output of the slicer, and it needs to have a sampling clock in order to do that. In order to give the best rejection of noise and jitter, the clock edges which operate the sampler must be in the centre of the eyes.

As has been stated, a separate clock is not practicable in recording or transmission. A fixed-frequency clock at the receiver is of no use as even if it was sufficiently stable, it would not know what phase to run at.

The only way in which the sampling clock can be obtained is to use a phase-locked loop to regenerate it from the clock content of the self-clocking channel-coded waveform. In phase-locked loops, the voltage-controlled oscillator is driven by a phase error measured between the output and some reference, such that the output eventually has the same frequency as the reference. If a divider is placed between the VCO and the phase comparator, as in section 2.9, the VCO frequency can be made to be a multiple of the reference. This also has the effect of making the loop more heavily damped. If a channel-coded waveform is used as a reference to a PLL, the loop will be able to make a phase comparison whenever a transition arrives and will run at the channel bit rate. When there are several detents between transitions, the loop will *flywheel* at the last known frequency and phase until it can rephase at a subsequent transition. Thus a continuous clock is re-created from the clock content of the channel waveform. In a recorder, if the speed of the medium should change, the PLL will change frequency to follow. Once the loop is locked, clock edges will be phased with the average phase of the jittering edges of the input waveform. If, for example, rising edges of the clock are phased to input transitions, then falling edges will be in the centre of the eyes. If these edges are used to clock the sampling process, the

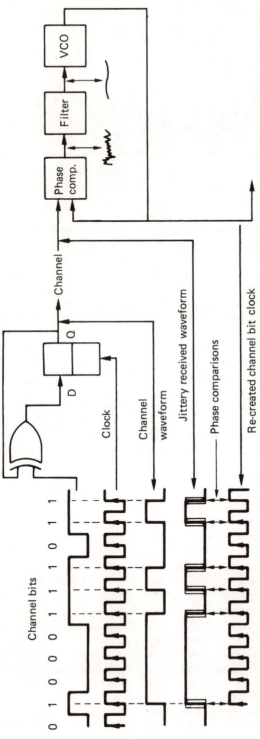

Figure 6.21 The clocking system when channel coding is used. The encoder clock runs at the channel bit rate, and any transitions in the channel must coincide with encoder clock edges. The reason for doing this is that, at the data separator, the PLL can lock to the edges of the channel signal, which represents an intermittent clock, and turn it into a continuous clock. The jitter in the edges of the channel signal causes noise in the phase error of the PLL, but the damping acts as a filter and the PLL runs at the average phase of the channel bits, rejecting the jitter.

maximum jitter and noise can be rejected. The output of the slicer when sampled by the PLL edge at the centre of an eye is the value of a channel bit. Figure 6.21 shows the complete clocking system of a channel code from encoder to data separator.

Clearly, data cannot be separated if the PLL is not locked, but it cannot be locked until it has seen transitions for a reasonable period. In recorders, which have discontinuous recorded blocks to allow editing, the solution is to precede each data block with a pattern of transitions whose sole purpose is to provide a timing reference for synchronizing the phase-locked loop. This pattern is known as a preamble. In interfaces, the transmission can be continuous and there is no difficulty remaining in lock indefinitely. There will simply be a short delay on first applying the signal before the receiver locks to it.

One potential problem area which is frequently overlooked is to ensure that the VCO in the receiving PLL is correctly centred. If it is not, it will be running with a static phase error and will not sample the received waveform at the centre of the eyes. The sampled bits will be more prone to noise and jitter errors. VCO centring can simply be checked by displaying the control voltage. This should not change significantly when the input is momentarily interrupted.

6.10 Channel coding

It is not practicable simply to serialize raw data in a shift register for the purpose of recording or for transmission except over relatively short distances. Practical systems require the use of a modulation scheme, known as a channel code, which expresses the data as waveforms which are self-clocking in order to reject jitter, separate the received bits and to avoid skew on separate clock lines. The coded waveforms should further be DC-free or nearly so to enable slicing in the presence of losses and have a narrower spectrum than the raw data both for economy and to make equalization easier.

Jitter causes uncertainty about the time at which a particular event occurred. The frequency response of the channel then places an overall limit on the spacing of events in the channel. Particular emphasis must be placed on the interplay of bandwidth, jitter and noise, which will be shown here to be the key to the design of a successful channel code.

Figure 6.22 shows that a channel coder is necessary prior to the record stage, and that a decoder, known as a data separator, is necessary after the replay stage. The output of the channel coder is generally a logic-level signal which contains a 'high' state when a transition is to be generated. The waveform generator produces the transitions in a signal whose level and impedance is suitable for driving the medium or channel. The signal may be bipolar or unipolar as appropriate.

Some codes eliminate DC entirely, which is advantageous for cable transmission, optical media and rotary head recording. Some codes can reduce the channel bandwidth needed by lowering the upper spectral limit. This permits higher linear density, usually at the expense of jitter rejection. Other codes narrow the spectrum, by raising the lower limit. A code with a narrow spectrum has a number of advantages. The reduction in asymmetry will reduce peak shift and data separators can lock more readily because the range of frequencies in the code is smaller. In theory the narrower the spectrum, the less noise will be suffered, but this is only achieved if filtering is employed. Filters can easily cause phase errors which will nullify any gain.

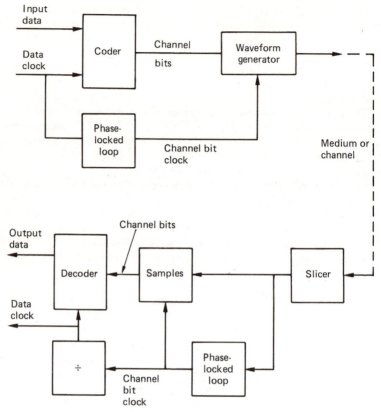

Figure 6.22 The major components of a channel coding system. See text for details.

A convenient definition of a channel code (for there are certainly others) is 'A method of modulating real data such that they can be reliably received despite the shortcomings of a real channel, while making maximum economic use of the channel capacity'. The basic time periods of a channel-coded waveform are called positions or detents, in which the transmitted voltage will be reversed or stay the same. The symbol used for the units of channel time is T_d.

One of the fundamental parameters of a channel code is the density ratio (DR). One definition of density ratio is that it is the worst-case ratio of the number of data bits recorded to the number of transitions in the channel. It can also be thought of as the ratio between the Nyquist rate of the data (one-half the bit rate) and the frequency response required in the channel. The storage density of data recorders has steadily increased due to improvements in medium and transducer technology, but modern storage densities are also a function of improvements in channel coding.

As jitter is such an important issue in digital recording and transmission, a parameter has been introduced to quantify the ability of a channel code to reject time instability. This parameter, the jitter margin, also known as the window margin or phase margin (T_w), is defined as the permitted range of time over which a transition can still be received correctly, divided by the data bit-cell period (T).

Since equalization is often difficult in practice, a code which has a large jitter margin will sometimes be used because it resists the effects of inter-symbol interference well. Such a code may achieve a better performance in practice than a code with a higher density ratio but poor jitter performance.

A more realistic comparison of code performance will be obtained by taking into account both density ratio and jitter margin. This is the purpose of the figure of merit (FoM), which is defined as DR × T_w.

6.11 Simple codes

In the Non-Return to Zero (NRZ) code shown in Figure 6.23(a), the record current does not cease between bits, but flows at all times in one direction or the other dependent on the state of the bit to be recorded. This results in a replay pulse only when the data bits change from state to another. As a result, if one pulse was missed, the subsequent bits would be inverted. This was avoided by adapting the coding such that the record current would change state or invert whenever a data one occurred, leading to the term Non-Return to Zero Invert or NRZI shown in Figure 6.23b). In NRZI a replay pulse occurs whenever there is a data one. Clearly, neither NRZ or NRZI are self-clocking, but require a separate clock track. Skew between tracks can only be avoided by working at low density and so the system cannot be used directly for digital video. However, virtually all the codes used for magnetic recording are based on the principle of reversing the record current to produce a transition.

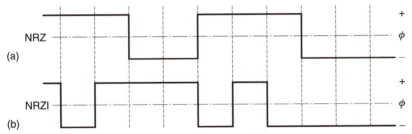

Figure 6.23 In the NRZ code (a) a missing replay pulse inverts every following bit. This was overcome in the NRZI code (b) which reverses write current on a data one.

The FM code, also known as Manchester code or bi-phase mark code, shown in Figure 6.24(a) was the first practical self-clocking binary code and it is suitable for both transmission and recording. It is DC-free and very easy to encode and decode. It is the code specified for the AES/EBU digital audio interconnect standard. In the field of recording it remains in use today only where density is not of prime importance, for example in SMPTE/EBU timecode for professional audio and video recorders.

In FM there is always a transition at the bit-cell boundary which acts as a clock. For a data one, there is an additional transition at the bit-cell centre. Figure 6.24(a) shows that each data bit can be represented by two channel bits. For a data zero, they will be 10, and for a data one they will be 11. Since the first bit is always one, it conveys no information, and is responsible for the density ratio of only one-half. Since there can be two transitions for each data bit, the jitter

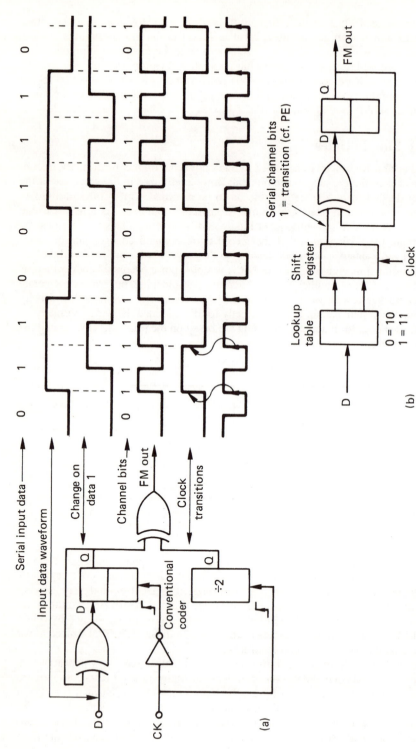

Figure 6.24 FM encoding. At (a) are the FM waveform and the channel bits which may be used to describe transitions in it. The FM coder is shown at (b).

margin can only be half a bit, and the resulting FoM is only 0.25. The high clock content of FM does, however, mean that data recovery is possible over a wide range of speeds; hence the use of timecode. The lowest frequency in FM is due to a stream of zeros and is equal to half the bit rate. The highest frequency is due to a stream of ones, and is equal to the bit rate. Thus the fundamentals of FM are within a band of one octave. Effective equalization is generally possible over such a band. FM is not polarity-conscious and can be inverted without changing the data.

Figure 6.24(b) shows how an FM coder works. Data words are loaded into the input shift register which is clocked at the data bit rate. Each data bit is converted to two channel bits in the codebook or look-up table. These channel bits are loaded into the output register. The output register is clocked twice as fast as the input register because there are twice as many channel bits as data bits. The ratio of the two clocks is called the code rate, in this case it is a rate one-half code. Ones in the serial channel bit output represent transitions whereas zeros represent no change. The channel bits are fed to the waveform generator which is a one-bit delay, clocked at the channel bit rate, and an exclusive-OR gate. This changes state when a channel bit one is input. The result is a coded FM waveform where there is always a transition at the beginning of the data bit period, and a second optional transition whose presence indicates a one.

In modified frequency modulation (MFM) also known as Miller code,[5] the highly redundant clock content of FM was reduced by the use of a phase-locked loop in the receiver which could flywheel over missing clock transitions. This technique is implicit in all the more advanced codes. Figure 6.25(a) shows that the bit-cell centre transition on a data one was retained, but the bit-cell boundary transition is now required only between successive zeros. There are still two channel bits for every data bit, but adjacent channel bits will never be one, doubling the minimum time between transitions, and giving a DR of 1. Clearly, the coding of the current bit is now influenced by the preceding bit. The maximum number of prior bits which affect the current bit is known as the

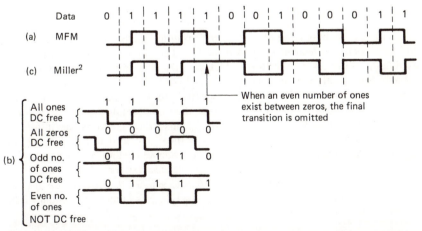

Figure 6.25 MFM or Miller code is generated as shown here. The minimum transition spacing is twice that of FM or PE. MFM is not always DC-free as shown at (b). This can be overcome by the modification of (c) which results in the Miller^2 code.

constraint length L_c, measured in data-bit periods. For MFM $L_c = T$. Another way of considering the constraint length is that it assesses the number of data bits which may be corrupted if the receiver misplaces one transition. If L_c is long, all errors will be burst errors.

MFM doubled the density ratio compared to FM and PE without changing the jitter performance; thus the FoM also doubles, becoming 0.5. It was adopted for many rigid disks at the time of its development, and remains in use on double-density floppy disks. It is not, however, DC-free. Figure 6.25(b) shows how MFM can have DC content under certain conditions.

The Miller[2] code is derived from MFM, and Figure 6.25(c) shows that the DC content is eliminated by a slight increase in complexity.[6,7] Wherever an even number of ones occurs between zeros, the transition at the last one is omitted. This creates two additional, longer run lengths and increases the T_{max} of the code. The decoder can detect these longer run lengths in order to re-insert the suppressed ones. The FoM of Miller[2] is 0.5 as for MFM.

6.12 Group codes

Further improvements in coding rely on converting patterns of real data to patterns of channel bits with more desirable characteristics using a conversion table known as a codebook. If a data symbol of m bits is considered, it can have 2^m different combinations. As it is intended to discard undesirable patterns to improve the code, it follows that the number of channel bits n must be greater than m. The number of patterns which can be discarded is:

$$2^n - 2^m$$

One name for the principle is group code recording (GCR), and an important parameter is the code rate, defined as:

$$R = \frac{m}{n}$$

It will be evident that the jitter margin T_w is numerically equal to the code rate, and so a code rate near to unity is desirable. The choice of patterns which are used in the codebook will be those which give the desired balance between clock content, bandwidth and DC content.

Figure 6.26 shows that the upper spectral limit can be made to be some fraction of the channel bit rate according to the minimum distance between ones in the channel bits. This is known as T_{min}, also referred to as the minimum transition parameter M and in both cases is measured in data bits T. It can be obtained by multiplying the number of channel detent periods between transitions by the code rate. Unfortunately, codes are measured by the number of consecutive zeros in the channel bits, given the symbol d, which is always one less than the number of detent periods. In fact T_{min} is numerically equal to the density ratio.

$$T_{min} = M = DR = \frac{(d + 1) \times m}{n}$$

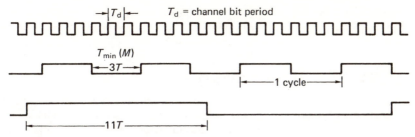

Figure 6.26 A channel code can control its spectrum by placing limits on T_{min} (M) and T_{max} which define upper and lower frequencies. The ratio of T_{max}/T_{min} determines the asymmetry of waveform and predicts DC content and peak shift. Example shown is EFM.

It will be evident that choosing a low code rate could increase the density ratio, but it will impair the jitter margin. The figure of merit is:

$$\text{FoM} = \text{DR} \times T_w = \frac{(d + 1) \times m^2}{n^2}$$

since $T_w = m/n$

Figure 6.26 also shows that the lower spectral limit is influenced by the maximum distance between transitions T_{max}. This is also obtained by multiplying the maximum number of detent periods between transitions by the code rate. Again, codes are measured by the maximum number of zeros between channel ones, k, and so:

$$T_{max} = \frac{(k + 1) \times m}{n}$$

and the maximum/minimum ratio P is:

$$p = \frac{(k + 1)}{(d + 1)}$$

The length of time between channel transitions is known as the *run length*. Another name for this class is the run-length-limited (RLL) codes.[8] Since m data bits are considered as one symbol, the constraint length L_c will be increased in RLL codes to at least m. It is, however, possible for a code to have run-length limits without it being a group code.

In practice, the junction of two adjacent channel symbols may violate run-length limits, and it may be necessary to create a further codebook of symbol size $2n$ which converts violating code pairs to acceptable patterns. This is known as merging and follows the golden rule that the substitute $2n$ symbol must finish with a pattern which eliminates the possibility of a subsequent violation. These patterns must also differ from all other symbols.

Substitution may also be used to different degrees in the same nominal code in order to allow a choice of maximum run length, e.g. 3PM. The maximum number of symbols involved in a substitution is denoted by r. There are many RLL codes and the parameters d, k, m, n, and r are a way of comparing them.

Group codes are used extensively in recording and transmission. DVTRs and magnetic disks use group codes optimized for jitter rejection whereas optical disks use group codes optimized for density ratio.

6.13 Randomizing and encryption

Randomizing is not a channel code, but a technique which can be used in conjunction with almost any channel code. It is widely used in digital audio and video broadcasting and in a number of recording and transmission formats. The randomizing system is arranged outside the channel coder. Figure 6.27 shows that, at the encoder, a pseudo-random sequence is added modulo-2 to the serial data. This process makes the signal spectrum in the channel more uniform, drastically reduces T_{max} and reduces DC content. At the receiver the transitions are converted back to a serial bitstream to which the same pseudo-random sequence is again added modulo-2. As a result, the random signal cancels itself out to leave only the serial data, provided that the two pseudo-random sequences are synchronized to bit accuracy.

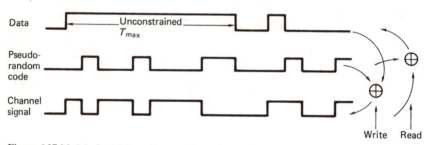

Figure 6.27 Modulo-2 addition with a pseudo-random code removes unconstrained runs in real data. Identical process must be provided on replay.

Many channel codes, especially group codes, display pattern sensitivity because some waveforms are more sensitive to peak shift distortion than others. Pattern sensitivity is only a problem if a sustained series of sensitive symbols needs to be recorded. Randomizing ensures that this cannot happen because it breaks up any regularity or repetition in the data. The data randomizing is performed by using the exclusive-OR function of the data and a pseudo-random sequence as the input to the channel coder. On replay the same sequence is generated, synchronized to bit accuracy, and the exclusive-OR of the replay bitstream and the sequence is the original data.

The generation of randomizing polynomials was described in section 2.8. Clearly, the sync pattern cannot be randomized, since this causes a Catch–22 situation where it is not possible to synchronize the sequence for replay until the sync pattern is read, but it is not possible to read the sync pattern until the sequence is synchronized!

In recorders, the randomizing is block based, since this matches the block structure on the medium. Where there is no obvious block structure, convolutional or endless randomizing can be used. In convolutional randomizing, the signal sent down the channel is the serial data waveform which has been convolved with the impulse response of a digital filter. On reception the signal is deconvolved to restore the original data.

Convolutional randomizing is used in the serial digital interface (SDI) which carries 4:2:2 sampled video. Figure 6.28(a) shows that the filter is an infinite impulse response (IIR) filter which has recursive paths from the output back to the input. As it is a one-bit filter its output cannot decay, and once excited, it runs indefinitely. The filter is followed by a transition generator which consists of a one-bit delay and an exclusive-OR gate. An input 1 results in an output transition on the next clock edge. An input 0 results in no transition.

A result of the infinite impulse response of the filter is that frequent transitions are generated in the channel which result in sufficient clock content for the phase-locked loop in the receiver.

Transitions are converted back to 1s by a differentiator in the receiver. This consists of a one-bit delay with an exclusive-OR gate comparing the input and the output. When a transition passes through the delay, the input and the output will be different and the gate outputs a 1 which enters the deconvolution circuit.

Figure 6.28(b) shows that in the deconvolution circuit a data bit is simply the exclusive-OR of a number of channel bits at a fixed spacing. The deconvolution is implemented with a shift register having the exclusive-OR gates connected in a reverse pattern to that in the encoder. The same effect as block randomizing is obtained, in that long runs are broken up and the DC content is reduced, but it has the advantage over block randomizing that no synchronizing is required to remove the randomizing, although it will still be necessary for deserialization. Clearly, the system will take a few clock periods to produce valid data after commencement of transmission, but this is no problem on a permanent wired connection where the transmission is continuous.

In a randomized transmission, if the receiver is not able to re-create the pseudo-random sequence, the data cannot be decoded. This can be used as the basis for encryption in which only authorized users can decode transmitted data. In an encryption system, the goal is security whereas in a channel-coding system the goal is simplicity. Channel coders use pseudo-random sequences because these are economical to create using feedback shift registers. However, there are a limited number of pseudo-random sequences and it would be too easy to try them all until the correct one was found. Encryption systems use the same processes, but the key sequence which is added to the data at the encoder is truly random. This makes it much harder for unauthorized parties to access the data. Only a receiver in possession of the correct sequence can decode the channel signal. If the sequence is made long enough, the probability of stumbling across the sequence by trial and error can be made sufficiently small. Security systems of this kind can be compromised if the delivery of the key to the authorized user is intercepted.

6.14 Partial response

It has been stated that a magnetic head acts as a transversal filter, because it has two poles which scan the medium at different times. In addition the output is differentiated, so that the head may be thought of as a (1 min D) impulse response system, where D is the delay which is a function of the tape speed and gap size. It is this delay which results in intersymbol interference. Conventional

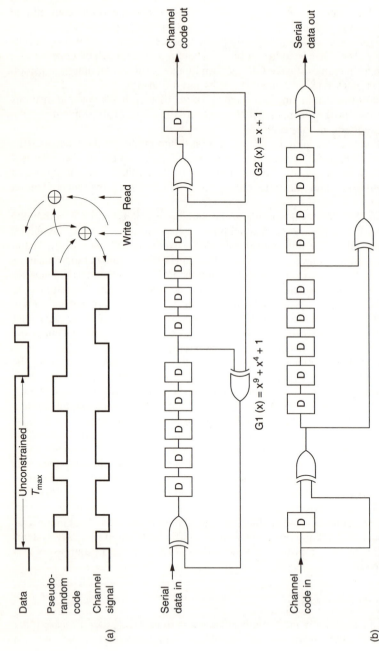

Figure 6.28 (a) Modulo-2 addition with a pseudo-random code removes unconstrained runs in real data. Identical process must be provided on replay. (b) Convolutional randomizing encoder, at top, transmits exclusive-OR of three bits at a fixed spacing in the data. One bit delay, far right, produces channel transitions from data ones. Decoder, below, has opposing one bit delay to return from transitions to data levels, followed by an opposing shift register which exactly reverses the coding process.

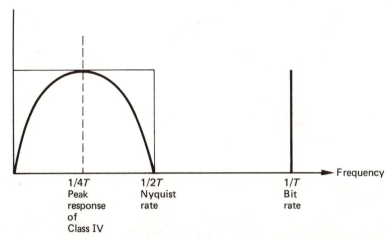

1/4T
Peak
response
of
Class IV

1/2T
Nyquist
rate

1/T
Bit
rate

Frequency

Figure 6.29 Class IV response has spectral nulls at DC and the Nyquist rate, giving a noise advantage, since magnetic replay signal is weak at both frequencies in a high-density channel.

equalizers attempt to oppose this effect, and succeed in raising the noise level in the process of making the frequency response linear. Figure 6.29 shows that the frequency response necessary to pass data with insignificant peak shift is a bandwidth of half the bit rate, which is the Nyquist rate. In Class IV partial response, the frequency response of the system is made to have nulls at DC and at the bit rate. Such a frequency response is particularly advantageous for rotary head recorders as it is DC-free and the low-frequency content is minimal, hence the use in Digital Betacam. The required response is achieved by an overall impulse response of $(1 - D^2)$ where D is now the bit period. There are a number of ways in which this can be done.

If the head gap is made equal to one bit, the $(1 - D)$ head response may be converted to the desired response by the use of a $(1 + D)$ filter, as in Figure 6.30(a).[9] Alternatively, a head of unspecified gapwidth may be connected to an integrator, and equalized flat to reproduce the record current waveform before being fed to a $(1 - D^2)$ filter as in Figure 6.30(b).[10]

The result of both of these techniques is a ternary signal. The eye pattern has two sets of eyes as in Figure 6.30(c).[11] When slicing such a signal, a smaller amount of noise will cause an error than in the binary case.

The treatment of the signal thus far represents an equalization technique, and not a channel code. However, to take full advantage of Class IV partial response, suitable precoding is necessary prior to recording, which does then constitute a channel-coding technique. This precoding is shown in Figure 6.31(a). Data are added modulo-2 to themselves with a two-bit delay. The effect of this precoding is that the outer levels of the ternary signals, which represent data ones, alternate in polarity on all odd bits and on all even bits. This is because the precoder acts like two interleaved one-bit delay circuits, as in Figure 6.31(b). As this alternation of polarity is a form of redundancy, it can be used to recover the 3 dB SNR loss encountered in slicing a ternary eye pattern (see Figure 6.32(a)).

Viterbi decoding[12] can be used for this purpose. In Viterbi decoding, each channel bit is not sliced individually; the slicing decision is made in the context of

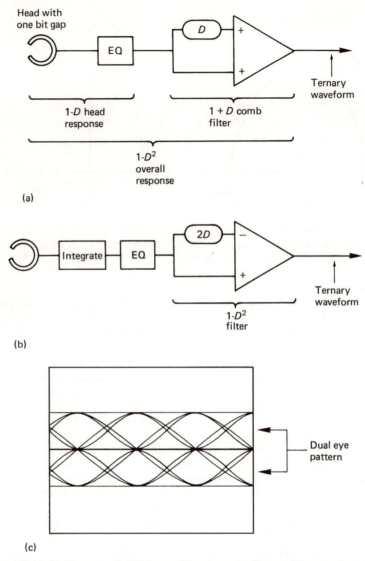

Figure 6.30 (a), (b) Two ways of obtaining partial response. (c) Characteristic eye pattern of ternary signal.

adjacent decisions. Figure 6.32(b) shows a replay waveform which is so noisy that, at the decision point, the signal voltage crosses the centre of the eye, and the slicer alone cannot tell whether the correct decision is an inner or an outer level. In this case, the decoder essentially allows both decisions to stand, in order to see what happens. A symbol representing indecision is output. It will be seen from the figure that as subsequent bits are received, one of these decisions will result in an absurd situation, which indicates that the other decision was the right one. The decoder can then locate the undecided symbol and set it to the correct value.

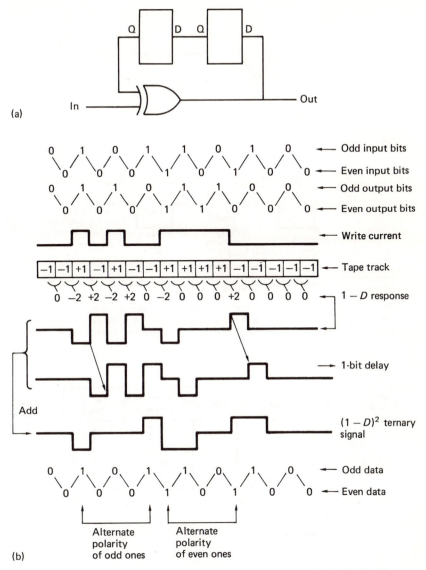

Figure 6.31 Class IV precoding at (a) causes redundancy in replay signal as derived in (b).

Viterbi decoding requires more information about the signal voltage than a simple binary slicer can discern. Figure 6.33 shows that the replay waveform is sampled and quantized so that it can be processed in digital logic. The sampling rate is obtained from the embedded clock content of the replay waveform. The digital Viterbi processing logic must be able to operate at high speed to handle serial signals from a DVTR head. Its application in Digital Betacam is eased somewhat by the adoption of compression which reduces the data rate at the heads by a factor of two.

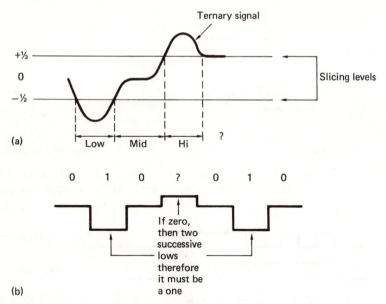

(a)

(b)

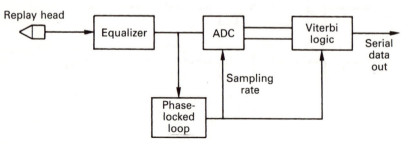

Figure 6.32 (a) A ternary signal suffers a noise penalty because there are two slicing levels. (b) The redundancy is used to determine the bit value in the presence of noise. Here the pulse height has been reduced to make it ambiguous 1/0, but only 1 is valid as zero violates the redundancy rules.

Figure 6.33 A Viterbi decoder is implemented in the digital domain by sampling the replay waveform with a clock locked to the embedded clock of the channel code.

Clearly a ternary signal having a dual eye pattern is more sensitive than a binary signal, and it is important to keep the maximum run length T_{max} small in order to have accurate AGC. The use of pseudo-random coding along with partial response equalization and precoding is a logical combination.[13]

There is then considerable overlap between the channel code and the error-correction system. Viterbi decoding is primarily applicable to channels with random errors due to Gaussian statistics, and they cannot cope with burst errors. In a head-noise-limited system, however, the use of a Viterbi detector could increase the power of an separate burst error-correction system by relieving it of the need to correct random errors due to noise. The error-correction system could then concentrate on correcting burst errors unimpaired.

6.15 Synchronizing

Once the PLL in the data separator has locked to the clock content of the transmission, a serial channel bitstream and a channel bit clock will emerge from the sampler. In a group code, it is essential to know where a group of channel bits begins in order to assemble groups for decoding to data bit groups. In a randomizing system it is equally vital to know at what point in the serial data stream the words or samples commence. In serial transmission and in recording, channel bit groups or randomized data words are sent one after the other, one bit at a time, with no spaces in between, so that although the designer knows that a data block contains, say, 128 bytes, the receiver simply finds 1024 bits in a row. If the exact position of the first bit is not known, then it is not possible to put all the bits in the right places in the right bytes; a process known as deserializing. The effect of sync slippage is devastating, because a one-bit disparity between the bit count and the bitstream will corrupt every symbol in the block.

The synchronization of the data separator and the synchronization to the block format are two distinct problems, which are often solved by the same sync pattern. Deserializing requires a shift register which is fed with serial data and read out once per word. The sync detector is simply a set of logic gates which are arranged to recognize a specific pattern in the register. The sync pattern is either identical for every block or has a restricted number of versions and it will be recognized by the replay circuitry and used to reset the bit count through the block. Then by counting channel bits and dividing by the group size, groups can be deserialized and decoded to data groups. In a randomized system, the pseudo-random sequence generator is also reset. Then counting derandomized bits from the sync pattern and dividing by the wordlength enables the replay circuitry to deserialize the data words.

Even if a specific code were excluded from the recorded data so that it could be used for synchronizing, this cannot ensure that the same pattern cannot be falsely created at the junction between two allowable data words. Figure 6.34 shows how false synchronizing can occur due to concatenation. It is thus not practical to use a bit pattern which is a data code value in a simple synchronizing recognizer. The problem is overcome in some synchronous systems by using the fact that sync patterns occur exactly once per block and therefore contain redundancy. If the pattern is seen by the recognizer at block rate, a genuine sync condition exists. Sync patterns seen at other times must be false. Such systems

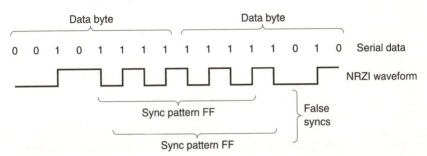

Figure 6.34 Concatenation of two words can result in the accidental generation of a word which is reserved for synchronizing.

take a few milliseconds before sync is achieved, but once achieved it should not be lost unless the transmission is interrupted.

In run-length-limited codes false syncs are not a problem. The sync pattern is no longer a data bit pattern but is a specific waveform. If the sync waveform contains run lengths which violate the normal coding limits, there is no way that these run lengths can occur in encoded data, nor any possibility that they will be interpreted as data. They can, however, be readily detected by the replay circuitry.

In a group code there are many more combinations of channel bits than there are combinations of data bits. Thus after all data bit patterns have been allocated group patterns, there are still many unused group patterns which cannot occur in the data. With care, group patterns can be found which cannot occur due to the concatenation of any pair of groups representing data. These are then unique and can be used for synchronizing.

6.16 Basic error correction

There are many different types of recording and transmission channel and consequently there will be many different mechanisms which may result in errors. Bit errors in video cause 'sparkles' in the picture whose effect depends upon the significance of the affected bit. Errors in compressed data are more serious as they may cause the decoder to lose sync.

In magnetic recording, data can be corrupted by mechanical problems such as media dropout and poor tracking or head contact, or Gaussian thermal noise in replay circuits and heads. In optical recording, contamination of the medium interrupts the light beam. When group codes are used, a single defect in a group changes the group symbol and may cause errors up to the size of the group. Single-bit errors are therefore less common in group-coded channels. Inside equipment, data are conveyed on short wires and the noise environment is under the designer's control. With suitable design techniques, errors can be made effectively negligible whereas in communication systems, there is considerably less control of the electromagnetic environment.

Irrespective of the cause, all these mechanisms cause one of two effects. There are large isolated corruptions, called error bursts, where numerous bits are corrupted all together in an area which is otherwise error-free, and there are random errors affecting single bits or symbols. Whatever the mechanism, the result will be that the received data will not be exactly the same as those sent. In binary the discrete bits will be each either right or wrong. If a binary digit is known to be wrong, it is only necessary to invert its state and then it must be right. Thus error correction itself is trivial; the hard part is working out *which* bits need correcting.

There are a number of terms which have idiomatic meanings in error correction. The raw BER (bit error rate) is the error rate of the medium, whereas the residual or uncorrected BER is the rate at which the error-correction system fails to detect or miscorrects errors. In practical digital systems, the residual BER is negligibly small. If the error correction is turned off, the two figures become the same.

Error correction works by adding some bits to the data which are calculated from the data. This creates an entity called a codeword which spans a greater length of time than one bit alone. The statistics of noise means that whilst one bit may be lost in a codeword, the loss of the rest of the codeword because of noise is highly improbable. As will be described later in this chapter, codewords are designed to be able to correct totally a finite number of corrupted bits. The greater the timespan over which the coding is performed, or, on a recording medium, the greater area over which the coding is performed, the greater will be the reliability achieved, although this does mean that an encoding delay will be experienced on recording, and a similar or greater decoding delay on reproduction.

Shannon[14] disclosed that a message can be sent to any desired degree of accuracy provided that it is spread over a sufficient timespan. Engineers have to compromise, because an infinite coding delay in the recovery of an error-free signal is not acceptable. Digital interfaces such as SDI (see Chapter 9) do not employ error correction because the build-up of coding delays in large production systems is unacceptable.

If error correction is necessary as a practical matter, it is then only a small step to put it to maximum use. All error correction depends on adding bits to the original message, and this, of course, increases the number of bits to be recorded, although it does not increase the information recorded. It might be imagined that error correction is going to reduce storage capacity, because space has to be found for all the extra bits. Nothing could be further from the truth. Once an error-correction system is used, the signal-to-noise ratio of the channel can be reduced, because the raised BER of the channel will be overcome by the error-correction system. Reduction of the SNR by 3 dB in a magnetic track can be achieved by halving the track width, provided that the system is not dominated by head or preamplifier noise. This doubles the recording density, making the storage of the additional bits needed for error correction a trivial matter. By a similar argument, the power of a digital transmitter can be reduced if error correction is used. In short, error correction is not a nuisance to be tolerated; it is a vital tool needed to maximize the efficiency of storage devices and transmission. Convergent systems would not be economically viable without it.

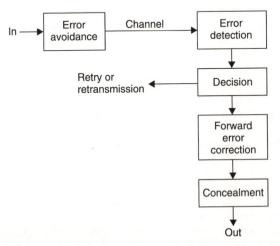

Figure 6.35 Error-handling strategies can be divided into avoiding errors, detecting errors and deciding what to do about them. Some possibilities are shown here. Of all these the detection is the most critical, as nothing can be done if the error is not detected.

Figure 6.35 shows the broad subdivisions of error handling. The first stage might be called error avoidance and includes such measures as creating bad block files on hard disks or using verified media. Properly terminating network cabling is also in this category. Placing the audio blocks near to the centre of the tape in DVTRs is a further example. The data pass through the channel, which causes whatever corruptions it feels like. On receipt of the data the occurrence of errors is first detected, and this process must be extremely reliable, as it does not matter how effective the correction or how good the concealment algorithm, if it is not known that they are necessary! The detection of an error then results in a course of action being decided.

In the case of a file transfer, real-time operation is not required. If a disk drive detects a read error a retry is easy as the disk is turning at several thousand rpm and will quickly re-present the data. An error due to a dust particle may not occur on the next revolution. A packet in error in a network will result in a retransmission. Many magnetic tape systems have *read after write*. During recording, offtape data are immediately checked for errors. If an error is detected, the tape may abort the recording, reverse to the beginning of the current block and erase it. The data from that block may then be recorded further down the tape. This is the recording equivalent of a retransmission in a communications system.

In many cases of digital video or audio replay a retry or retransmission is not possible because the data are required in real time. In this case the solution is to encode the message using a system which is sufficiently powerful to correct the errors in real time. These are called forward error-correcting schemes (FEC). The term 'forward' implies that the transmitter does not need to take any action in the case of an error; the receiver will perform the correction.

6.17 Concealment by interpolation

There are some practical differences between data recording for video and the computer data recording application. Although video or audio recorders seldom have time for retries, they have the advantage that there is a certain amount of redundancy in the information conveyed. Thus if an error cannot be corrected, then it can be concealed. If a sample is lost, it is possible to obtain an approximation to it by interpolating between samples in the vicinity of the missing one. Clearly, concealment of any kind cannot be used with computer instructions or compressed data, although concealment can be applied after compressed signals have been decoded.

If there is too much corruption for concealment, the only course in video is repeat the previous field or frame in a freeze as it is unlikely that the corrupt picture is watchable. In audio the equivalent is muting.

In general, if use is to be made of concealment on replay, the data must generally be reordered or shuffled prior to recording. To take a simple example, odd-numbered samples are recorded in a different area of the medium from even-numbered samples. On playback, if a gross error occurs on the medium, depending on its position, the result will be either corrupted odd samples or corrupted even samples, but it is most unlikely that both will be lost. Interpolation is then possible if the power of the correction system is exceeded. In practice the shuffle employed in digital video recorders is two-dimensional and rather more complex. Further details can be found in Chapter 8. The concealment technique described here is only suitable for PCM recording. If compression has been employed, different concealment techniques will be needed.

It should be stressed that corrected data are undistinguishable from the original and thus there can be no visible or audible artifacts. In contrast, concealment is only an approximation to the original information and could be detectable. In practical equipment, concealment occurs infrequently unless there is a defect requiring attention, and its presence is difficult to see.

6.18 Parity

The error-detection and error-correction processes are closely related and will be dealt with together here. The actual correction of an error is simplified tremendously by the adoption of binary. As there are only two symbols, 0 and 1, it is enough to know that a symbol is wrong, and the correct value is obvious. Figure 6.36 shows a minimal circuit required for correction once the bit in error has been identified. The XOR (exclusive-OR) gate shows up extensively in error correction and the figure also shows the truth table. One way of remembering the characteristics of this useful device is that there will be an output when the inputs are different. Inspection of the truth table will show that there is an even number of ones in each row (zero is an even number) and so the device could also be called an even parity gate. The XOR gate is also a adder in modulo-2.

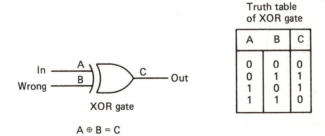

Truth table
of XOR gate

A	B	C
0	0	0
0	1	1
1	0	1
1	1	0

$A \oplus B = C$

Figure 6.36 Once the position of the error is identified, the correction process in binary is easy.

Parity is a fundamental concept in error detection. In Figure 6.37, the example is given of a four-bit data word which is to be protected. If an extra bit is added to the word which is calculated in such a way that the total number of ones in the five-bit word is even, this property can be tested on receipt. The generation of the parity bit can be performed by a number of the ubiquitous XOR gates configured into what is known as a parity tree. In the figure, if a bit is corrupted, the received message will be seen no longer to have an even number of ones. If two bits are corrupted, the failure will be undetected. This example can be used to introduce much of the terminology of error correction. The extra bit added to the message carries no information of its own, since it is calculated from the other bits. It is therefore called a *redundant* bit.

The addition of the redundant bit gives the message a special property, i.e. the number of ones is even. A message having some special property *irrespective of the actual data content* is called a *codeword*. All error correction relies on adding

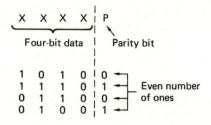

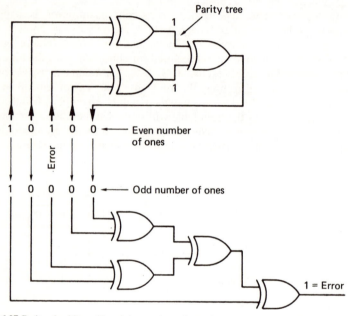

Figure 6.37 Parity checking adds up the number of ones in a word using, in this example, parity trees. One error bit and odd numbers of errors are detected. Even numbers of errors cannot be detected.

redundancy to real data to form codewords for transmission. If any corruption occurs, the intention is that the received message will not have the special property; in other words, if the received message is not a codeword there has definitely been an error. The receiver can check for the special property without any prior knowledge of the data content. Thus the same check can be made on all received data. If the received message is a codeword, there probably has not been an error. The word 'probably' must be used because the figure shows that two bits in error will cause the received message to be a codeword, which cannot be discerned from an error-free message.

If it is known that generally the only failure mechanism in the channel in question is loss of a single bit, it is *assumed* that receipt of a codeword means that there has been no error. If there is a probability of two error bits, that becomes

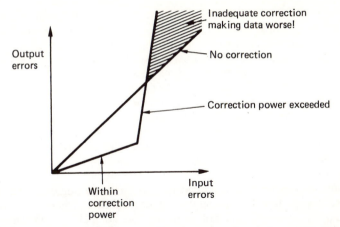

Figure 6.38 An error-correction system can only reduce errors at normal error rates at the expense of increasing errors at higher rates. It is most important to keep a system working to the left of the knee in the graph.

very nearly the probability of failing to detect an error, since all odd numbers of errors will be detected, and a four-bit error is much less likely. It is paramount in all error-correction systems that the protection used should be appropriate for the probability of errors to be encountered. An inadequate error-correction system is actually worse than not having any correction. Error correction works by trading probabilities. Error-free performance with a certain error rate is achieved at the expense of performance at higher error rates. Figure 6.38 shows the effect of an error-correction system on the residual BER for a given raw BER. It will be seen that there is a characteristic knee in the graph. If the expected raw BER has been misjudged, the consequences can be disastrous. Another result demonstrated by the example is that we can only guarantee to detect the same number of bits in error as there are redundant bits.

6.19 Block and convolutional codes

Figure 6.39(a) shows a strategy known as a crossword code, or product code. The data are formed into a two-dimensional array, in which each location can be a single bit or a multi-bit symbol. Parity is then generated on both rows and columns. If a single bit or symbol fails, one row parity check and one column parity check will fail, and the failure can be located at the intersection of the two failing checks. Although two symbols in error confuse this simple scheme, using more complex coding in a two-dimensional structure is very powerful, and further examples will be given throughout this chapter.

The example of Figure 6.39(a) assembles the data to be coded into a block of finite size and then each codeword is calculated by taking a different set of symbols. This should be contrasted with the operation of the circuit of Figure 6.39(b). Here the data are not in a block, but form an endless stream. A shift register allows four symbols to be available simultaneously to the encoder. The action of the encoder depends upon the delays. When symbol 3 emerges from the first delay, it will be added (modulo-2) to symbol 6. When this sum emerges from the second delay, it will be added to symbol 9 and so on. The codeword produced is shown in Figure 6.39(c) where it will be seen to be bent such that it has a vertical section and a diagonal section. Four symbols later the next codeword will be created one column further over in the data.

This is a convolutional code because the coder always takes parity on the same pattern of symbols which is convolved with the data stream on an endless basis. Figure 6.39(c) also shows that if an error occurs, it can be located because it will cause parity errors in two codewords. The error will be on the diagonal part of one codeword and on the vertical part of the other so that it can be located uniquely at the intersection and corrected by parity.

Comparison with the block code of Figure 6.39(a) will show that the convolutional code needs less redundancy for the same single-symbol location and correction performance as only a single redundant symbol is required for every four data symbols. Convolutional codes are computed on an endless basis which makes them inconvenient in recording applications where editing is anticipated. Here the block code is more appropriate as it allows edit gaps to be created between codes. In the case of uncorrectable errors, the convolutional principle causes the syndromes to be affected for some time afterwards and results in miscorrections of symbols which were not actually in error. This is called error propagation and is a characteristic of convolutional codes. Recording media tend to produce somewhat variant error statistics because media defects and mechanical problems cause errors which do not fit the classical additive noise channel. Convolutional codes can easily be taken beyond their correcting power if used with real recording media.

In transmission and broadcasting, the error statistics are more stable and the editing requirement is absent. As a result, convolutional codes tend to be used in digital broadcasting as will be seen in Chapter 9.

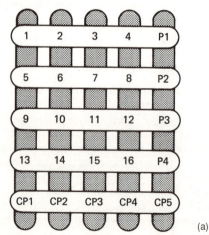

Figure 6.39 A block code is shown in (a). Each location in the block can be a bit or a word. Horizontal parity checks are made by adding P1, P2, etc., and cross-parity or vertical checks are made by adding CP1, CP2, etc. Any symbol in error will be at the intersection of the two failing codewords.

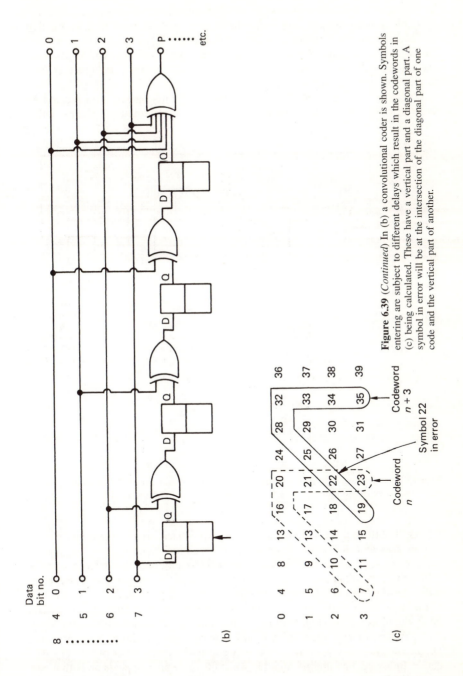

(b)

(c)

Figure 6.39 (*Continued*) In (b) a convolutional coder is shown. Symbols entering are subject to different delays which result in the codewords in (c) being calculated. These have a vertical part and a diagonal part. A symbol in error will be at the intersection of the diagonal part of one code and the vertical part of another.

6.20 Cyclic codes

In digital recording applications, the data are stored serially on a track, and it is desirable to use relatively large data blocks to reduce the amount of the medium devoted to preambles, addressing and synchronizing. The principle of codewords having a special characteristic will still be employed, but they will be generated and checked algorithmically by equations. The syndrome will then be converted to the bit(s) in error by solving equations.

Where data can be accessed serially, simple circuitry can be used because the same gate will be used for many XOR operations. The circuit of Figure 6.40 is

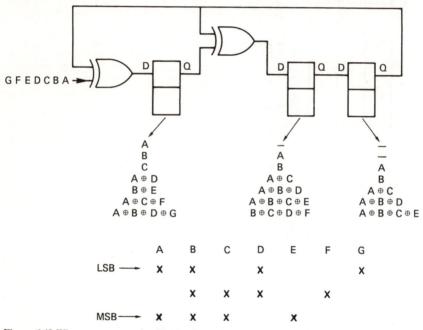

Figure 6.40 When seven successive bits A–G are clocked into this circuit, the contents of the three latches are shown for each clock. The final result is a parity-check matrix.

a kind of shift register, but with a particular feedback arrangement which leads it to be known as a twisted-ring counter. If seven message bits A–G are applied serially to this circuit, and each one of them is clocked, the outcome can be followed in the diagram. As bit A is presented and the system is clocked, bit A will enter the left-hand latch. When bits B and C are presented, A moves across to the right. Both XOR gates will have A on the upper input from the right-hand latch, the left one has D on the lower input and the right one has B on the lower input. When clocked, the left latch will thus be loaded with the XOR of A and D, and the right one with the XOR of A and B. The remainder of the sequence can be followed, bearing in mind that when the same term appears on both inputs of an XOR gate, it goes out, as the exclusive-OR of something with itself is nothing. At the end of the process, the latches contain three different expressions.

Essentially, the circuit makes three parity checks through the message, leaving the result of each in the three stages of the register. In the figure, these expressions have been used to draw up a check matrix. The significance of these steps can now be explained.

The bits A B C and D are four data bits, and the bits E F and G are redundancy. When the redundancy is calculated, bit E is chosen so that there are an even number of ones in bits A B C and E; bit F is chosen such that the same applies to bits B C D and F, and similarly for bit G. Thus the four data bits and the three check bits form a seven-bit codeword. If there is no error in the codeword, when it is fed into the circuit shown, the result of each of the three parity checks will be zero and every stage of the shift register will be cleared. As the register has eight possible states, and one of them is the error-free condition, then there are seven remaining states, hence the seven-bit codeword. If a bit in the codeword is corrupted, there will be a non-zero result. For example, if bit D fails, the check on bits A B D and G will fail, and a one will appear in the left-hand latch. The check on bits B C D F will also fail, and the centre latch will set. The check on bits A B C E will not fail, because D is not involved in it, making the right-hand bit zero. There will be a syndrome of 110 in the register, and this will be seen from the check matrix to correspond to an error in bit D. Whichever bit fails, there will be a different three-bit syndrome which uniquely identifies the failed bit. As there are only three latches, there can be eight different syndromes. One of these is zero, which is the error-free condition, and so there are seven remaining error syndromes. The length of the codeword cannot exceed seven bits, or there would not be enough syndromes to correct all the bits. This can also be made to tie in with the generation of the check matrix. If fourteen bits, A to N, were fed into the circuit shown, the result would be that the check matrix repeated twice, and if a syndrome of 101 were to result, it could not be determined whether bit D or bit K failed. Because the check repeats every seven bits, the code is said to be a cyclic redundancy check (CRC) code.

It has been seen that the circuit shown makes a matrix check on a received word to determine if there has been an error, but the same circuit can also be used to generate the check bits. To visualize how this is done, examine what happens if only the data bits A B C and D are known, and the check bits E F and G are set to zero. If this message, ABCD000, is fed into the circuit, the left-hand latch will afterwards contain the XOR of A B C and zero, which is, of course, what E should be. The centre latch will contain the XOR of B C D and zero, which is what F should be and so on. This process is not quite ideal, however, because it is necessary to wait for three clock periods after entering the data before the check bits are available. Where the data are simultaneously being recorded and fed into the encoder, the delay would prevent the check bits being easily added to the end of the data stream. This problem can be overcome by slightly modifying the encoder circuit as shown in Figure 6.41. By moving the position of the input to the right, the operation of the circuit is advanced so that the check bits are ready after only four clocks. The process can be followed in the diagram for the four data bits A B C and D. On the first clock, bit A enters the left two latches, whereas on the second clock, bit B will appear on the upper input of the left XOR gate, with bit A on the lower input, causing the centre latch to load the XOR of A and B and so on.

The way in which the cyclic codes work has been described in engineering terms, but it can be described mathematically if analysis is contemplated.

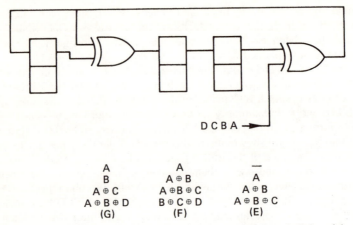

D C B A →

A	A	$\overline{A}$
B	A ⊕ B	A ⊕ B
A ⊕ C	A ⊕ B ⊕ C	A ⊕ B ⊕ C
A ⊕ B ⊕ D	B ⊕ C ⊕ D	
(G)	(F)	(E)

Figure 6.41 By moving the insertion point three places to the right, the calculation of the check bits is completed in only four clock periods and they can follow the data immediately. This is equivalent to premultiplying the data by x^3.

Just as the position of a decimal digit in a number determines the power of ten (whether that digit means one, ten or a hundred), the position of a binary digit determines the power of two (whether it means one, two or four). It is possible to rewrite a binary number so that it is expressed as a list of powers of two. For example, the binary number 1101 means $8 + 4 + 1$, and can be written:

$$2^3 + 2^2 + 2^0$$

In fact, much of the theory of error correction applies to symbols in number bases other than 2, so that the number can also be written more generally as

$$x^3 + x^2 + 1 \ (2^0 = 1)$$

which also looks much more impressive. This expression, containing as it does various powers, is, of course, a polynomial, and the circuit of Figure 6.40 which has been seen to construct a parity-check matrix on a codeword can also be described as calculating the remainder due to dividing the input by a polynomial using modulo-2 arithmetic. In modulo-2 there are no borrows or carries, and addition and subtraction are replaced by the XOR function, which makes hardware implementation very easy. In Figure 6.42 it will be seen that the circuit of Figure 6.40 actually divides the codeword by a polynomial which is

$$x^3 + x + 1 \ \text{or} \ 1011$$

This can be deduced from the fact that the right-hand bit is fed into two lower-order stages of the register at once. Once all the bits of the message have been clocked in, the circuit contains the remainder. In mathematical terms, the special property of a codeword is that it is a polynomial which yields a remainder of zero when divided by the generating polynomial. The receiver will make this division, and the result should be zero in the error-free case. Thus the codeword itself

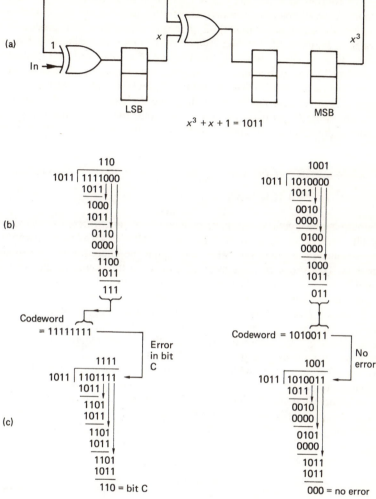

Figure 6.42 Circuit of Figure 6.40 divides by $x^3 + x + 1$ to find remainder. At (b) this is used to calculate check bits. At (c) right, zero syndrome, no error.

disappears from the division. If an error has occurred it is considered that this is due to an error polynomial which has been added to the codeword polynomial. If a codeword divided by the check polynomial is zero, a non-zero syndrome must represent the error polynomial divided by the check polynomial. Thus if the syndrome is multiplied by the check polynomial, the latter will be cancelled out and the result will be the error polynomial. If this is added modulo-2 to the received word, it will cancel out the error and leave the corrected data.

Some examples of modulo-2 division are given in Figure 6.42 which can be compared with the parallel computation of parity checks according to the matrix of Figure 6.40.

The process of generating the codeword from the original data can also be described mathematically. If a codeword has to give zero remainder when divided, it follows that the data can be converted to a codeword by adding the remainder when the data are divided. Generally speaking, the remainder would have to be subtracted, but in modulo-2 there is no distinction. This process is also illustrated in Figure 6.42. The four data bits have three zeros placed on the right-hand end, to make the wordlength equal to that of a codeword, and this word is then divided by the polynomial to calculate the remainder. The remainder is added to the zero-extended data to form a codeword. The modified circuit of Figure 6.41 can be described as premultiplying the data by x^3 before dividing.

CRC codes are of primary importance for detecting errors, and several have been standardized for use in digital communications. The most common of these are:

$$x^{16} + x^{15} + x^2 + 1 \text{ (CRC-16)}$$

$$x^{16} + x^{12} + x^5 + 1 \text{ (CRC-CCITT)}$$

The sixteen-bit cyclic codes have codewords of length $2^{16}-1$ or 65 535 bits long. This may be too long for the application. Another problem with very long codes is that with a given raw BER, the longer the code, the more errors will occur in it. There may be enough errors to exceed the power of the code. The solution in both cases is to shorten or *puncture* the code. Figure 6.43 shows that in a punctured code, only the end of the codeword is used, and the data and redundancy are preceded by a string of zeros. It is not necessary to record these zeros, and, of course, errors cannot occur in them. Implementing a punctured code is easy. If a CRC generator starts with the register cleared and is fed with serial zeros, it will not change its state. Thus it is not necessary to provide the zeros, encoding can begin with the first data bit. In the same way, the leading zeros need not be provided during playback. The only precaution needed is that if a syndrome calculates the location of an error, this will be from the beginning of the codeword not from the beginning of the data. Where codes are used for detection only, this is of no consequence.

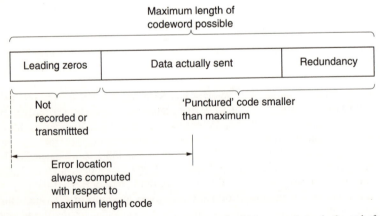

Figure 6.43 Codewords are often shortened, or punctured, which means that only the end of the codeword is actually transmitted. The only precaution to be taken when puncturing codes is that the computed position of an error will be from the beginning of the codeword, not from the beginning of the message.

6.21 Introduction to the Reed–Solomon codes

The Reed–Solomon codes (Irving Reed and Gustave Solomon) are inherently burst correcting[15] because they work on multi-bit symbols rather than individual bits. The R–S codes are also extremely flexible in use. One code may be used both to detect and correct errors and the number of bursts which are correctable can be chosen at the design stage by the amount of redundancy. A further advantage of the R–S codes is that they can be used in conjunction with a separate error-detection mechanism in which case they perform the correction only by erasure. R–S codes operate at the theoretical limit of correcting efficiency. In other words, no more efficient code can be found.

In the simple CRC system described in section 6.20, the effect of the error is detected by ensuring that the codeword can be divided by a polynomial. The CRC codeword was created by adding a redundant symbol to the data. In the Reed–Solomon codes, several errors can be isolated by ensuring that the codeword will divide by a number of polynomials. Clearly, if the codeword must divide by, say, two polynomials, it must have two redundant symbols. This is the minimum case of an R–S code. On receiving an R–S-coded message there will be two syndromes following the division. In the error-free case, these will both be zero. If both are not zero, there is an error.

It has been stated that the effect of an error is to add an error polynomial to the message polynomial. The number of terms in the error polynomial is the same as the number of errors in the codeword. The codeword divides to zero and the syndromes are a function of the error only. There are two syndromes and two equations. By solving these simultaneous equations it is possible to obtain two unknowns. One of these is the position of the error, known as the *locator* and the other is the error bit pattern, known as the *corrector*. As the locator is the same size as the code symbol, the length of the codeword is determined by the size of the symbol. A symbol size of eight bits is commonly used because it fits in conveniently with both sixteen-bit audio samples and byte-oriented computers. An eight-bit syndrome results in a locator of the same wordlength. Eight bits have 2^8 combinations, but one of these is the error-free condition, and so the locator can specify one of only 255 symbols. As each symbol contains eight bits, the codeword will be $255 \times 8 = 2040$ bits long.

As further examples, five-bit symbols could be used to form a codeword 31 symbols long, and three-bit symbols would form a codeword seven symbols long. This latter size is small enough to permit some worked examples, and will

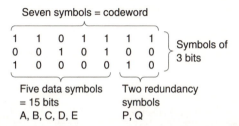

Seven symbols = codeword

```
1   1   0   1   1   1   1
0   0   1   0   1   0   0      Symbols of
1   0   0   0   0   1   0      3 bits
```

Five data symbols   Two redundancy
= 15 bits   symbols
A, B, C, D, E   P, Q

Figure 6.44 A Reed–Solomon codeword. As the symbols are of three bits, there can only be eight possible syndrome values. One of these is all zeros, the error-free case, and so it is only possible to point to seven errors: hence the codeword length of seven symbols. Two of these are redundant, leaving five data symbols.

be used further here. Figure 6.44 shows that in the seven-symbol codeword, five symbols of three bits each, A–E, are the data, and P and Q are the two redundant symbols. This simple example will locate and correct a single symbol in error. It does not matter, however, how many bits in the symbol are in error.

The two check symbols are solutions to the following equations:

$$A \oplus B \oplus C \oplus D \oplus E \oplus P \oplus Q = 0 \ (\oplus = \text{XOR symbol})$$

$$a^7 A \oplus a^6 B \oplus a^5 C \oplus a^4 D \oplus a^3 E \oplus a^2 P \oplus aQ = 0$$

where a is a constant. The original data A–E followed by the redundancy P and Q pass through the channel.

The receiver makes two checks on the message to see if it is a codeword. This is done by calculating syndromes using the following expressions, where the (') implies the received symbol which is not necessarily correct:

$$S_0 = A' \oplus B' \oplus C' \oplus D' \oplus E' \ P' \oplus Q'$$

(This is in fact a simple parity check.)

$$S_1 = a^7 A' \oplus a^6 B' \oplus a^5 C' \oplus a^4 D' \oplus a^3 E' \oplus a^2 P' \oplus aQ'$$

If two syndromes of all zeros are not obtained, there has been an error. The information carried in the syndromes will be used to correct the error. For the purpose of illustration, let it be considered that D' has been corrupted before moving to the general case. D' can be considered to be the result of adding an error of value E to the original value D such that $D' = D \oplus E$.

As $A \oplus B \oplus C \oplus D \oplus E \oplus P \oplus Q = 0$

then $A \oplus B \oplus C \oplus (D \oplus E) \oplus E \oplus P \oplus Q = E = S_0$

As $D' = D \oplus E$

then $D = D' \oplus E = D' \oplus S_0$

Thus the value of the corrector is known immediately because it is the same as the parity syndrome S_0. The corrected data symbol is obtained simply by adding S_0 to the incorrect symbol.

At this stage, however, the corrupted symbol has not yet been identified, but this is equally straightforward:

As $a^7 A \oplus a^6 B \oplus a^5 C \oplus a^4 D \oplus a^3 E \oplus a^2 P \oplus aQ = 0$

then:

$$a^7 A \oplus a^6 B \oplus a^5 C \oplus a^4 (D \oplus E) \oplus a^3 E \oplus a^2 P \oplus aQ = a^4 E = S_1$$

Thus the syndrome S_1 is the error bit pattern E, but it has been raised to a power of a which is a function of the position of the error symbol in the block. If the position of the error is in symbol k, then k is the locator value and:

$$S_0 \times a^k = S_1$$

Hence:

$$a^k = \frac{S_1}{S_0}$$

The value of k can be found by multiplying S_0 by various powers of a until the product is the same as S_1. Then the power of a necessary is equal to k. The use of the descending powers of a in the codeword calculation is now clear because the error is then multiplied by a different power of a dependent upon its position, known as the locator, because it gives the position of the error. The process of finding the error position by experiment is known as a Chien search.

Whilst the expressions above show that the values of P and Q are such that the two syndrome expressions sum to zero, it is not yet clear how P and Q are calculated from the data. Expressions for P and Q can be found by solving the two R–S equations simultaneously. This has been done in Appendix 6.1. The following expressions must be used to calculate P and Q from the data in order to satisfy the codeword equations. These are:

$$P = a^6\,A \oplus a B \oplus a^2\,C \oplus a^5\,D \oplus a^3\,E$$

$$Q = a^2\,A \oplus a^3\,B \oplus a^6\,C \oplus a^4\,D \oplus aE$$

In both the calculation of the redundancy shown here and the calculation of the corrector and the locator it is necessary to perform numerous multiplications and raising to powers. This appears to present a formidable calculation problem at both the encoder and the decoder. This would be the case if the calculations involved were conventionally executed. However, the calculations can be simplified by using logarithms. Instead of multiplying two numbers, their logarithms are added. In order to find the cube of a number, its logarithm is added three times. Division is performed by subtracting the logarithms. Thus all the manipulations necessary can be achieved with addition or subtraction, which is straightforward in logic circuits.

The success of this approach depends upon simple implementation of log tables. As was seen in section 2.26, raising a constant, a, known as the *primitive element* to successively higher powers in modulo-2 gives rise to a Galois field. Each element of the field represents a different power n of a. It is a fundamental of the R–S codes that all the symbols used for data, redundancy and syndromes are considered to be elements of a Galois field. The number of bits in the symbol determines the size of the Galois field, and hence the number of symbols in the codeword.

In Figure 6.45, the binary values of the elements are shown alongside the power of a they represent. In the R–S codes, symbols are no longer considered simply as binary numbers, but also as equivalent powers of a. In Reed–Solomon coding and decoding, each symbol will be multiplied by some power of a. Thus if the symbol is also known as a power of a it is only necessary to add the two powers. For example, if it is necessary to multiply the data symbol 100 by a^3, the calculation proceeds as follows, referring to Figure 6.45.

$$100 = a^2 \text{ so } 100 \times a^3 = a^{(2\,+\,3)} = a^5 = 111$$

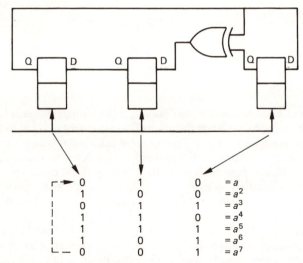

Figure 6.45 The bit patterns of a Galois field expressed as powers of the primitive element a. This diagram can be used as a form of log table in order to multiply binary numbers. Instead of an actual multiplication, the appropriate powers of a are simply added.

Note that the results of a Galois multiplication are quite different from binary multiplication. Because all products must be elements of the field, sums of powers which exceed seven wrap around by having seven subtracted. For example:

$$a^5 \times a^6 = a^{11} = a^4 = 110$$

Figure 6.46 shows some examples of circuits which will perform this kind of multiplication. Note that they require a minimum amount of logic.

Figure 6.47 given an example of the Reed–Solomon encoding process. The Galois field shown in Figure 6.45 has been used, having the primitive element a = 010. At the beginning of the calculation of P, the symbol A is multiplied by a^6. This is done by converting A to a power of a. According to Figure 6.45, 101 = a^6 and so the product will be $a^{(6 + 6)} = a^{12} = a^5 = 111$. In the same way, B is multiplied by a, and so on, and the products are added modulo-2. A similar process is used to calculate Q.

Figure 6.48 shows a circuit which can calculate P or Q. The symbols A–E are presented in succession, and the circuit is clocked for each one. On the first clock, a^6A is stored in the left-hand latch. If B is now provided at the input, the second GF multiplier produces aB and this is added to the output of the first latch and when clocked will be stored in the second latch which now contains a^6A + aB. The process continues in this fashion until the complete expression for P is available in the right-hand latch. The intermediate contents of the right-hand latch are ignored.

The entire codeword now exists, and can be recorded or transmitted. Figure 6.47 also demonstrates that the codeword satisfies the checking equations. The modulo-2 sum of the seven symbols, S_0, is 000 because each column has an even number of ones. The calculation of S_1 requires multiplication by descending

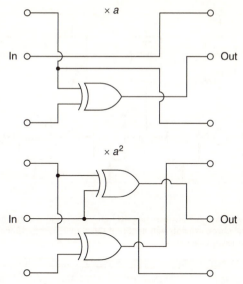

Figure 6.46 Some examples of GF multiplier circuits.

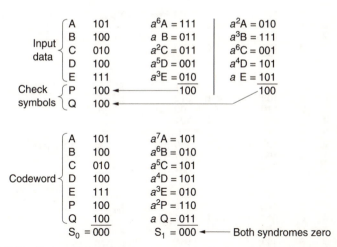

Figure 6.47 Five data symbols A–E are used as terms in the generator polynomials derived in Appendix 6.1 to calculate two redundant symbols P and Q. An example is shown at the top. Below is the result of using the codeword symbols A–Q as terms in the checking polynomials. As there is no error, both syndromes are zero.

powers of a. The modulo-2 sum of the products is again zero. These calculations confirm that the redundancy calculation was properly carried out.

Figure 6.49 gives three examples of error correction based on this codeword. The erroneous symbol is marked with a dash. As there has been an error, the syndromes S_0 and S_1 will not be zero.

Figure 6.50 shows circuits suitable for parallel calculation of the two syndromes at the receiver. The S_0 circuit is a simple parity checker which

A, B, C, D, E

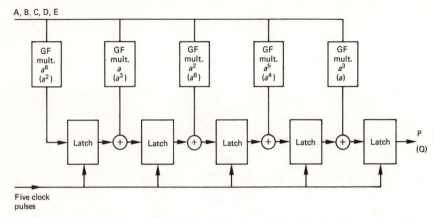

Five clock
pulses

Figure 6.48 If the five data symbols of Figure 6.47 are supplied to this circuit in sequence, after
five clocks, one of the check symbols will appear at the output. Terms without brackets will
calculate P, bracketed terms calculate Q.

7	A	101	$a^7 A = 101$	$\dfrac{S_1}{S_0} = \dfrac{a^4}{1} = a^4$
6	B	100	$a^6 B = 010$	
5	C	010	$a^5 C = 101$	
4	D'	101	$a^4 D' = 011$	$k = 4$
3	E	111	$a^3 E = 010$	
2	P	100	$a^2 P = 110$	$D' + S_0 = 101 + 001$
1	Q	100	$a\ Q = 011$	$D = 100$
	$S_0 = \overline{001}$		$S_1 = \overline{110}$	

7	A	101	$a^7 A = 101$	$\dfrac{S_1}{S_0} = \dfrac{1}{a^2} = \dfrac{1}{a^2} \times \dfrac{a^5}{a^5} = a^5$
6	B	100	$a^6 B = 010$	
5	C'	110	$a^5 C = 100$	
4	D	100	$a^4 D = 101$	$k = 5$
3	E	111	$a^3 E = 010$	
2	P	100	$a^2 P = 110$	$C' + S_0 = 110 + 100$
1	Q	100	$a\ Q = 011$	$C = 010$
	$S_0 = \overline{100}$		$S_1 = \overline{001}$	

7	A'	111	$a^7 A = 111$	$\dfrac{S_1}{S_0} = \dfrac{a}{a} = 001 = a^7$
6	B	100	$a^6 B = 010$	
5	C	010	$a^5 C = 101$	
4	D	100	$a^4 D = 101$	$k = 7$
3	E	111	$a^3 E = 010$	
2	P	100	$a^2 P = 110$	$A' + S_0 = 111 + 010$
1	Q	100	$a\ Q = 011$	$A = 101$
	$S_0 = \overline{010}$		$S_1 = \overline{010}$	

Figure 6.49 Three examples of error location and correction. The number of bits in error in a
symbol is irrelevant; if all three were wrong, S_0 would be 111, but correction is still possible.

accumulates the modulo-2 sum of all symbols fed to it. The S_1 circuit is more
subtle, because it contains a Galois field (GF) multiplier in a feedback loop, such
that early symbols fed in are raised to higher powers than later symbols because
they have been recirculated through the GF multiplier more often. It is possible
to compare the operation of these circuits with the example of Figure 6.49 and
with subsequent examples to confirm that the same results are obtained.

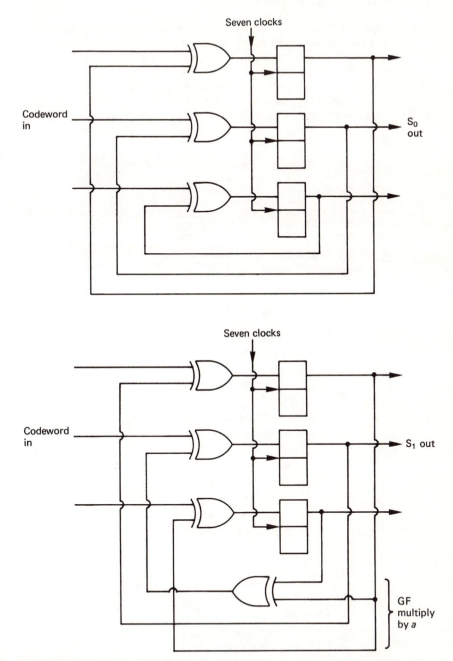

Figure 6.50 Circuits for parallel calculation of syndromes S_0, S_1. S_0 is a simple parity check. S_1 has a GF multiplication by a in the feedback, so that A is multiplied by a^7, B is multiplied by a^6, etc., and all are summed to give S_1.

6.22 Correction by erasure

In the examples of Figure 6.49, two redundant symbols P and Q have been used to locate and correct one error symbol. If the positions of errors are known by some separate mechanism (see product codes, section 6.24) the locator need not be calculated. The simultaneous equations may instead be solved for two correctors. In this case the number of symbols which can be corrected is equal to the number of redundant symbols. In Figure 6.51(a) two errors have taken place, and it is known that they are in symbols C and D. Since S_0 is a simple parity

$$
\begin{array}{lll}
A & 101 & a'A = \qquad 101 \\
B & 100 & a^6 B = \qquad 010 \\
(C \oplus E_C) & 001 & a^5 (C \oplus E_D) \quad 111 \\
(D \oplus E_D) & 010 & a^4 (D \oplus E_D) \quad 111 \\
E & 111 & a^3 E = \qquad 010 \\
P & 100 & a^2 P = \qquad 110 \\
Q & \underline{100} & a\ Q = \qquad \underline{011} \\
S_1 \quad = & 101 & S_1 = \qquad 000
\end{array}
$$

$$S_0 = E_C \oplus E_D \qquad S_1 = a^5 E_C \oplus a^4 E_D$$

$$S_1 = a^5 E_C \oplus a^4 (S_0 \oplus E_C)$$

$$= a^5 E_C \ a^4 S_0 \oplus a^4 E_C$$

$$\therefore E_C = \frac{S_1 \oplus a^4 S_0}{a^5 \oplus a^4} = \frac{000 \oplus 011}{001} = 011$$

$$C = (C \oplus E_C) \oplus E_C = 001 \oplus 011 = \underline{010}$$

$$S_1 = a^5 (S_0 \oplus E_D \oplus a^4 E_D$$

$$= a^5 S_0 \oplus a^5 E_D \oplus a^4 E_D$$

$$\therefore E_D = \frac{S_1 \oplus a^5 S_0}{a^5 \oplus a^4} = \frac{000 \oplus 110}{001} = 110$$

$$D = (D \oplus E_D) + E_D = 010 \oplus 110 = \underline{100} \qquad (a)$$

$$
\begin{array}{lll}
A & 101 & a^7 A = 101 \\
B & 100 & a^6 B = 010 \quad S_0 = C \oplus D \\
C & \underline{000} & a^5 C = \underline{000} \\
D & \underline{000} & a^4 D = \underline{000} \quad S_1 = a^5 C \oplus a^4 D \\
E & 111 & a^3 E = 010 \\
P & 100 & a^2 P = 110 \\
Q & \underline{100} & a\ Q = \underline{011} \\
S_0 & = 100 & S_1 = 000
\end{array}
$$

$$S_1 = a^5 S_0 \oplus a^5 D \oplus a^4 D = a^5 S_0 \ D$$

$$\therefore D = S_1 \oplus a^5 S_0 = 000 \oplus 100 = \underline{100}$$

$$S_1 = a^5 C \oplus a^4 C \oplus a^4 S_0 = C \oplus a^4 S_0$$

$$\therefore C = S_1 \oplus a^4 S_0 = 000 \oplus 010 = \underline{010}$$

(b)

Figure 6.51 If the location of errors is known, then the syndromes are a known function of the two errors as shown in (a). It is, however, much simpler to set the incorrect symbols to zero, i.e. to *erase* them as in (b). Then the syndromes are a function of the wanted symbols and correction is easier.

check, it will reflect the modulo-2 sum of the two errors. Hence $S_1 = EC \oplus ED$.

The two errors will have been multiplied by different powers in S_1, such that:

$$S_1 = a^5\ EC \oplus a^4\ ED$$

These two equations can be solved, as shown in the figure, to find EC and ED, and the correct value of the symbols will be obtained by adding these correctors to the erroneous values. It is, however, easier to set the values of the symbols in error to zero. In this way the nature of the error is rendered irrelevant and it does not enter the calculation. This setting of symbols to zero gives rise to the term erasure. In this case,

$$S_0 = C \oplus D$$
$$S_1 = a^5\ C + a^4\ D$$

Erasing the symbols in error makes the errors equal to the correct symbol values and these are found more simply as shown in Figure 6.51(b)

Practical systems will be designed to correct more symbols in error than in the simple examples given here. If it is proposed to correct by erasure an arbitrary number of symbols in error given by t, the codeword must be divisible by t different polynomials. Alternatively, if the errors must be located and corrected, $2t$ polynomials will be needed. These will be of the form $(x + a^n)$ where n takes all values up to t or $2t$. a is the primitive element discussed in section 2.26.

Where four symbols are to be corrected by erasure, or two symbols are to be located and corrected, four redundant symbols are necessary, and the codeword polynomial must then be divisible by

$$(x + a^0)\ (x + a^1)\ (x + a^2)\ (x + a^3)$$

Upon receipt of the message, four syndromes must be calculated, and the four correctors or the two error patterns and their positions are determined by solving four simultaneous equations. This generally requires an iterative procedure, and a number of algorithms have been developed for the purpose.[16–18] Modern DVTR formats use eight-bit R–S codes and erasure extensively. The primitive polynomial commonly used with GF(256) is:

$$x^8 + x^4 + x^3 + x^2 + 1$$

The codeword will be 255 bytes long but will often be shortened by puncturing. The larger Galois fields require less redundancy, but the computational problem increases. LSI chips have been developed specifically for R–S decoding in many high-volume formats.

6.23 Interleaving

The concept of bit interleaving was introduced in connection with a single-bit correcting code to allow it to correct small bursts. With burst-correcting codes such as Reed–Solomon, bit interleave is unnecessary. In most channels, particularly high-density recording channels used for digital video or audio, the burst size may be many bytes rather than bits, and to rely on a code alone to correct such errors would require a lot of redundancy. The solution in this case is to employ symbol interleaving, as shown in Figure 6.52. Several codewords are encoded from input data, but these are not recorded in the order they were input, but are physically reordered in the channel, so that a real burst error is split into smaller bursts in several codewords. The size of the burst seen by each codeword is now determined primarily by the parameters of the interleave, and Figure 6.53 shows that the probability of occurrence of bursts with respect to the burst length in a given codeword is modified. The number of bits in the interleave word can be made equal to the burst-correcting ability of the code in the knowledge that it will be exceeded only very infrequently.

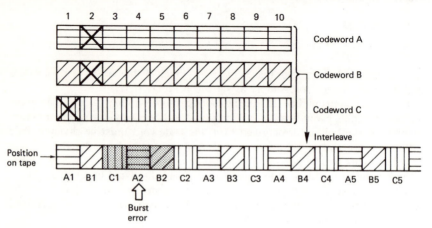

Figure 6.52 The interleave controls the size of burst errors in individual codewords.

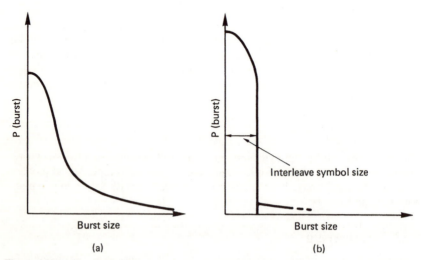

Figure 6.53 (a) The distribution of burst sizes might look like this. (b) Following interleave, the burst size within a codeword is controlled to that of the interleave symbol size, except for gross errors which have low probability.

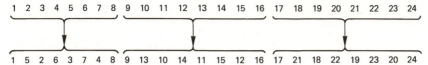

Figure 6.54 In block interleaving, data are scrambled within blocks which are themselves in the correct order.

There are a number of different ways in which interleaving can be performed. Figure 6.54 shows that in block interleaving, words are reordered within blocks which are themselves in the correct order. This approach is attractive for rotary-head recorders, because the scanning process naturally divides the tape up into blocks. The block interleave is achieved by writing samples into a memory in sequential address locations from a counter, and reading the memory with non-sequential addresses from a sequencer. The effect is to convert a one-dimensional sequence of samples into a two-dimensional structure having rows and columns.

The alternative to block interleaving is convolutional interleaving where the interleave process is endless. In Figure 6.55 symbols are assembled into short blocks and then delayed by an amount proportional to the position in the block. It will be seen from the figure that the delays have the effect of shearing the symbols so that columns on the left side of the diagram become diagonals on the right. When the columns on the right are read, the convolutional interleave will be obtained. Convolutional interleave works well in transmission applications such as DVB where there is no natural track break. Convolutional interleave has the advantage of requiring less memory to implement than a block code. This is because a block code requires the entire block to be written into the memory before it can be read, whereas a convolutional code requires only enough memory to cause the required delays.

6.24 Product codes

In the presence of burst errors alone, the system of interleaving works very well, but it is known that in most practical channels there are also uncorrelated errors of a few bits due to noise. Figure 6.56 shows an interleaving system where a dropout-induced burst error has occurred which is at the maximum correctable size. All three codewords involved are working at their limit of one symbol. A random error due to noise in the vicinity of a burst error will cause the correction power of the code to be exceeded. Thus a random error of a single bit causes a further entire symbol to fail. This is a weakness of an interleave solely designed to handle dropout-induced bursts. Practical high-density equipment must address the problem of noise-induced or random errors and burst errors occurring at the same time. This is done by forming codewords both before and after the interleave process. In block interleaving, this results in a *product code*, whereas in the case of convolutional interleave the result is called *cross-interleaving*.

Figure 6.57 shows that in a product code the redundancy calculated first and checked last is called the outer code, and the redundancy calculated second and checked first is called the inner code. The inner code is formed along tracks on the medium. Random errors due to noise are corrected by the inner code and do not impair the burst-correcting power of the outer code. Burst errors are declared

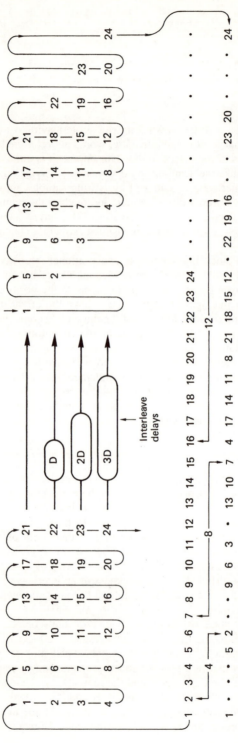

Figure 6.55 In convolutional interleaving, samples are formed into a rectangular array, which is sheared by subjecting each row to a different delay. The sheared array is read in vertical columns to provide the interleaved output. In this example, samples will be found at 4, 8 and 12 places away from their original order.

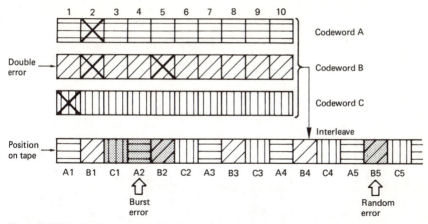

Figure 6.56 The interleave system falls down when a random error occurs adjacent to a burst.

uncorrectable by the inner code which flags the bad samples on the way into the de-interleave memory. The outer code reads the error flags in order to correct the flagged symbols by erasure. The error flags are also known as erasure flags. As it does not have to compute the error locations, the outer code needs half as much redundancy for the same correction power. Thus the inner code redundancy does not raise the code overhead. The combination of codewords with interleaving in several dimensions yields an error-protection strategy which is truly synergistic, in that the end result is more powerful than the sum of the parts. Needless to say, the technique is used extensively in modern storage formats.

Appendix 6.1 Calculation of Reed–Solomon generator polynomials

For a Reed–Solomon codeword over $GF(2^3)$, there will be seven three-bit symbols. For location and correction of one symbol, there must be two redundant symbols P and Q, leaving A–E for data.

The following expressions must be true, where a is the primitive element of $x^3 \oplus x \oplus 1$ and $\oplus$ is XOR throughout:

$$A \oplus B \oplus C \oplus D \oplus E \oplus P \oplus Q = 0 \tag{1}$$

$$a^7 A \oplus a^6 B \oplus a^5 C \oplus a^4 D \oplus a^3 E \oplus a^2 P \oplus aQ = 0 \tag{2}$$

Dividing equation (2) by a:

$$a^6 A \oplus a^5 B \oplus a^4 C \oplus a^3 D \oplus a^2 E \oplus aP \oplus Q = 0$$

$$= A \oplus B \oplus C \oplus D \oplus E \oplus P \oplus Q$$

Cancelling Q, and collecting terms:

$$(a^6 \oplus 1)A \oplus (a^5 \oplus 1)B \oplus (a^4 \oplus 1)C \oplus (a^3 \oplus 1)D \oplus (a^2 \oplus 1)E$$

$$= (a \oplus 1)P$$

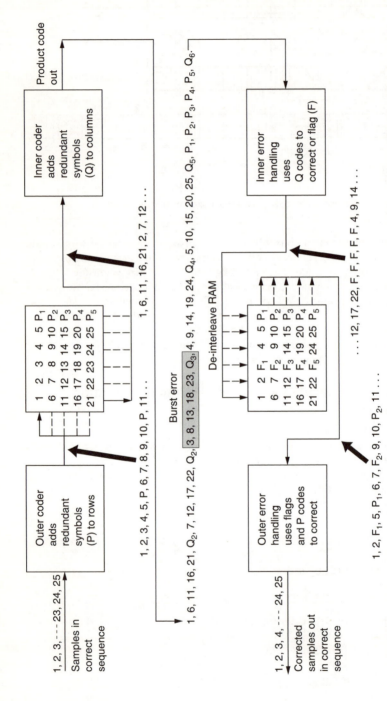

Figure 6.57 In addition to the redundancy P on rows, inner redundancy Q is also generated on columns. On replay, the Q code checker will pass on flags F if it finds an error too large to handle itself. The flags pass through the de-interleave process and are used by the outer error correction to identify which symbol in the row needs correcting with P redundancy. The concept of crossing two codes in this way is called a product code.

Using section 2.26 to calculate $(a^n + 1)$, e.g. $a^6 + 1 = 101 + 001 = 100 = a^2$:

$$a^2A \oplus a^4B \oplus a^5C \oplus aD \oplus a^6E = a^3P$$
$$a^6A \oplus aB \oplus a^2C \oplus a^5D \oplus a^3E = P$$

Multiplying equation (1) by a^2 and equating to equation (2):

$$a^2A \oplus a^2B \oplus a^2C \oplus a^2D \oplus a^2E \oplus a^2P \oplus a^2Q = 0$$
$$= a^7A \oplus a^6B \oplus a^5C \oplus a^4D \oplus a^3E \oplus a^2P \oplus aQ$$

Cancelling terms a^2P and collecting terms (remember $a^2 \oplus a^2 = 0$):

$$(a^7 \oplus a^2)A \oplus (a^6 \oplus a^2)B \oplus (a^5 \oplus a^2)C \oplus (a^4 \oplus a^2)D \oplus$$
$$(a^3 \oplus a^2)E = (a^2 \oplus a)Q$$

Adding powers according to section 2.26, e.g.

$$a^7 \oplus a^2 = 001 \oplus 100 = 101 = a^6:$$
$$a^6A \oplus B \oplus a^3C \oplus aD \oplus a^5E = a^4Q$$
$$a^2A \oplus a^3B \oplus a^6C \oplus a^4D \oplus aE = Q$$

References

1. Deeley, E.M., Integrating and differentiating channels in digital tape recording. *Radio Electron. Eng.*, **56**, 169–173 (1986)
2. Mee, C.D., *The Physics of Magnetic Recording*, Amsterdam and New York: Elsevier–North-Holland Publishing (1978)
3. Jacoby, G.V., Signal equalization in digital magnetic recording. *IEEE Trans. Magn.*, **MAG–11**, 302–305 (1975)
4. Schneider, R.C., An improved pulse-slimming method for magnetic recording. *IEEE Trans. Magn.*, **MAG–11**, 1240–1241 (1975)
5. Miller, A., US Patent. No.3 108 261
6. Mallinson, J.C. and Miller, J.W., Optimum codes for digital magnetic recording. *Radio and Electron. Eng.*, **47**, 172–176 (1977)
7. Miller, J.W., DC-free encoding for data transmission system. US Patent 4 027 335 (1977)
8. Tang, D.T., Run-length-limited codes. IEEE International Symposium on Information Theory (1969)
9. Yokoyama, K., Digital video tape recorder. *NHK Technical Monograph*, No.31 (March 1982)
10. Coleman, C.H. *et al.*, High data rate magnetic recording in a single channel. *J. IERE*, **55**, 229–236 (1985)
11. Kobayashi, H., Application of partial response channel coding to magnetic recording systems. *IBM J. Res. Dev.*, **14**, 368–375 (1970)
12. Forney, G.D., JR, The Viterbi algorithm, *Proc. IEEE*, **61**, 268–278 (1973)
13. Wood, R.W. and Petersen, D.A., Viterbi detection of Class IV partial response on a magnetic recording channel. *IEEE Trans. Commun.*, **34**, 454–461 (1968)
14. Shannon, C.E., A mathematical theory of communication. *Bell System Tech. J.*, **27**, 379 (1948)
15. Reed, I.S. and Solomon, G., Polynomial codes over certain finite fields. *J. Soc. Indust. Appl. Math.*, **8**, 300–304 (1960)
16. Berlekamp, E.R., *Algebraic Coding Theory*. New York: McGraw-Hill (1967). Reprint edition: Laguna Hills, CA: Aegean Park Press (1983)
17. Sugiyama, Y. *et al.*, An erasures and errors decoding algorithm for Goppa codes. *IEEE Trans. Inf. Theory*, **IT–22** (1976)
18. Peterson, W.W. and Weldon, E.J., *Error Correcting Codes*, 2nd edn, Cambridge MA: MIT Press (1972)

Chapter 7

Disks in digital video

7.1 Types of disk

Disk drives came into being as random-access file-storage devices for digital computers. The explosion in the growth of personal computers has fuelled demand for low-cost high-density disk drives and the rapid access offered is increasingly finding applications in digital video. After lengthy development, optical disks are also emerging in digital video applications.

Figure 7.1 shows that, in a disk drive, the data are recorded on a circular track. In hard-disk drives, the disk rotates at several thousand rev/min so that the head-to-disk speed is of the order of 100 miles per hour. At this speed no contact can be tolerated, and the head flies on a boundary layer of air turning with the disk at a height measured in microinches. The longest time it is necessary to wait to access a given data block is a few milliseconds. To increase the storage capacity of the drive without a proportional increase in cost, many concentric tracks are recorded on the disk surface, and the head is mounted on a positioner which can rapidly bring the head to any desired track. Such a machine is termed a moving-head disk drive. An increase in capacity could be obtained by assembling many disks on a common spindle to make a disk pack. The small size of magnetic heads allows the disks to be placed close together. If the positioner is designed so that it can remove the heads away from the disk completely, it can be exchanged. The exchangeable-pack moving-head disk drive became the standard for mainframe and minicomputers for a long time.

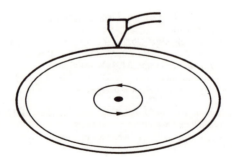

Figure 7.1 The rotating store concept. Data on the rotating circular track are repeatedly presented to the head.

312

Later came the so-called Winchester technology disks, where the disk and positioner formed a compact sealed unit which allowed increased storage capacity but precluded exchange of the disk pack alone.

Disk drive development has been phenomenally rapid. The first flying-head disks were about 3 feet across. Subsequently disk sizes of 14, 8, $5\frac{1}{4}$, $3\frac{1}{2}$ and $1\frac{7}{8}$ inches were developed. Despite the reduction in size, the storage capacity is not compromised because the recording density has increased and continues to increase. In fact there is an advantage in making a drive smaller because the moving parts are then lighter and travel a shorter distance, improving access time.

There are numerous types of optical disk, which have different characteristics. The basic principles of optical disk readout are introduced in section 7.10. Optical disks fall into three broad groups which can usefully be compared.

1 The Compact Disc: its data derivative CD-ROM and the later DVD are examples of a read-only laser disk, which is designed for mass duplication by stamping. They cannot be recorded.
2 Some laser disks can be recorded, but once recorded they cannot be edited or erased because some permanent mechanical or chemical change has been made. These are usually referred to as write-once-read-many (WORM) disks.
3 Erasable optical disks have essentially the same characteristic as magnetic disks, in that new and different recordings can be made in the same track indefinitely, but there is usually a separate erase cycle needed before a new recording can be made since overwrite is not always possible.

Figure 7.2 introduces the essential subsystems of a disk drive which will be discussed here. Magnetic drives and optical drives are similar in that both have a spindle drive mechanism to revolve the disk, and a positioner to give radial access across the disk surface. In the optical drive, the positioner has to carry a collection of lasers, lenses, prisms, gratings and so on, and will be rather larger than a magnetic head. The heavier pickup cannot be accelerated as fast as a magnetic-drive positioner, and access time is slower. A large number of pickups on one positioner makes matters worse. For this reason and because of the larger

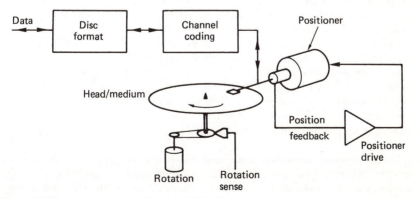

Figure 7.2 The main subsystems of a typical disk drive.

spacing needed between the disks, multi-platter optical disks are uncommon. Instead 'juke box' mechanisms have been developed to allow a large library of optical disks to be mechanically accessed by one or more drives. Access time is sometimes reduced by having more than one positioner per disk; a technique adopted rarely in magnetic drives. A penalty of the very small track pitch possible in laser disks, which gives the enormous storage capacity, is that very accurate track following is needed, and it takes some time to lock onto a track. For this reason tracks on laser disks are usually made as a continuous spiral, rather than the concentric rings of magnetic disks. In this way, a continuous data transfer involves no more than track following once the beginning of the file is located.

Rigid disks are made from aluminium alloy. Magnetic-oxide types use an aluminium oxide substrate, or undercoat, giving a flat surface to which the oxide binder can adhere. Later metallic disks having higher coercivity are electroplated with the magnetic medium. In both cases the surface finish must be extremely good owing to the very small flying height of the head. As the head-to-disk speed and recording density are functions of track radius, the data are confined to the outer areas of the disks to minimize the change in these parameters. As a result, the centre of the pack is often an empty well. In fixed (i.e. non-interchangeable) disks the drive motor is often installed in the centre well.

The information layer of optical disks may be made of a variety of substances, depending on the working principle. This layer is invariably protected beneath a thick transparent layer of glass or polycarbonate.

Exchangeable optical and magnetic disks are usually fitted in protective cartridges. These have various shutters which retract on insertion in the drive to allow access by the drive spindle and heads. Removable packs usually seat on a taper to ensure concentricity and are held to the spindle by a permanent magnet. A lever mechanism may be incorporated into the cartridge to assist their removal.

7.2 Magnetic disks

In all technologies there are specialist terms, and those relating to magnetic disks will be explained here. Figure 7.3 shows a typical multiplatter magnetic disk pack in conceptual form. Given a particular set of coordinates (cylinder, head, sector), known as a disk physical address, one unique data block is defined. A common block capacity is 512 bytes. The subdivision into sectors is sometimes omitted for special applications. A disk drive can be randomly accessed, because any block address can follow any other, but unlike a RAM, at each address a large block of data is stored, rather than a single word.

Magnetic disk drives permanently sacrifice storage density in order to offer rapid access. The use of a flying head with a deliberate air gap between it and the medium is necessary because of the high medium speed, but this causes a severe separation loss which restricts the linear density available. The air gap must be accurately maintained, and consequently the head is of low mass and is mounted flexibly.

The aerohydrodynamic part of the head is known as the slipper; it is designed to provide lift from the boundary layer which changes rapidly with changes in flying height. It is not initially obvious that the difficulty with disk heads is not making them fly, but making them fly close enough to the disk surface. The boundary layer

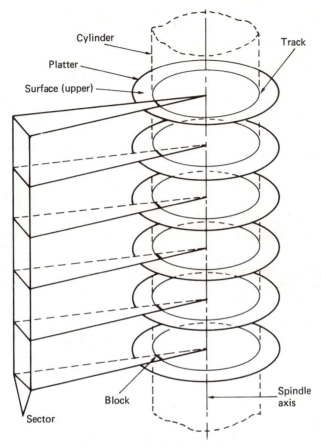

Figure 7.3 Disk terminology. Surface: one side of a platter. Track: path described on a surface by a fixed head. Cylinder: imaginary shape intersecting all surfaces at tracks of the same radius. Sector: angular subdivision of pack. Block: that part of a track within one sector. Each block has a unique cylinder, head and sector address.

travelling at the disk surface has the same speed as the disk, but as height increases, it slows down due to drag from the surrounding air. As the lift is a function of relative air speed, the closer the slipper comes to the disk, the greater the lift will be. The slipper is therefore mounted at the end of a rigid cantilever sprung towards the medium. The force with which the head is pressed towards the disk by the spring is equal to the lift at the designed flying height. Because of the spring, the head may rise and fall over small warps in the disk. It would be virtually impossible to manufacture disks flat enough to dispense with this feature. As the slipper negotiates a warp it will pitch and roll in addition to rising and falling, but it must be prevented from yawing, as this would cause an azimuth error. Downthrust is applied to the aerodynamic centre by a spherical thrust button, and the required degrees of freedom are supplied by a thin flexible gimbal. The slipper has to bleed away surplus air in order to approach close enough to the disk, and holes or grooves are usually provided for this purpose in the same way that pinch rollers on some tape decks have grooves to prevent tape slip.

In exchangeable-pack drives, there will be a ramp on the side of the cantilever which engages a fixed block when the heads are retracted in order to lift them away from the disk surface.

Figure 7.4 shows how disk heads are made. The magnetic circuit of disk heads was originally assembled from discrete magnetic elements. As the gap and flying height became smaller to increase linear recording density, the slipper was made from ferrite, and became part of the magnetic circuit. This was completed by a small C-shaped ferrite piece which carried the coil. Ferrite heads were restricted in the coercivity of disk they could write without saturating. In thin-film heads, the magnetic circuit and coil are both formed by deposition on a substrate which becomes the rear of the slipper.

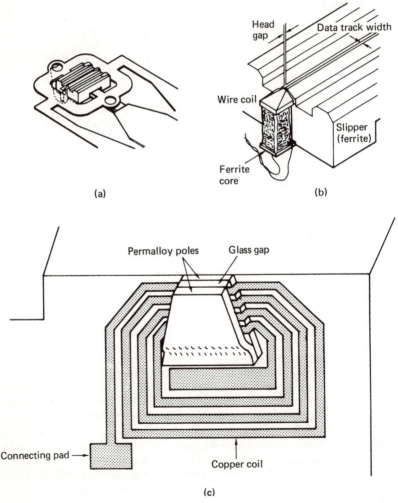

Figure 7.4 (a) Winchester head construction showing large air bleed grooves. (b) Close-up of slipper showing magnetic circuit on trailing edge. (c) Thin film head is fabricated on the end of the slipper using microcircuit technology.

In a moving-head device it is not practicable to position separate erase, record and playback heads accurately. Erase is by overwriting, and reading and writing are carried out by the same head. The presence of the air film causes severe separation loss, and peak shift distortion is a major problem. The flying height of the head varies with the radius of the disk track, and it is difficult to provide accurate equalization of the replay channel because of this. The write current is often controlled as a function of track radius so that the changing reluctance of the air gap does not change the resulting record flux. Automatic gain control (AGC) is used on replay to compensate for changes in signal amplitude from the head.

Equalization may be used on recording in the form of precompensation, which moves recorded transitions in such a way as to oppose the effects of peak shift in addition to any replay equalization used.

Early disks used FM coding, which was easy to decode, but had a poor density ratio. The invention of MFM revolutionized hard disks, and further progress led to run-length-limited codes such as 2/3 and 2/7 which had a high density ratio without sacrificing the large jitter window necessary to reject peak shift distortion. Partial response is also suited to disks.

Typical drives have several heads, but with the exception of special-purpose parallel-transfer machines, only one head will be active at any one time, which means that the read and write circuitry can be shared between the heads. The read channel usually incorporates AGC, which will be overridden by the control logic between data blocks in order to search for address marks, which are short unmodulated areas of track. As a block preamble is entered, the AGC will be enabled to allow a rapid gain adjustment.

7.3 Accessing the blocks

The servo system required to move the heads rapidly between tracks, and yet hold them in place accurately for data transfer, is a fascinating and complex piece of engineering. In exchangeable-pack drives, the disk positioner moves on a straight axis which passes through the spindle. Motive power is generally by moving-coil drive, because of the small moving mass which this technique permits.

When a drive is track-following, it is said to be detented, in fine mode or in linear mode depending on the manufacturer. When a drive is seeking from one track to another, it can be described as being in coarse mode or velocity mode. These are the two major operating modes of the servo.

Moving-coil actuators do not naturally detent and require power to stay on-track. The servo system needs positional feedback of some kind. The purpose of the feedback will be one or more of the following:

1 to count the number of cylinders crossed during a seek
2 to generate a signal proportional to carriage velocity
3 to generate a position error proportional to the distance from the centre of the desired track

Magnetic and optical drives obtain these feedback signals in different ways. Many positioners incorporate a tacho which may be a magnetic moving-coil type or its complementary equivalent, the moving-magnet type. Both generate a voltage proportional to velocity, and can give no positional information.

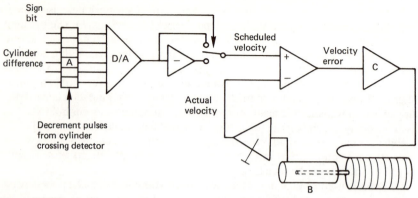

Figure 7.5 Control of carriage velocity by cylinder difference. The cylinder difference is loaded into the difference counter A. A digital-to-analog convertor generates an analog voltage from the cylinder difference, known as the scheduled velocity. This is compared with the actual velocity from the transducer B in order to generate the velocity error which drives the servo amplifier C.

A seek is a process where the positioner moves from one cylinder to another. The speed with which a seek can be completed is a major factor in determining the access time of the drive. The main parameter controlling the carriage during a seek is the cylinder difference, which is obtained by subtracting the current cylinder address from the desired cylinder address. The cylinder difference will be a signed binary number representing the number of cylinders to be crossed to reach the target, direction being indicated by the sign. The cylinder difference is loaded into a counter which is decremented each time a cylinder is crossed. The counter drives a DAC which generates an analog voltage proportional to the cylinder difference. As Figure 7.5 shows, this voltage, known as the scheduled velocity, is compared with the output of the carriage-velocity tacho. Any difference between the two results in a velocity error which drives the carriage to cancel the error. As the carriage approaches the target cylinder, the cylinder difference becomes smaller, with the result that the run-in to the target is critically damped to eliminate overshoot.

Figure 7.6(a) shows graphs of scheduled velocity, actual velocity and motor current with respect to cylinder difference during a seek. In the first half of the seek, the actual velocity is less than the scheduled velocity, causing a large velocity error which saturates the amplifier and provides maximum carriage acceleration. In the second half of the graphs, the scheduled velocity is falling below the actual velocity, generating a negative velocity error which drives a reverse current through the motor to slow the carriage down. The scheduled deceleration slope clearly cannot be steeper than the saturated acceleration slope. Areas A and B on the graph will be about equal, as the kinetic energy put into the carriage has to be taken out. The current through the motor is continuous, and would result in a heating problem, so to counter this, the DAC is made non-linear so that above a certain cylinder difference no increase in scheduled velocity will occur. This results in the graph of Figure 7.6(b). The actual velocity graph is called a velocity profile. It consists of three regions: acceleration, where the system is saturated; a constant velocity plateau, where the only power needed is to overcome friction; and the scheduled run-in to the desired cylinder. Dissipation is only significant in the first and last regions.

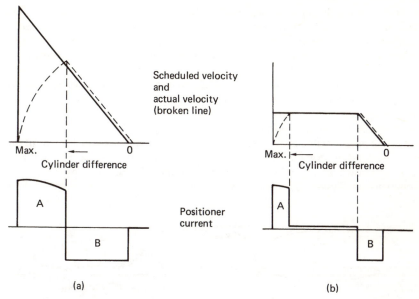

Figure 7.6 In the simple arrangement at (a) the dissipation in the positioner is continuous, causing a heating problem. The effect of limiting the scheduled velocity above a certain cylinder difference is apparent in (b) where heavy positioner current only flows during acceleration and deceleration. During the plateau of the velocity profile, only enough current to overcome friction is necessary. The curvature of the acceleration slope is due to the back EMF of the positioner motor.

The track-following accuracy of a drive positioner will be impaired if there is bearing runout, and so the spindle bearings are made to a high degree of precision.

In order to control reading and writing, the drive control circuitry needs to know which cylinder the heads are on, and which sector is currently under the head. Sector information used to be obtained from a sensor which detects holes or slots cut in the hub of the disk. Modern drives will obtain this information from the disk surface as will be seen. The result is that a sector counter in the control logic remains in step with the physical rotation of the disk. The desired sector address is loaded into a register, which is compared with the sector counter. When the two match, the desired sector has been found. This process is referred to as a search, and usually takes place after a seek. Having found the correct physical place on the disk, the next step is to read the header associated with the data block to confirm that the disk address contained there is the same as the desired address.

7.4 Servo-surface disks

One of the major problems to be overcome in the development of high-density disk drives was that of keeping the heads on-track despite changes of temperature. The very narrow tracks used in digital recording have similar dimensions to the amount a disk will expand as it warms up. The cantilevers and the drive base all expand and contract, conspiring with thermal drift in the cylinder transducer to limit track pitch. The breakthrough in disk density came with the introduction of the servo-surface drive. The position error in a servo-surface drive is derived from a head reading the disk itself. This virtually eliminates thermal effects on head positioning and allows great increases in storage density.

In a multiplatter drive, one surface of the pack holds servo information which is read by the servo head. In a ten-platter pack this means that 5 per cent of the medium area is lost, but this is unimportant since the increase in density allowed is enormous. Using one side of a single-platter cartridge for servo information would be unacceptable as it represents 50 per cent of the medium area, so in this case the servo information can be interleaved with sectors on the data surfaces. This is known as an embedded-servo technique. These two approaches are contrasted in Figure 7.7.

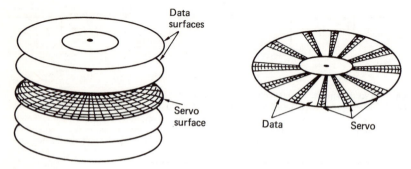

Figure 7.7 In a multiplatter disk pack, one surface is dedicated to servo information. In a single platter, the servo information is embedded in the data on the same surfaces.

The servo surface is written at the time of disk pack manufacture, and the disk drive can only read it. Writing the servo surface has nothing to do with disk formatting, which affects the data storage areas only. As there are exactly the same number of pulses on every track on the servo surface, it is possible to describe the rotational position of the disk simply by counting them. All that is needed is an unique pattern of missing pulses once per revolution to act as an index point, and the sector transducer can also be eliminated.

The advantage of deriving the sector count from the servo surface is that the number of sectors on the disk can be varied. Any number of sectors can be accommodated by feeding the pulse signal through a programmable divider, so the same disk and drive can be used in numerous different applications.

7.5 Winchester technology

In order to offer extremely high capacity per spindle, which reduces the cost per bit, a disk drive must have very narrow tracks placed close together, and must use very short recorded wavelengths, which implies that the flying height of the heads must be small. The so-called Winchester technology is one approach to high storage density. The technology was developed by IBM, and the name came about because the model number of the development drive was the same as that of the famous rifle.

Reduction in flying height magnifies the problem of providing a contaminant-free environment. A conventional disk is well protected whilst inside the drive, but outside the drive the effects of contamination become intolerable.

In exchangeable-pack drives, there is a real limit to the track pitch that can be achieved because of the difficulty or cost of engineering head-alignment mechanisms to make the necessary minute adjustments to give interchange compatibility.

The essence of Winchester technology is that each disk pack has its own set of read/write and servo heads, with an integral positioner. The whole is protected by a dust-free enclosure, and the unit is referred to as a head disk assembly, or HDA.

As the HDA contains its own heads, compatibility problems do not exist, and no head alignment is necessary or provided for. It is thus possible to reduce track pitch considerably compared with exchangeable pack drives. The sealed environment ensures complete cleanliness which permits a reduction in flying height without loss of reliability, and hence leads to an increased linear density. If the rotational speed is maintained, this can also result in an increase in data transfer rate. The HDA is completely sealed, but some have a small filtered port to equalize pressure.

An exchangeable-pack drive must retract the heads to facilitate pack removal. With Winchester technology this is not necessary. An area of the disk surface is reserved as a landing strip for the heads. The disk surface is lubricated, and the heads are designed to withstand landing and take-off without damage. Winchester heads have very large air-bleed grooves to allow low flying height with a much smaller downthrust from the cantilever, and so they exert less force on the disk surface during contact. When the term 'parking' is used in the context of Winchester technology, it refers to the positioning of the heads over the landing area.

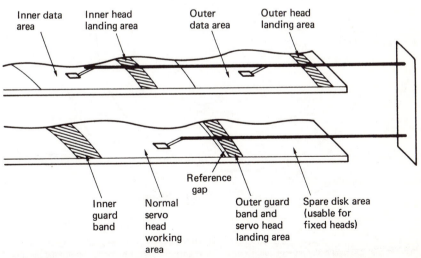

Figure 7.8 When more than one head is used per surface, the positioner still only requires one servo head. This is often arranged to be equidistant from the read/write heads for thermal stability.

Disk rotation must be started and stopped quickly to minimize the length of time the heads slide over the medium. This is conveniently achieved with a servo-controlled brushless motor which has dynamic braking ability. A major advantage of contact start/stop is that more than one head can be used on each surface if retraction is not needed. This leads to two gains: first, the travel of the positioner is reduced in proportion to the number of heads per surface, reducing access time; and, second, more data can be transferred at a given detented carriage position before a seek to the next cylinder becomes necessary. This increases the speed of long transfers. Figure 7.8 illustrates the relationships of the heads in such a system.

Figure 7.9 shows that rotary positioners are feasible in Winchester drives; they cannot be used in exchangeable-pack drives because of interchange problems. There are some advantages to a rotary positioner. It can be placed in the corner of a compact HDA allowing smaller overall size. The manufacturing cost will be less than a linear positioner because fewer bearings and precision bars are needed. Significantly, a rotary positioner can be made faster since its inertia is smaller. With a linear positioner all parts move at the same speed. In a rotary positioner, only the heads move at full speed, as the parts closer to the shaft must move more slowly. The principle of many rotary positioners is exactly that of a moving-coil ammeter, where current is converted directly into torque.

One characteristic of rotary positioners is that there is a component of windage on the heads which tends to pull the positioner in towards the spindle. Windage can be overcome in rotary positioners by feeding the current cylinder address to a ROM which sends a code to a DAC. This produces an offset voltage which is

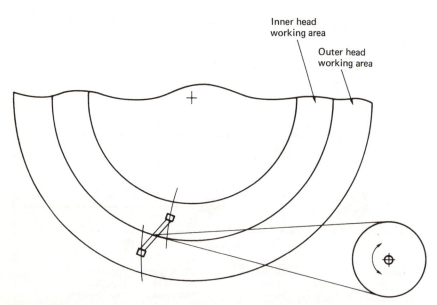

Figure 7.9 A rotary positioner with two heads per surface. The tolerances involved in the spacing between the heads and the axis of rotation mean that each arm records data in an unique position. Those data can only be read back by the same heads, which rules out the use of a rotary positioner in exchangeable-pack drives. In a head disk assembly the problem of compatibility does not arise.

fed to the positioner driver to generate a torque that balances the windage whatever the position of the heads.

When extremely small track spacing is contemplated, it cannot be assumed that all the heads will track the servo head due to temperature gradients. In this case the embedded-servo approach must be used, where each head has its own alignment patterns. The servo surface is often retained in such drives to allow coarse positioning, velocity feedback and index and write-clock generation, in addition to locating the guard bands for landing the heads.

Winchester drives have been made with massive capacity, but the problem of backup is then magnified, and the general trend has been for the physical size of the drive to come down as the storage density increases in order to improve access time and to facilitate the construction of storage arrays (see section 7.8). Very small Winchester disk drives are now available which plug into standard integrated circuit sockets. These are competing with RAM for memory applications where non-volatility is important.

7.6 The disk controller

A disk controller is a unit which is interposed between the drives and the rest of the system. It consists of two main parts; that which issues control signals to and obtains status from the drives, and that which handles the data to be stored and retrieved. Both parts are synchronized by the control sequencer. The essentials of a disk controller are determined by the characteristics of drives and the functions needed, and so they do not vary greatly. It is desirable for economic reasons to use a commercially available disk controller intended for computers. Such controllers are adequate for still store applications, but cannot support the data rate required for real-time moving video unless data reduction is employed. Disk drives are generally built to interface to a standard controller interface, such as the SCSI bus. The disk controller will then be a unit which interfaces the drive bus to the host computer system.

The execution of a function by a disk subsystem requires a complex series of steps, and decisions must be made between the steps to decide what the next will be. There is a parallel with computation, where the function is the equivalent of an instruction, and the sequencer steps needed are the equivalent of the microinstructions needed to execute the instruction. The major failing in this analogy is that the sequence in a disk drive must be accurately synchronized to the rotation of the disk.

Most disk controllers use direct memory access, which means that they have the ability to transfer disk data in and out of the associated memory without the assistance of the processor. In order to cause a file transfer, the disk controller must be told the physical disk address (cylinder, sector, track), the physical memory address where the file begins, the size of the file and the direction of transfer (read or write). The controller will then position the disk heads, address the memory, and transfer the samples. One disk transfer may consist of many contiguous disk blocks, and the controller will automatically increment the disk-address registers as each block is completed. As the disk turns, the sector address increases until the end of the track is reached. The track or head address will then be incremented and the sector address reset so that transfer continues at the beginning of the next track. This process continues until all the heads have been used in turn. In this case both the head address and sector address will be reset, and the cylinder address will be incremented, which causes a seek. A seek which takes place because of a data transfer is called an implied seek, because it is not necessary formally to instruct the system to perform it. As disk drives are block-structured devices, and the error correction is codeword-based, the controller will always complete a block even if the size of the file is less than a whole number of blocks. This is done by packing the last block with zeros.

The status system allows the controller to find out about the operation of the drive, both as a feedback mechanism for the control process, and to handle any errors. Upon completion of a function, it is the status system which interrupts the control processor to tell it that another function can be undertaken.

In a system where there are several drives connected to the controller via a common bus, it is possible for non data-transfer functions such as seeks to take place in some drives simultaneously with a data transfer in another.

Before a data transfer can take place, the selected drive must physically access the desired block, and confirm this by reading the block header. Following a seek to the required cylinder, the positioner will confirm that the heads are on track and settled. The desired head will be selected, and then a search for the correct sector begins. This is done by comparing the desired sector with the current sector register, which is typically incremented by dividing down servo-surface pulses. When the two counts are equal, the head is about to enter the desired block. Figure 7.10 shows the structure of a typical magnetic disk track. In between blocks are placed address marks, which are areas without transitions which the read circuits can detect. Following detection of the address mark, the sequencer is roughly synchronized to begin handling the block. As the block is entered, the data separator locks to the preamble, and in due course the sync pattern will be found. This sets to zero a counter which divides the data-bit rate by eight, allowing the serial recording to be correctly assembled into bytes, and also allowing the sequencer to count the position of the head through the block in order to perform all the necessary steps at the right time.

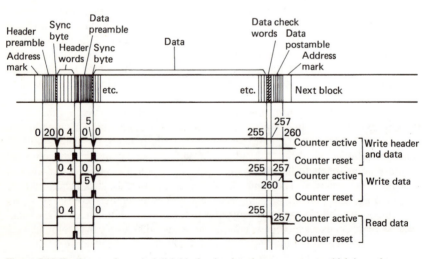

Figure 7.10 The format of a typical disk block related to the count process which is used to establish where in the block the head is at any time. During a read the count is derived from the actual data read, but during a write, the count is derived from the write clock.

The first header word is usually the cylinder address, and this is compared with the contents of the desired cylinder register. The second header word will contain the sector and track address of the block, and these will also be compared with the desired addresses. There may also be bad-block flags and/or defect-skipping information. At the end of the header is a CRCC which will be used to ensure that the header was read correctly. Figure 7.11 shows a flowchart of the position verification, after which a data transfer can proceed. The header reading is completely automatic. The only time it is necessary formally to command a header to be read is when checking that a disk has been formatted correctly.

During the read of a data block, the sequencer is employed again. The sync pattern at the beginning of the data is detected as before, following which the actual data arrive. These bits are converted to byte or sample parallel, and sent to the memory by DMA. When the sequencer has counted the last data-byte off the track, the redundancy for the error-correction system will be following.

During a write function, the header-check function will also take place as it is perhaps even more important not to write in the wrong place on a disk. Once the header has been checked and found to be correct, the write process for the associated data block can begin. The preambles, sync pattern, data block, redundancy and postamble have all to be written contiguously. This is taken care

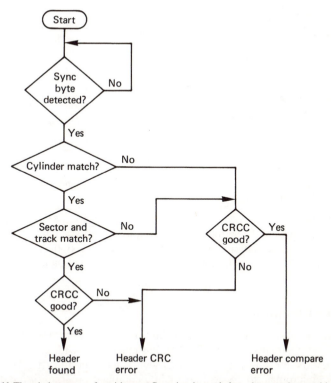

Figure 7.11 The vital process of position confirmation is carried out in accordance with the above flowchart. The appropriate words from the header are compared in turn with the contents of the disk-address registers in the subsystem. Only if the correct header has been found and read properly will the data transfer take place.

of by the sequencer, which is obtaining timing information from the servo surface to lock the block structure to the angular position of the disk. This should be contrasted with the read function, where the timing comes directly from the data.

When video samples are fed into a disk-based system, from a digital interface or from an A/D convertor, they will be placed in a buffer memory, from which the disk controller will read them by DMA. The continuous-input sample stream will be split up into disk blocks for disk storage.

The disk transfers must, by definition, be intermittent, because there are headers between contiguous sectors. Once all the sectors on a particular cylinder have been used, it will be necessary to seek to the next cylinder, which will cause a further interruption to the data transfer. If a bad block is encountered, the sequence will be interrupted until it has passed. The instantaneous data rate of a parallel transfer drive is made higher than the continuous video data rate, so that there is time for the positioner to move whilst the video output is supplied from the FIFO memory. In replay, the drive controller attempts to keep the FIFO as full as possible by issuing a read command as soon as one block space appears in the FIFO. This allows the maximum time for a seek to take place before reading must resume. Figure 7.12 shows the action of the FIFO. Whilst recording, the drive

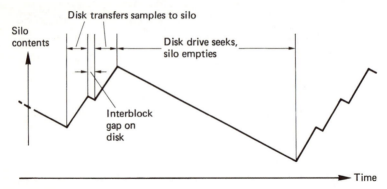

Figure 7.12 During a video replay sequence, silo is constantly emptied to provide samples, and is refilled in blocks by the drive.

controller attempts to keep the FIFO as empty as possible by issuing write commands as soon as a block of data is present. In this way the amount of time available to seek is maximized in the presence of a continuous video sample input.

7.7 Defect handling

The protection of data recorded on disks differs considerably from the approach used on other media in digital video. This has much to do with the intolerance of data processors to errors when compared with video data. In particular, it is not possible to interpolate to conceal errors in a computer program or a data file.

In the same way that magnetic tape is subject to dropouts, magnetic disks suffer from surface defects whose effect is to corrupt data. The shorter wavelengths employed as disk densities increase are affected more by a given size of defect. Attempting to make a perfect disk is subject to a law of diminishing returns, and eventually a state is reached where it becomes more cost-effective to invest in a defect-handling system.

In the construction of bad-block files, a brand-new disk is tested by the operating system. Known patterns are written everywhere on the disk, and these are read back and verified. Following this the system gives the disk a volume name, and creates on it a directory structure which keeps records of the position and size of every file subsequently written. The physical disk address of every block which fails to verify is allocated to a file which has an entry in the disk directory. In this way, when genuine data files come to be written, the bad blocks appear to the system to be in use storing a fictitious file, and no attempt will be made to write there. Some disks have dedicated tracks where defect information can be written during manufacture or by subsequent verification programs, and these permit a speedy construction of the system bad-block file.

7.8 RAID arrays

Whilst the MTBF of a disk drive is very high, it is a simple matter of statistics that when a large number of drives is assembled in a system the time between failures becomes shorter. Disk drives are sealed units and the disks cannot be removed if there is an electronic failure. Even if this were possible the system cannot usually afford downtime whilst such a data recovery takes place.

Consequently any system in which the data are valuable must take steps to ensure data integrity. This is commonly done using RAID (redundant array of inexpensive disks) technology. Figure 7.13 shows that in a RAID array data blocks are spread across a number of drives.

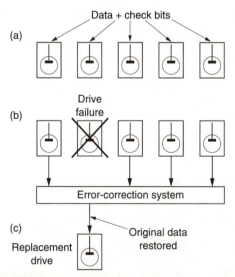

Figure 7.13 In RAID technology, data and redundancy are spread over a number of drives (a). In the case of a drive failure (b) the error-correction system can correct for the loss and continue operation. When the drive is replaced (c) the data can be rewritten so that the system can then survive a further failure.

An error-correcting check symbol (typically Reed–Solomon) is stored on a redundant drive. The error-correction is powerful enough to fully correct any error in the block due to a single failed drive. In RAID arrays the drives are designed to be hot-plugged (replaced without removing power) so if a drive fails it is simply physically replaced with a new one. The error-correction system will rewrite the drive with the data which was lost with the failed unit.

When a large number of disk drives are arrayed together, it is necessary and desirable to spread files across all the drives in a RAID array. Whilst this ensures data integrity, it also means that the data transfer rate is multiplied by the number of drives sharing the data. This means that the data transfer rate can be extremely high and new approaches are necessary to move the data in and out of the disk system.

7.9 Disk servers

The disk controller will automatically divide files up into blocks of the appropriate size for recording. If any partial blocks are left over, these will be zero stuffed. Consequently disk stores are not constrained to files of a particular size. Unlike a DVTR which always stores the same amount of data per field, a disk system can store a different amount of data for each field if needs be.

This means that disks are not standards dependent. A disk system can mix 4:4:4, 4:2:2 and 4:2:0 files and it doesn't care whether the video is interlaced or not or compressed or not. It can mix 525- and 625-line files and it can mix 4:3 and 16:9 aspect ratios. This an advantage in news systems where compression is used. If a given compression scheme is used at the time of recording e.g. DVCPRO, the video can remain in the compressed data domain when it is loaded onto the disk system for editing. This avoids concatenation of codecs which is generally bad news in compressed systems.

One of the happy consequences of the move to disk drives in production is that the actual picture format used need no longer be fixed. With computer graphics and broadcast video visibly merging, interlace may well be doomed. In the near future it will be possible to use non-interlaced HD cameras, and downconvert to a non-interlaced intermediate-resolution production format.

As production units such as mixers, character generators, paint systems and DVEs become increasingly software driven, such a format is much easier to adopt than in the days of analog where the functionality was frozen into the circuitry. Following production the intermediate format can be converted to any present or future emission standard.

7.10 Optical disk principles

In order to record MO disks or replay any optical disk, a source of monochromatic light is required. The light source must have low noise otherwise the variations in intensity due to the noise of the source will mask the variations due to reading the disk. The requirement for a low-noise monochromatic light source is economically met using a semiconductor laser.

In the LED, the light produced is incoherent or noisy. In the laser, the ends of the semiconductor are optically flat mirrors, which produce an optically resonant cavity. One photon can bounce to and fro, exciting others in synchronism, to produce coherent light. This is known as Light Amplification by Stimulated

Emission of Radiation, mercifully abbreviated to LASER, and can result in a runaway condition, where all available energy is used up in one flash. In injection lasers, an equilibrium is reached between energy input and light output, allowing continuous operation with a clean output. The equilibrium is delicate, and such devices are usually fed from a current source. To avoid runaway when temperature change disturbs the equilibrium, a photosensor is often fed back to the current source. Such lasers have a finite life, and become steadily less efficient. The feedback will maintain output, and it is possible to anticipate the failure of the laser by monitoring the drive voltage needed to give the correct output.

Many rerecordable or eraseable optical disks rely on magneto-optics. The storage medium is magnetic, but the writing mechanism is the heat produced by light from a laser; hence the term 'thermomagneto-optics'. The advantage of this writing mechanism is that there is no physical contact between the writing head and the medium. The distance can be several millimetres, some of which is taken up with a protective layer to prevent corrosion. Originally, this layer was glass, but engineering plastics have now taken over.

The laser beam will supply a relatively high power for writing, since it is supplying heat energy. For reading, the laser power is reduced, such that it cannot heat the medium past the Curie temperature, and it is left on continuously.

Whatever the type of disk being read, it must be illuminated by the laser beam. Some of the light reflected back from the disk re-enters the aperture of the objective lens. The pickup must be capable of separating the reflected light from the incident light. When playing prerecorded disks such as CDs or DVDs, the phase contrast readout process results in a variation of intensity of the light returning to the pickup. When playing MO disks, the intensity does not change, but the magnetic recording on the disk rotates the plane of polarization one way or the other depending on the direction of the vertical magnetization. Figure 7.14(a) shows that a polarizing prism is required to linearly polarize the light from the laser on its way to the disk. Light returning from the disk has had its plane of polarization rotated by approximately 1 degree. This is an extremely small rotation. Figure 7.14(b) shows that the returning rotated light can be considered to be composed of two orthogonal components. R_x is the component which is in the same plane as the illumination and is called the ordinary component and R_y in the component due to the Kerr effect rotation and is known as the magneto-optic component. A polarizing beam splitter mounted squarely would reflect the magneto-optic component R_y very well because it is at right-angles to the transmission plane of the prism, but the ordinary component would pass straight on in the direction of the laser. By rotating the prism slightly a small amount of the ordinary component is also reflected. Figure 7.14(c) shows that when combined with the magneto-optic component, the angle of rotation has increased. Detecting this rotation requires a further polarizing prism or analyser as shown. The prism is twisted such that the transmission plane is at 45° to the planes of R_x and R_y. Thus with an unmagnetized disk, half of the light is transmitted by the prism and half is reflected. If the magnetic field of the disk turns the plane of polarization towards the transmission plane of the prism, more light is transmitted and less is reflected. Conversely if the plane of polarization is rotated away from the transmission plane, less light is transmitted and more is reflected. If two sensors are used, one for transmitted light and one for reflected light, the difference between the two sensor outputs will be a waveform

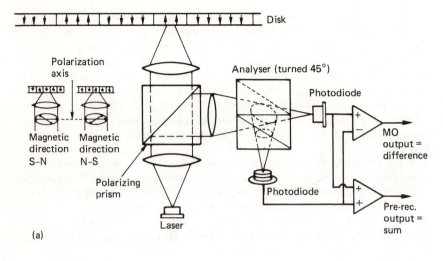

(a)

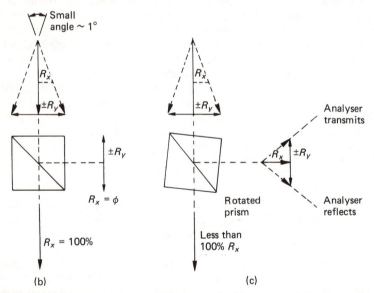

(b)   (c)

Figure 7.14 A pickup suitable for the replay of magneto-optic disks must respond to very small rotations of the plane of polarization.

representing the angle of polarization and thus the recording on the disk. This differential analyser eliminates common-mode noise in the reflected beam.

High-density recording implies short wavelengths. Using a laser focused on the disk from a distance allows short-wavelength recordings to be played back without physical contact, whereas conventional magnetic recording requires intimate contact and implies a wear mechanism, the need for periodic cleaning, and susceptibility to contamination.

The information layer is read through the thickness of the disk; this approach causes the readout beam to enter and leave the disk surface through the largest possible area. Despite the minute spot size of about 1 micrometre diameter, light enters and leaves through a 1 mm diameter circle. As a result, surface debris has to be three orders of magnitude larger than the readout spot before the beam is obscured. This approach has the further advantage in MO drives that the magnetic head, on the opposite side to the laser pickup, is then closer to the magnetic layer in the disk.

7.11 Focus and tracking systems

The frequency response of the laser pickup and the amount of crosstalk are both a function of the spot size and care must be taken to keep the beam focused on the information layer. If the spot on the disk becomes too large, it will be unable to discern the smaller features of the track, and can also be affected by the adjacent track. Disk warp and thickness irregularities will cause focal-plane movement beyond the depth of focus of the optical system, and a focus servo system will be needed. The depth of field is related to the numerical aperture, which is defined, and the accuracy of the servo must be sufficient to keep the focal plane within that depth, which is typically ±1 mm.

The track pitch of a typical optical disk is of the order of a micrometre, and this is much smaller than the accuracy to which the player chuck or the disk centre hole can be made; on a typical player, runout will swing several tracks past a fixed pickup. The non-contact readout means that there is no inherent mechanical guidance of the pickup and a suitable servo system must be provided.

The focus servo moves a lens along the optical axis in order to keep the spot in focus. Since dynamic focus-changes are largely due to warps, the focus system must have a frequency response in excess of the rotational speed. A moving-coil actuator is often used owing to the small moving mass which this permits. Figure 7.15 shows that a cylindrical magnet assembly almost identical to that of a loudspeaker can be used, coaxial with the light beam. Alternatively a moving magnet design can be used. A rare-earth magnet allows a sufficiently strong magnetic field without excessive weight.

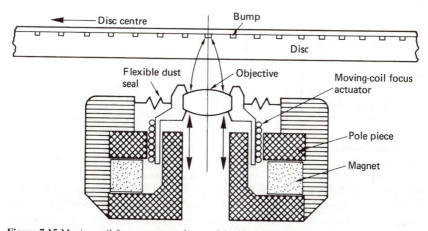

Figure 7.15 Moving-coil-focus servo can be coaxial with the light beam as shown.

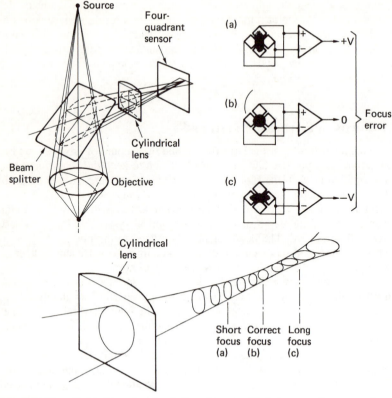

Figure 7.16 The cylindrical lens focus method produces an elliptical spot on the sensor whose aspect ratio is detected by a four-quadrant sensor to produce a focus error.

A focus-error system is necessary to drive the lens. There are a number of ways in which this can be derived, the most common of which will be described here.

In Figure 7.16 a cylindrical lens is installed between the beam splitter and the photosensor. The effect of this lens is that the beam has no focal point on the sensor. In one plane, the cylindrical lens appears parallel-sided, and has negligible effect on the focal length of the main system, whereas in the other plane, the lens shortens the focal length. The image will be an ellipse whose aspect ratio changes as a function of the state of focus. Between the two foci, the image will be circular. The aspect ratio of the ellipse, and hence the focus error, can be found by dividing the sensor into quadrants. When these are connected as shown, the focus-error signal is generated. The data readout signal is the sum of the quadrant outputs.

Figure 7.17 shows the knife-edge method of determining focus. A split sensor is also required. At (a) the focal point is coincident with the knife-edge, so it has little effect on the beam. At (b) the focal point is to the right of the knife-edge, and rising rays are interrupted, reducing the output of the upper sensor. At (c) the focal point is to the left of the knife-edge, and descending rays are interrupted, reducing the output of the lower sensor. The focus error is derived by comparing

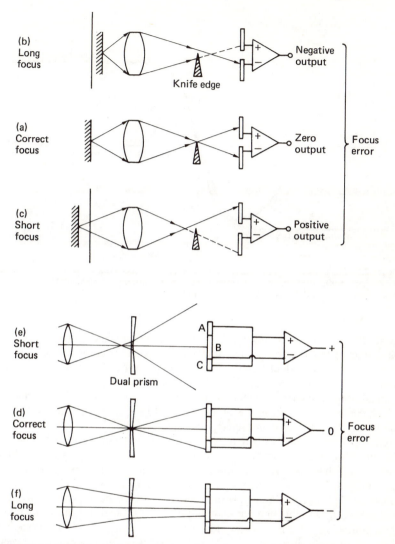

Figure 7.17 (a)-(c) Knife-edge focus-method requires only two sensors, but is critically dependent on knife-edge position. (d)-(f) Twin-prism method requires three sensors (A, B, C), where focus error is (A + C) – B. Prism alignment reduces sensitivity without causing focus error.

the outputs of the two halves of the sensor. A drawback of the knife-edge system is that the lateral position of the knife-edge is critical, and adjustment is necessary. To overcome this problem, the knife edge can be replaced by a pair of prisms, as shown in Figure 7.17(d)-(f). Mechanical tolerances then only affect the sensitivity, without causing a focus offset.

The cylindrical lens method is compared with the knife-edge/prism method in Figure 7.18, which shows that the cylindrical lens method has a much smaller capture range. A focus-search mechanism will be required, which moves the focus servo over its entire travel, looking for a zero crossing. At this time the

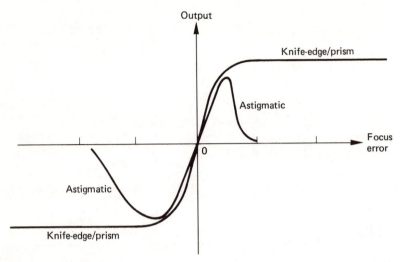

Figure 7.18 Comparison of capture range of knife-edge/prism method and astigmatic (cylindrical lens) system. Knife edge may have range of 1 mm, whereas astigmatic may only have a range of 40 micrometres, requiring a focus-search mechanism.

feedback loop will be completed, and the sensor will remain on the linear part of its characteristic. The spiral track of CD and DVD starts at the inside and works outwards. This was deliberately arranged because there is less vertical runout near the hub, and initial focusing will be easier.

In addition to the track runout mentioned above, there are further mechanisms which cause tracking error. A warped disk will not present its surface at 90° to the beam, but will constantly change the angle of incidence during two whole cycles per revolution. Owing to the change of refractive index at the disk surface, the tilt will change the apparent position of the track to the pickup, and Figure 7.19 shows that this makes it appear wavy. Warp also results in coma of the readout spot. The disk format specifies a maximum warp amplitude to keep these effects under control. Finally, vibrations induced in the player from outside, particularly in portable and automotive players, will tend to disturb tracking. A track-following servo is necessary to keep the spot centralized on the track in the presence of these difficulties. There are several ways in which a tracking error can be derived.

In the three-spot method, two additional light beams are focused on the disk track, one offset to each side of the track centre-line. Figure 7.20 shows that, as one side spot moves away from the track into the mirror area, there is less destructive interference and more reflection. This causes the average amplitude of the side spots to change differentially with tracking error. The laser head contains a diffraction grating which produces the side spots, and two extra photosensors onto which the reflections of the side spots will fall. The side spots feed a differential amplifier, which has a low-pass filter to reject the channel-code information and retain the average brightness difference. Some players use a delay line in one of the side-spot signals whose period is equal to the time taken for the disk to travel between the side spots. This helps the differential amplifier to cancel the channel code.

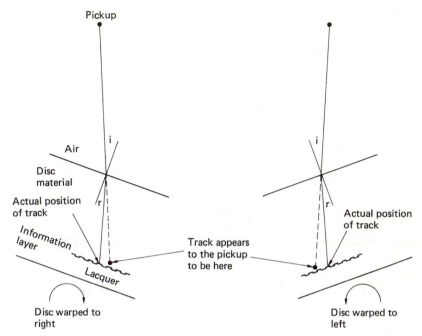

Pickup

i

Air

Disc material

Actual position of track

r

Information layer

Lacquer

Track appears to the pickup to be here

Disc warped to right

i

r

Actual position of track

Disc warped to left

Figure 7.19 Owing to refraction, the angle of incidence (*i*) is greater than the angle of refraction (*r*). Disk warp causes the apparent position of the track (dotted line) to move, requiring the tracking servo to correct.

The side spots are generated as follows. When a wavefront reaches an aperture which is small compared to the wavelength, the aperture acts as a point source, and the process of diffraction can be observed as a spherical wavefront leaving the aperture as in Figure 7.21. Where the wavefront passes through a regular structure, known as a diffraction grating, light on the far side will form new wavefronts wherever radiation is in phase, and Figure 7.22 shows that these will be at an angle to the normal depending on the spacing of the structure and the wavelength of the light. A diffraction grating illuminated by white light will produce a dispersed spectrum at each side of the normal. To obtain a fixed angle of diffraction, monochromatic light is necessary.

The alternative approach to tracking-error detection is to analyse the diffraction pattern of the reflected beam. The effect of an off-centre spot is to rotate the radial diffraction pattern about an axis along the track. Figure 7.23 shows that, if a split sensor is used, one half will see greater modulation than the other when off-track. Such a system may be prone to develop an offset due either to drift or to contamination of the optics, although the capture range is large. A further tracking mechanism is often added to obviate the need for periodic adjustment. Figure 7.24 shows that in this dither-based system, a sinusoidal drive is fed to the tracking servo, causing a radial oscillation of spot position of about ±50 nm. This results in modulation of the envelope of the readout signal, which can be synchronously detected to obtain the sense of the error. The dither can be produced by vibrating a mirror in the light path, which enables a high frequency to be used, or by oscillating the whole pickup at a lower frequency.

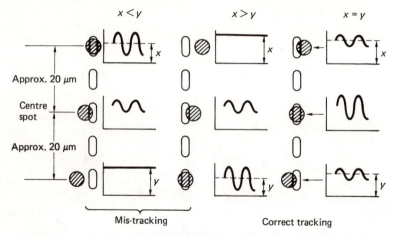

Figure 7.20 Three-spot method of producing tracking error compares average level of side-spot signals. Side spots are produced by a diffraction grating and require their own sensors.

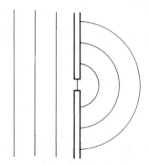

Figure 7.21 Diffraction as a plane wave reaches a small aperture.

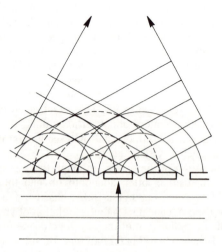

Figure 7.22 In a diffraction grating, constructive interference can take place at more than one angle for a single wavelength.

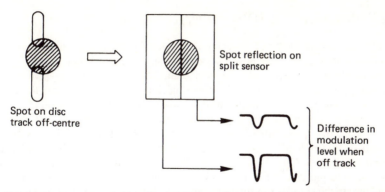

Figure 7.23 Split-sensor method of producing tracking error focuses image of spot onto sensor. One side of spot will have more modulation when off track.

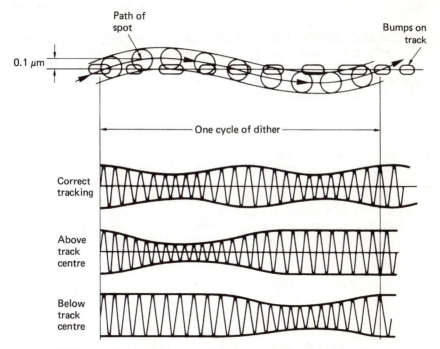

Figure 7.24 Dither applied to readout spot modulates the readout envelope. A tracking error can be derived.

In prerecorded disks there is obviously a track to follow, but in recordable disks provision has to be made for track following during the first recording of a blank disk. This is typically done by pressing the tracks in the form of continuous grooves. The grooves may be produced with a lateral wobble so that the wobble frequency can be used to measure the speed of the track during recording.

7.12 Structure of a DVD player

Figure 7.25 shows the block diagram of a typical DVD player, and illustrates the essential components. The most natural division within the block diagram is into the control/servo system and the data path. The control system provides the interface between the user and the servo mechanisms, and performs the logical interlocking required for safety and the correct sequence of operation.

The servo systems include any power-operated loading drawer and chucking mechanism, the spindle-drive servo, and the focus and tracking servos already described. Power loading is usually implemented on players where the disk is placed in a drawer. Once the drawer has been pulled into the machine, the disk is lowered onto the drive spindle, and clamped at the centre, a process known as chucking. In the simpler top-loading machines, the disk is placed on the spindle by hand, and the clamp is attached to the lid so that it operates as the lid is closed.

The lid or drawer mechanisms have a safety switch which prevents the laser operating if the machine is open. This is to ensure that there can be no conceivable hazard to the user. In actuality there is very little hazard in a DVD pickup. This is because the beam is focused a few millimetres away from the objective lens, and beyond the focal point the beam diverges and the intensity falls rapidly. It is almost impossible to position the eye at the focal point when the pickup is mounted in the player, but it would be foolhardy to attempt to disprove this.

The data path consists of the data separator, the de-interleaving and error-correction process followed by a RAM buffer which supplies the MPEG decoder. The data separator converts the EFMplus readout waveform into data. Following data separation the error-correction and de-interleave processes take place. Because of the interleave system, there are two opportunities for correction, first, using the inner code prior to de-interleaving, and second, using the outer code after de-interleaving. In Chapter 6 it was shown that interleaving is designed to spread the effects of burst errors among many different codewords, so that the errors in each are reduced. However, the process can be impaired if a small random error, due perhaps to an imperfection in manufacture, occurs close to a burst error caused by surface contamination. The function of the inner

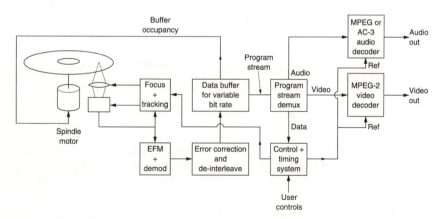

Figure 7.25 A DVD player's essential parts. See text for details.

redundancy is to correct single-symbol errors, so that the power of interleaving to handle bursts is undiminished, and to generate error flags for the outer system when a gross error is encountered.

The EFMplus coding is a group code which means that a small defect which changes one channel pattern into another could have corrupted up to eight data bits. In the worst case, if the small defect is on the boundary between two channel patterns, two successive bytes could be corrupted. However, the final odd/even interleave on encoding ensures that the two bytes damaged will be in different inner codewords; thus a random error can never corrupt two bytes in one inner codeword, and random errors are therefore always correctable.

The de-interleave process is achieved by writing sequentially into a memory and reading out using a sequencer. The outer decoder will then correct any burst errors in the data. As MPEG data are very sensitive to error the error-correction performance has to be extremely good.

Following the de-interleave and outer error-correction process an MPEG program stream (see Chapter 5) emerges. Some of the program stream data will be video, some will be audio and this will be routed to the appropriate decoder. It is a fundamental concept of DVD that the bit rate of this program stream is not fixed, but can vary with the difficulty of the program material in order to maintain consistent image quality. The bit rate is changed by changing the speed of the disk. However, there is a complication because the disk uses constant linear velocity rather than constant angular velocity. It is not possible to obtain a particular bit rate with a fixed spindle speed.

The solution is to use a RAM buffer between the transport and the MPEG decoders. The RAM is addressed by counters which are arranged to overflow, giving the memory a ring structure as described in Section 1.7. Writing into the memory is done using clocks derived from the disk whose frequency rises and falls with runout, whereas reading is done by the decoder which, for each picture, will take as much data as is required from the buffer.

The buffer will only function properly if the two addresses are kept apart. This implies that the amount of data read from the disk over the long term must equal the amount of data used by the MPEG decoders. This is done by analysing the address relationship of the buffer. If the disk is turning too fast, the write address will move towards the read address; if the disk is turning too slowly, the write address moves away from the read address. Subtraction of the two addresses produces an error signal which can be fed to the spindle motor.

The speed of the motor is unimportant. The important factor is that the data rate needed by the decoder is correct, and the system will drive the spindle at whatever speed is necessary so that the buffer neither underflows nor overflows.

The MPEG decoder will convert the compressed elementary streams into PCM video and audio and place the pictures and audio blocks into RAM. These will be read out of RAM whenever the time stamps recorded with each picture or audio block match the state of a time stamp counter. If bidirectional coding is used, the RAM readout sequence will convert the recorded picture sequence back to the real-time sequence. The time stamp counter is derived from a crystal oscillator in the player which is divided down to provide the 90 kHz time stamp clock.

As a result, the frame rate at which the disk was mastered will be replicated as the pictures are read from RAM. Once a picture buffer is read out, this will trigger the decoder to decode another picture. It will read data from the buffer until this has been completed and thus indirectly influence the disk speed.

Owing to the use of constant linear velocity, the disk speed will be wrong if the pickup is suddenly made to jump to a different radius using manual search controls. This may force the data separator out of lock, or cause a buffer overflow and the decoder may freeze briefly until this has been remedied.

The control system of a DVD player is inevitably microprocessor-based, and as such does not differ greatly in hardware terms from any other microprocessor-controlled device. Operator controls will simply interface to processor input ports and the various servo systems will be enabled or overridden by output ports. Software, or more correctly firmware, connects the two. The necessary controls are Play and Eject, with the addition in most players of at least Pause and some buttons which allow rapid skipping through the program material.

Although machines vary in detail, the flowchart of Figure 7.26 shows the logic flow of a simple player, from Start being pressed to pictures and sound emerging. At the beginning, the emphasis is on bringing the various servos into operation. Towards the end, the disk subcode is read in order to locate the beginning of the first section of the program material.

When track-following, the tracking-error feedback loop is closed, but for track crossing, in order to locate a piece of action, the loop is opened, and a microprocessor signal forces the laser head to move. The tracking error becomes an approximate sinusoid as tracks are crossed. The cycles of tracking error can be counted as feedback to determine when the correct number of tracks have been crossed. The 'mirror' signal obtained when the readout spot is half a track away from target is used to brake pickup motion and re-enable the track-following feedback.

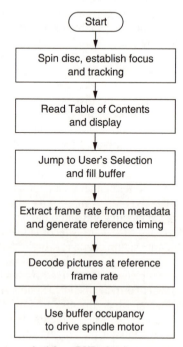

Figure 7.26 Simple processes required for a DVD player to operate.

7.13 Non-linear video editing

Non-linear editing takes advantage of the freedom to store digitized image data in any suitable medium and the signal-processing techniques developed in computation. The images may have originated on film or video. Recently images which have been synthesized by computer have been added. Although aesthetically film and video have traditionally had little in common, from a purely technological standpoint many of the necessary processes are similar.

In all types of editing the goal is the appropriate sequence of material at the appropriate time. In an ideal world the difficulty and cost involved in creating the perfect edited work are discounted. In practice there is economic pressure to speed up the editing process and to use cheaper media. Editors will not accept new technologies if they form an obstacle to the creative process, but if a new approach to editing takes nothing away, it will be considered. If something is added, such as freedom or flexibility, so much the better.

When there was only film or video tape editing, it did not need a qualifying name. Now that images are stored as data, alternative storage media have become available which allow editors to reach the same goal but using different techniques. Whilst digital VTR formats copy their analog predecessors and support field-accurate editing on the tape itself, in all other digital editing samples from various sources are brought from the storage media to various pages of RAM. The edit is viewed by selectively processing two (or more) sample streams retrieved from RAM. Thus the nature of the storage medium does not affect the form of the edit in any way except the amount of time needed to execute it.

Tapes only allow serial access to data, whereas disks and RAM allow random access and so can be much faster. Editing using random access storage devices is very powerful as the shuttling of tape reels is avoided. The technique is sometimes called non-linear editing. This is not a very helpful name, as in these systems the editing itself is performed in RAM in the same way as before. In fact it is only the time axis of the storage medium which is non-linear.

7.14 The structure of a workstation

Figure 7.27 shows the general arrangement of a hard disk-based workstation. The VDU in such devices has a screen which is a montage of many different signals, each of which appear in windows. In addition to the video windows there will be a number of alphanumeric and graphic display areas required by the control system. There will also be a cursor which can be positioned by a trackball or mouse. The screen is refreshed by a framestore which is read at the screen refresh rate. The framestore can be simultaneously written by various processes to produce a windowed image. In addition to the VDU, there may be a second screen which reproduces full-size images for preview purposes.

A master timing generator provides reference signals to synchronize the internal processes. This also produces an external reference to which source devices such as VTRs can lock. The timing generator may free-run in a stand-alone system, or genlock to station reference to allow playout to air.

Digital inputs and outputs are provided, along with optional convertors to allow working in an analog environment. In many workstations, compression is employed, and the appropriate coding and decoding logic will be required

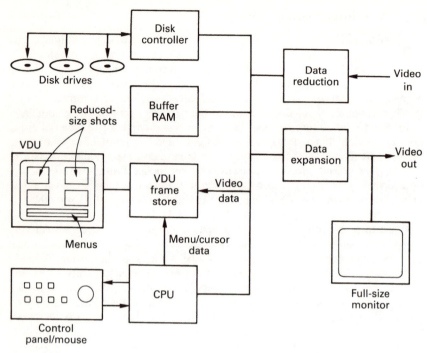

Figure 7.27 A hard-disk-based workstation. Note the screen which can display numerous clips at the same time.

adjacent to the inputs and outputs. With mild compresssion, the video output of the machine may be used directly for some purposes. This is known as on-line editing. Alternatively a high compression factor may be used, and the editor is then used only to create an edit decision list (EDL). This is known as off-line editing. The EDL is then used to control automatic editing of the full bandwidth source material, probably on tape.

Disk-based workstations fall into two categories depending on the relative emphasis of the vertical or horizontal aspects of the process. High-end post-production emphasizes the vertical aspect of the editing as a large number of layers may be used to create the output image. The length of such productions is generally quite short and so disk capacity is not an issue and data reduction will not be employed. In contrast, a general-purpose editor used for television program or film production will emphasize the horizontal aspect of the task. Extended recording ability will be needed, and data reduction is more likely.

The machine will be based around a high data rate bus, connecting the I/O, RAM, disk subsystem and the processor. If magnetic disks are used, these will be Winchester types, because they offer the largest capacity. Exchangeable magneto-optic disks may also be supported.

Before any editing can be performed, it is necessary to have source material on-line. If the source material exists on MO disks with the appropriate file structure, these may be used directly. Otherwise it will be necessary to input the material in real time and record it on magnetic disks via the compression system. In addition to recording the compressed source video, reduced versions of each field may also be recorded which are suitable for the screen windows.

Inputting the image data from film rushes requires telecine to disk transfer. Inputting from video tape requires dubbing. Both are time-consuming processes.

Time can be saved by involving the disk system at an early stage. In video systems, the disk system can record camera video and timecode alongside the VTRs. Editing can then begin as soon as shooting finishes. In film work, it is possible to use video-assisted cameras where a video camera runs from the film camera viewfinder. During filming, the video is recorded on disk and both record the same timecode. Once more, editing can begin as soon as shooting is finished.

7.15 Locating the edit point

Digital editors must simulate the 'rock and roll' process of edit-point location in VTRs or flatbeds where the tape or film is moved to and fro by the action of a shuttle knob, jog wheel or joystick. Whilst DVTRs with track-following systems can work in this way, disks cannot. Disk drives transfer data intermittently and not necessarily in real time. The solution is to transfer the recording in the area of the edit point to RAM in the editor. RAM access can take place at any speed or direction and the precise edit point can then be conveniently found by monitoring signals from the RAM. In a window-based display, a source recording is attributed to a particular window, and will be reproduced within that window, with timecode displayed adjacently.

Figure 7.28 shows how the area of the edit point is transferred to the memory. The source device is commanded to play, and the operator watches the replay in the selected window. The same samples are continuously written into a memory within the editor. This memory is addressed by a counter which repeatedly

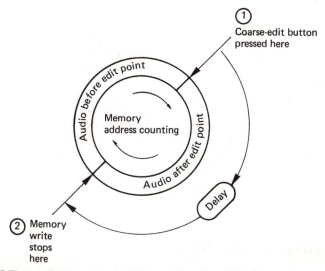

Figure 7.28 The use of a ring memory which overwrites allows storage of samples before and after the coarse edit point.

overflows to give the memory a ring-like structure rather like that of a timebase corrector, but somewhat larger. When the operator sees the rough area in which the edit is required, he will press a button. This action stops the memory writing, not immediately, but one half of the memory contents later. The effect is then that the memory contains an equal number of samples before and after the rough edit point. Once the recording is in the memory, it can be accessed at leisure, and the constraints of the source device play no further part in the edit-point location.

There are a number of ways in which the the memory can be read. If the field address in memory is supplied by a counter which is clocked at the appropriate rate, the edit area can be replayed at normal speed, or at some fraction of normal speed repeatedly. In order to simulate the analog method of finding an edit point, the operator is provided with a scrub wheel or rotor, and the memory field address will change at a rate proportional to the speed with which the rotor is turned, and in the same direction. Thus the recording can be seen forward or backward at any speed, and the effect is exactly that of turning the wheel on a flatbed or VTR.

If the position of the jog address pointer through the memory is compared with the addresses of the ends of the memory, it will be possible to anticipate that the pointer is about to reach the end of the memory. A disk transfer can be performed to fetch new data further up the time axis, so that it is possible to jog an indefinite distance along the source recording in a manner which is transparent to the user.

Samples which will be used to make the master recording need never pass through these processes; they are solely to assist in the location of the edit points.

The act of pressing the coarse edit-point button stores the timecode of the source at that point, which is frame-accurate. As the rotor is turned, the memory address is monitored, and used to update the timecode.

Before the edit can be performed, two edit points must be determined, the out-point at the end of the previously recorded signal, and the in-point at the beginning of the new signal. The second edit point can be determined by moving the cursor to a different screen window in which video from a different source is displayed. The jog wheel will now roll this material to locate the second edit point while the first source video remains frozen in the deselected window. The editor's microprocessor stores these in an EDL in order to control the automatic assemble process.

It is also possible to locate a rough edit point by typing in a previously noted timecode, and the image in the window will automatically jump to that time. In some systems, in addition to recording video and audio, there may also be text files locked to timecode which contain the dialog. Using these systems one can allocate a textual dialog display to a further window and scroll down the dialog or search for a key phrase as in a word processor. Unlike a word processor, the timecode pointer from the text access is used to jog the video window. As a result, an edit point can be located in the video if the actor's lines at the desired point are known.

7.16 Editing with disk drives

Using one or other of the above methods, an edit list can be made which contains an in-point, an out-point and a filename for each of the segments of video which

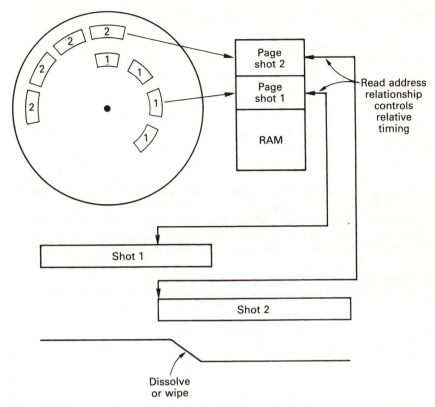

Figure 7.29 Sequence of events for a hard-disk edit. See text for details.

need to be assembled to make the final work, along with a timecode-referenced transition command and period for the vision mixer. This edit list will also be stored on the disk. When a preview of the edited work is required, the edit list is used to determine what files will be necessary and when, and this information drives the disk controller.

Figure 7.29 shows the events during an edit between two files. The edit list causes the relevant blocks from the first file to be transferred from disk to memory, and these will be read by the signal processor to produce the preview output. As the edit point approaches, the disk controller will also place blocks from the incoming file into the memory. In different areas of the memory there will be simultaneously the end of the outgoing recording and the beginning of the incoming recording. Before the edit point, only pixels from the outgoing recording are accessed, but as the transition begins, pixels from the incoming recording are also accessed, and for a time both data streams will be input to the vision mixer according to the transition period required. The output of the signal processor becomes the edited preview material, which can be checked for the required subjective effect. If necessary the in- or out-points can be trimmed, or the crossfade period changed, simply by modifying the edit-list file. The preview can be repeated as often as needed, until the desired effect is obtained. At this stage the edited work does not exist as a file, but is re-created each time by a further execution of the EDL. Thus a lengthy editing session need not fill up the disks.

It is important to realize that at no time during the edit process were the original files modified in any way. The editing was done solely by reading the files. The power of this approach is that if an edit list is created wrongly, the original recording is not damaged, and the problem can be put right simply by correcting the edit list. The advantage of a disk-based system for such work is that location of edit points, previews and reviews are all performed almost instantaneously, because of the random access of the disk. This can reduce the time taken to edit a program to a fraction of that needed with a tape machine.

During an edit, the disk controller has to provide data from two different files simultaneously, and so it has to work much harder than for a simple playback. If there are many close-spaced edits, the controller and drives may be hard-pressed to keep ahead of real time, especially if there are long transitions, because during a transition a vertical edit is taking place between two video signals and the source data rate is twice as great as during replay. A large buffer memory helps this situation because the drive can fill the memory with files before the edit actually begins, and thus the instantaneous sample rate can be met by allowing the memory to empty during disk-intensive periods.

Some drives rotate the sector addressing from one cylinder to the next so that the drive does not lose a revolution when it moves to the next cylinder. Disk-editor performance is usually specified in terms of peak editing activity which can be achieved, but with a recovery period between edits. If an unusually severe editing task is necessary where the drive just cannot access files fast enough, it will be necessary to rearrange the files on the disk surface so that files which will be needed at the same time are on nearby cylinders. An alternative is to spread the material between two or more drives so that overlapped seeks are possible.

Once the editing is finished, it will generally be necessary to transfer the edited material to form a contiguous recording so that the source files can make way for new work. In off-line editing, the source files already exist on tape or film and all that is needed is the EDL; the disk files can simply be erased. In on-line editing the disks hold original recordings and they will need to be backed up to tape if they will be required again. In large broadcast systems, the edited work can be broadcast directly from the disk file server. In smaller systems it will be necessary to output to some removable medium, since the Winchester drives in the editor have fixed media.

Chapter 8

Introduction to the digital VTR

8.1 History of DVTRs

The first production DVTR, launched in 1987, used the D-1 format which recorded colour difference data according to CCIR-601 on $\frac{3}{4}$-inch tape. D-1 was too early to take advantage of high-coercivity tapes and its recording density was quite low, leading to large cassettes and high running costs. As a result, D-1 found application only in high-end post-production suites. The D-2 format came next, but this was a composite digital format, handling conventional PAL and NTSC signals in digital form, and derived from a format developed by Ampex for an automated cart. machine. The choice of composite recording was intended to allow broadcasters directly to replace analog recorders with a digital machine. D-2 retained the cassette shell of D-1 but employed higher-coercivity tape and azimuth recording (see Chapter 6) to improve recording density and playing time. Early D-2 machines had no flying erase heads, and difficulties arose with audio edits. D-3 was designed by NHK, and put into production by Panasonic. This had twice the recording density of D-2; three times that of D-1. This permitted the use of $\frac{1}{2}$-inch tape, making a digital camcorder a possibility. D-3 used the same sampling structure as D-2 for its composite recordings. Coming later, D-3 had learned from earlier formats and had a more powerful error-correction strategy than earlier formats, particularly in the audio recording.

By this time the economics of VLSI chips had made compression in VTRs viable, and the first application was the Ampex DCT format which used approximately 2:1 compression so that component video could be recorded on an updated version of the $\frac{3}{4}$-inch cassettes and transports designed for D-2. When Sony were developing the Digital Betacam format, compatibility with the existing analog Betacam format was a priority. Digital Betacam uses the same cassette shells as the analog format, and certain models of the digital recorder can play existing analog tapes. Sony also adopted compression, but this was in order to allow the construction of a digital component VTR which offered sufficient playing time within the existing cassette dimensions.

The DV format was designed as a consumer digital device and the very small cassette allows highly miniaturized products. In the DV format relatively heavy compression is used to obtain a bit rate of 25 Mbits/s on tape. The results obtained from DV were so remarkable that professional equipment was developed from it under the name of DVCPRO. DVCPRO subsequently introduced higher bit rates of 50 and 100 Mbits/s.

The D-5 component format is backward compatible with D-3. The same cassettes are used and D-5 machines can play D-3 tapes. However, the tape speed is doubled in the component format in order to increase the bit rate to 360 Mbits/s. D-5 recorders have become very useful in the development of HDTV because with mild compression they can easily record HDTV formats. Some D-5 machines have also been made which can run at 700 Mbits/s.

The D-6 format is intended for high bit rate use in HDTV applications and uses a large number of heads in parallel. D-9 uses the compression algorithms of DVC to create a production-grade DVTR format using VHS-sized cassettes at lower cost than earlier formats.

The popularity of the DVTR has been eroded by the enormous growth of disk-based production equipment, but for archiving purposes and in portable applications the tape medium still offers advantages and will not go away readily. In the future recording technology will continue to advance and further tape formats will emerge with higher performance. These will be needed to support the emerging high-definition and electronic cinema applications.

8.2 The rotary-head tape transport

The high bit rate of digital video could be accommodated by a conventional tape deck having many parallel tracks, but each would need its own read/write electronics and the cost would be high. However, the main problem with such an approach is that the data rate is proportional to the tape speed. The provision of stunt modes such as still frame or picture in shuttle are difficult or impossible. The rotary-head recorder has the advantage that the spinning heads create a high head-to-tape speed offering a high bit rate recording with a small number of heads and without high linear tape speed. The head-to-tape speed is dominated by the rotational speed, and the linear tape speed can vary enormously without changing the frequencies produced by the head by very much. Whilst mechanically complex, the rotary-head transport has been raised to a high degree of refinement and offers the highest recording density and thus lowest cost per bit of all digital recorders.

Figure 8.1 shows that the tape is led around a rotating drum in a helix such that the entrance and exit heights are different. As a result, the rotating heads cross the tape at an angle and record a series of slanting tracks. The rotating heads turn at a speed which is locked to the video field rate so that a whole number of tracks results in each input field. Time compression can be used so that the switch from one track to the next falls within a gap between data blocks. Clearly, the slant tracks can only be played back properly if linear tape motion is controlled in some way. This is the job of the linear control track which carries a pulse corresponding to every slant track. The control track is played back in order to control the capstan. The breaking up of fields into several tracks is called 'segmentation' and it is used to keep the tracks reasonably short. The segments are invisibly reassembled in memory on replay to restore the original fields.

Figure 8.2 shows the important components of a rotary-head helical scan tape transport. There are four servo systems which must correctly interact to obtain all modes of operation: two reel servos, the drum servo and the capstan servo. The capstan and reel servos together move and tension the tape, and the drum servo moves the heads. For variable-speed operation a further servo system will be necessary to deflect the heads.

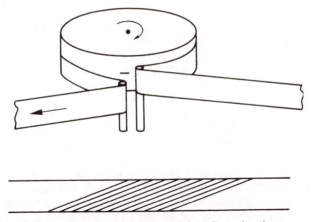

Figure 8.1 A rotary-head recorder. Medical scan records long diagonal tracks.

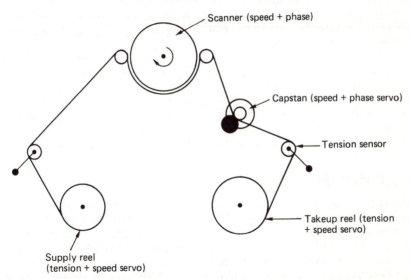

Figure 8.2 The four servos essential for proper operation of a helical-scan DVTR. Cassette-based units will also require loading and threading servos, and for variable speed a track-following servo will be necessary.

There are two approaches to capstan drive, those which use a pinch roller and those which do not. In a pinch roller drive, the tape is held against the capstan by pressure from a resilient roller which is normally pulled towards the capstan by a solenoid. The capstan only drives the tape over a narrow speed range, generally the range in which broadcastable pictures are required. Outside this range, the pinch roller retracts, the tape will be driven by reel motors alone, and the reel motors will need to change their operating mode; one bcomes a velocity servo whilst the other remains a tension servo.

In a pinch-rollerless transport, the tape is wrapped some way around a relatively large capstan, to give a good area of contact. The tape is always in contact with the capstan, irrespective of operating mode, and so the reel servos never need to change mode. A large capstan has to be used to give sufficient contact area, and to permit high shuttle speed without excessive motor rev/min. This means that at play speed it will be turning slowly, and must be accurately controlled and free from cogging. A multi-pole ironless rotor pancake-type brush motor is often used, or a sinusoidal drive brushless motor.

The simplest operating mode to consider is the first recording on a blank tape. In this mode, the capstan will rotate at constant speed, and drive the tape at the linear speed specified for the format. The drum must rotate at a precisely determined speed, so that the correct number of tracks per unit distance will be laid down on the tape. Since in a segmented recording each track will be a constant fraction of a television field, the drum speed must ultimately be determined by the incoming video signal to be recorded. To take the example of a 625-line Digital Betacam, having two record head pairs, six tracks or three segments will be necessary to record one field, and so the drum must make exactly one and a half complete revolutions in one field period, requiring it to run at 75 Hz. The phase of the drum rotation with respect to input video timing depends upon the time delay necessary to shuffle and interleave the video samples. This time will vary from a minimum of about one segment to more than a field depending on the format.

In order to obtain accurate tracking on replay, a phase comparison will be made between offtape control track pulses and pulses generated by the rotation of the drum. If the phase error between these is used to modify the capstan drive, the error can be eliminated, since the capstan drives the tape which produces the control track segment pulses. Eliminating this timing error results in the rotating heads following the tape tracks properly. Artificially delaying or advancing the reference pulses from the drum will result in a tracking adjustment.

8.3 Digital video cassettes

The D-1, D-2 and D-6 formats use the same mechanical parts and dimensions in their respective $\frac{3}{4}$-inch cassettes, even though the kind of tape and the track pattern are completely different. The Ampex DCT cassette is the same as a D-2 cassette. D-3 and D-5 use the same $\frac{1}{2}$-inch cassette, and Digital Betacam uses a $\frac{1}{2}$-inch cassette which is identical mechanically to the analog Betacam cassette, but uses different tape. The D-9 format uses a VHS cassette shell again with different tape.

The use of a cassette means that it is not as easy to provide a range of sizes as it is with open reels. Simply putting smaller reels in a cassette with the same hub spacing does not produce a significantly smaller cassette. The only solution is to specify different hub spacings for different sizes of cassette. This gives the best volumetric efficiency for storage, but it does mean that the transport must be able to reposition the reel drive motors if it is to play more than one size of cassette.

Digital Betacam offers two cassette sizes, whereas the other formats offer three. If the small, medium and large digital video cassettes are placed in a stack with their throats and tape guides in a vertical line, the centres of the hubs will be seen to fall on a pair of diagonal lines going outwards and backwards. This arrangement allows the reel motors to travel along a linear track in machines which accept more than one size. Figure 8.3 compares the sizes and capacities of the various digital cassettes.

		D-1	D-2	D-3	D-5	DCT	Digital Betacam	DVC
Track pitch (μm)		45	35	18	18	35	26	10
Tape speed (mm/s)		286.9	131.7	83.2	167.2	131.7	96.7	18.831
Play time (min)	S	14/11	32	64/50	32/25	32	40	60
	M	50/37	104	125/95	62/47	104	–	–
	L	101/75	208	245/185	123/92	208	124	270
Data rate (mbits/s)		216	142	142	288	113	126	25
Density (mbits/cm²)		4	5.8	13	13	5.8	10	40

Figure 8.3 Specifications of various digital tape formats compared.

8.4 DVTR block diagram

Figure 8.4(a) shows a representative block diagram of a full bit rate DVTR. Following the convertors will be the distribution of odd and even samples and a shuffle process for concealment purposes. An interleaved product code will be formed prior to the channel-coding stage which produces the recorded waveform. On replay the data separator decodes the channel code and the inner and outer codes perform correction as in section 6.24. Following the de-shuffle the data channels are recombined and any necessary concealment will take place.

Figure 8.4(b) shows the block diagram of a DVTR using compression. Data from the convertors is rearranged from the normal raster scan to sets of pixel blocks upon which the compression unit works. A common size is eight pixels horizontally by four vertically. The blocks are then shuffled for concealment purposes. The shuffled blocks are passed through the data reduction unit. The output of this is distributed and then assembled into product codes and channel coded as for a conventional recorder. On replay, data separation and error-correcting take place as before, but there is now a matching compression decoder which outputs pixel blocks. These are then de-shuffled prior to the error-concealment stage. As concealment is more difficult with pixel blocks, data from another field may be employed for concealment as well as data within the field.

The various DVTR formats largely employ the same processing stages, but there are considerable differences in the order in which these are applied. Distribution is shown in Figure 8.5(a). This is a process of sharing the input bit rate over two or more signal paths so that the bit rate recorded in each is reduced. The data are subsequently recombined on playback. Each signal path requires its own tape track and head. The parallel tracks which result form a *segment*.

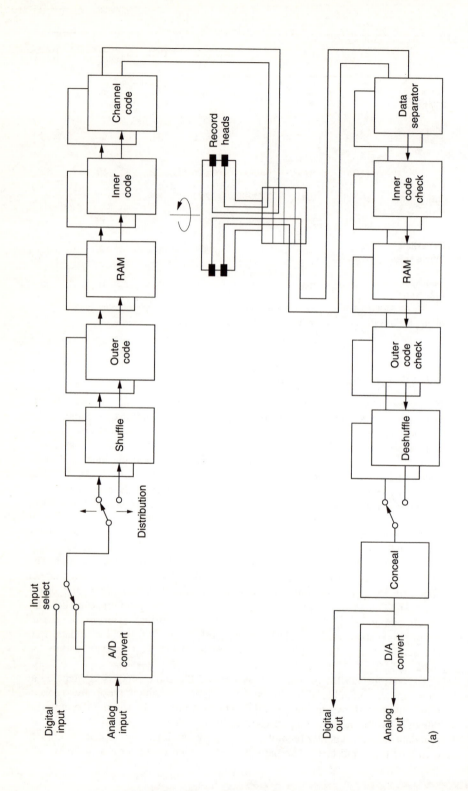

(a)

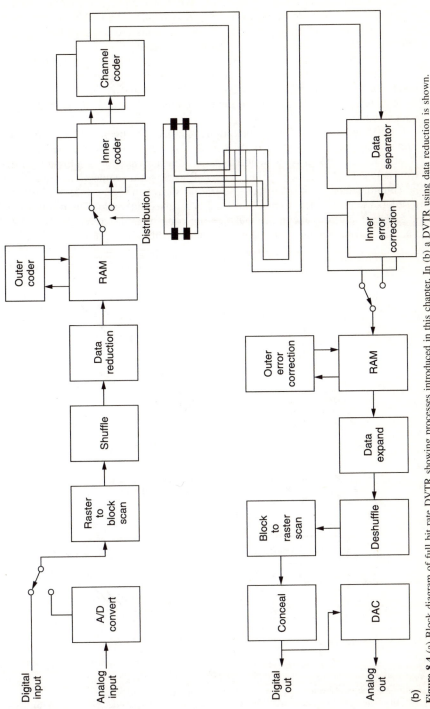

Figure 8.4 (a) Block diagram of full bit rate DVTR showing processes introduced in this chapter. In (b) a DVTR using data reduction is shown.

Segmentation is shown in Figure 8.5(b). This is the process of sharing the data resulting from one video field over several segments. The replay system must have some means to ensure that associated segments are reassembled into the original field. This is generally a function of the control track.

Figure 8.5(c) shows a product code. Data to be recorded are protected by two error-correcting codeword systems at right-angles; the inner code and the outer code (see Chapter 6). When it is working within its capacity the error-correction system returns corrupt data to their original value and its operation is undetectable.

If errors are too great for the correction system, concealment will be employed. Concealment is the *estimation* of missing data values from surviving data nearby. Nearby means data on vertical, horizontal or time axes as shown in Figure 8.5(d). Concealment relies upon distribution, as all tracks of a segment are unlikely to be simultaneously lost, and upon the *shuffle* shown in Figure 8.5(e). Shuffling reorders the pixels prior to recording and is reversed on replay. The result is that uncorrectable errors due to dropouts are not concentrated, but are spread out by the de-shuffle, making concealment easier. A different approach is required

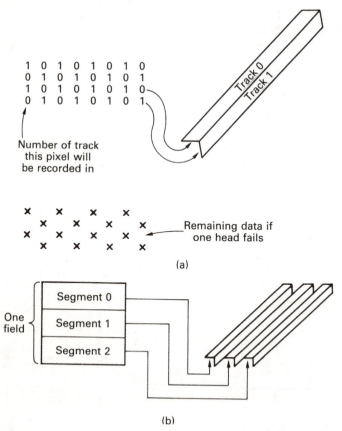

Figure 8.5 The fundamental stages of DVTR processing. At (a), distribution spreads data over more than one track to make concealment easier and to reduce the data rate per head. At (b) segmentation breaks video fields into manageable track lengths.

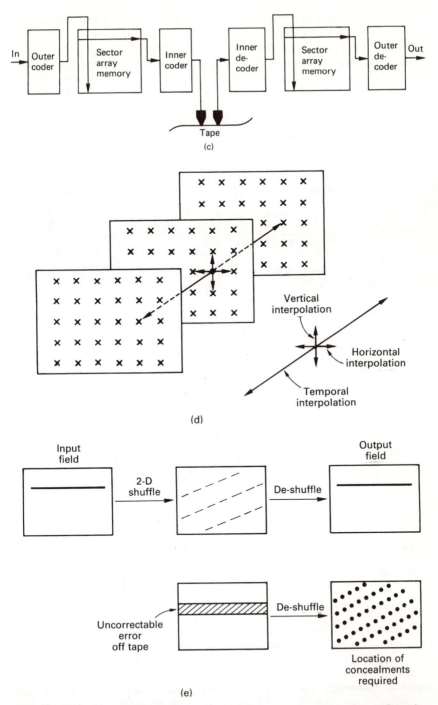

Figure 8.5 (*Continued*) Product codes (c) correct mixture of random and burst errors. Correction failure requires concealment which may be in three dimensions as shown in (d). Irregular shuffle (e) makes concealments less visible.

where compression is used because the data recorded are not pixels representing a point, but coefficients representing an area of the image. In this case it is the DCT blocks (typically eight pixels across) which must be shuffled.

There are two approaches to error correction in segmented recordings. In D-1 and D-2 the approach shown in Figure 8.6(a) is used. Here, following distribution the input field is segmented first, then each segment becomes an independent shuffled product code. This requires less RAM to implement, but it means that from an error-correction standpoint each tape track is self-contained and must deal alone with any errors encountered.

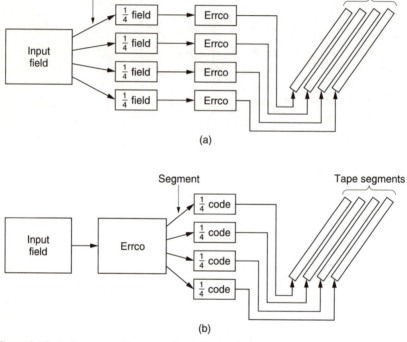

Figure 8.6 Early formats would segment data before producing product codes as in (a). Later formats perform product coding first, and then segment for recording as in (b). This gives more robust performance.

Later formats, beginning with D-3, use the approach shown in Figure 8.6(b). Here following distribution the entire field is used to produce one large shuffled product code in each channel. The product code is then segmented for recording on tape. Although more RAM is required to assemble the large product code, the result is that outer codewords on tape spread across several tracks and redundancy in one track can compensate for errors in another. The result is that size of a single burst error which can be fully corrected is increased. As the cost of RAM falls, this approach is becoming more common.

8.5 Operating modes of a DVTR

A simple recorder needs to do little more than record and play back, but the sophistication of modern production processes requires a great deal more flexibility. A production DVTR will need to support most if not all of the following:

- Offtape confidence replay must be available during recording.
- Timecode must be recorded and this must be playable at all tape speeds. Full remote control is required so that edit controllers can synchronize several machines via timecode.
- A high-quality video signal is required over a speed range of $-1 \times$ to $+3 \times$ normal speed. Audio recovery is required in order to locate edit points from dialogue.
- A picture of some kind is required over the whole shuttle speed range (typically $\pm 50 \times$ normal speed.
- Assembly and insert editing must be supported, and it must be possible to edit the audio and video channels independently.
- It must be possible slightly to change the replay speed in order to shorten or lengthen programs. Full audio and video quality must be available in this mode, known as tape speed override (TSO).
- For editing purposes, there is a requirement for the DVTR to be able to play back the tape with heads fitted in advance of the record head. The playback signal can be processed in some way and rerecorded in a single pass. This is known as preread, or read–modify–write operation.

8.6 Confidence replay

It is important to be quite certain that a recording is being made, and the only way of guaranteeing that this is so is actually to play the tape as it is being recorded. Extra heads are fitted to the revolving drum in such a way that they pass along the slant tracks directly behind the record heads. The drum must carry additional rotary transformers so that signals can simultaneously enter and leave the rotating head assembly. As can be seen in Figure 8.7, the input signal will be made

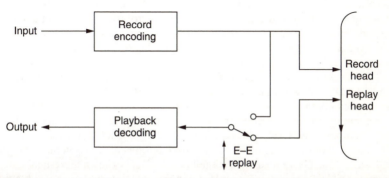

Figure 8.7 A professional DVTR uses confidence replay where the signal recorded is immediately played back. If the tape is not running, the heads are bypassed in E–E mode so that all of the circuitry can be checked.

available at the machine output if all is well. In analog machines it was traditional to assess the quality of the recording by watching a picture monitor connected to the confidence replay output during recording. With a digital machine this is not necessary, and instead the rate at which the replay channel performs error corrections should be monitored. In some machines the error rate is made available at an output socket so that remote or centralized data reliability logging can be used.

It will be seen from Figure 8.7 that when the machine is not running, a connection is made which bypasses the record and playback heads. The output signal in this mode has passed through every process in the machine except the actual tape/head system. This is known as E–E (electronics to electronics) mode, and is a good indication that the circuitry is functioning.

8.7 Colour framing

Composite video has a subcarrier added to the luminance in order to carry the colour difference signals. The frequency of this subcarrier is critical if it is to be invisible on monochrome TV sets, and as a result it does not have a whole number of cycles within a frame, but only returns to its starting phase once every two frames in NTSC and every four frames in PAL. These are known as colour-framing sequences. When playing back a composite recording, the offtape colour-frame sequence must be synchronized with the reference colour-frame sequence, otherwise composite replay signals cannot be mixed with signals from elsewhere in the facility. When editing composite recordings, the subcarrier phase must not be disturbed at the edit point and this puts constraints on the edit control process. In both cases the solution is to record the start of a colour-frame sequence in the control track of the tape. There is also a standardised algorithm linking the timecode with the colour-framing sequences.

In practice many component formats have colour framing so that they can give better results when working with decoded composite signals.

8.8 Timecode

Timecode is simply a label attached to each frame on the tape which contains the time at which it was recorded measured in hours, minutes, seconds and frames. There are two ways in which timecode data can be recorded. The first is to use a dedicated linear track, usually alongside the control track, in which there is one timecode entry for every tape frame. Such a linear track can easily be played back over a wide speed range by a stationary head. Timecode of this kind is known as LTC (linear timecode).

LTC clearly cannot be replayed when the tape is stopped or moving very slowly. In DVTRs with track-following heads, particularly those which support preread, head deflection may result in a frame being played which is not the one corresponding to the timecode from the stationary head. The player software needs to modify the LTC value as a function of the head deflection.

An alternative timecode is where the information is recorded in the video field itself, so that the above mismatch cannot occur. This is known as vertical interval timecode (VITC) because it is recorded in a line which is in the vertical blanking period. VITC has the advantage that it can be recovered with the tape stopped, but it cannot be recovered in shuttle because the rotary heads do not play entire tracks in this mode. DVTRs do not record the whole of the vertical blanking period, and if VITC is to be used, it must be inserted in a line which is within the recorded data area of the format concerned.

8.9 Picture in shuttle

A rotary-head recorder cannot follow the tape tracks properly when the tape is shuttled. Instead the heads cross the tracks at an angle and intermittently pick up short data blocks. Each of these blocks is an inner error-correcting codeword and this can be checked to see if the block was properly recovered. If this is the case, the data can be used to update a framestore which displays the shuttle picture. Clearly, the shuttle picture is a mosaic of parts of many fields. In addition to helping the concealment of errors, the shuffle process is beneficial to obtaining picture-in-shuttle. Owing to shuffle, a block recovered from the tape contains data from many places in the picture, and this gives a better result than if many pixels were available from one place in the picture. The twinkling effect seen in shuttle is due to the updating of individual pixels following de-shuffle.

When compression is used, the picture is processed in blocks, and these will be visible as mosaicing in the shuttle picture as the framestore is updated by the blocks.

In composite recorders, the subcarrier sequence is only preserved when playing at normal speed. In all other cases, extra *colour processing* is required to convert the disjointed replay signal into a continuous subcarrier once more.

8.10 Digital Betacam

Digital Betacam (DB) is a component format which accepts eight- or ten-bit 4:2:2 data with 720 luminance samples per active line and four channels of 48 kHz digital audio having up to twenty-bit wordlength. Video compression based on the discrete cosine transform is employed, with a compression factor of almost two to one (assuming eight-bit input). The audio data are uncompressed. The cassette shell of the half-inch analog Betacam format is retained, but contains 14 micrometre metal particle tape. The digital cassette contains an identification hole which allows the transport to identify the tape type. Unlike the other digital formats, only two cassette sizes are available. The large cassette offers 124 minutes of playing time; the small cassette plays for 40 minutes.

Owing to the trade-off between SNR and bandwidth which is a characteristic of digital recording, the tracks must be longer than in the analog Betacam format, but narrower. The drum diameter of the DB transport is 81.4 mm which is designed to produce tracks of the required length for digital recording. The helix angle of the digital drum is designed such that when an analog Betacam tape is driven past at the correct speed, the track angle is correct. Certain DB machines are fitted with analog heads which can trace the tracks of an analog tape. As the drum size is different, the analog playback signal is time compressed by about 9 per cent, but this is easily dealt with in the timebase-correction process.[1] The fixed heads are compatible with the analog Betacam positioning. The reverse compatibility is for playback only; the digital machine cannot record on analog cassettes.

Figure 8.8 shows the track pattern for 625/50 Betacam. The four digital audio channels are recorded in separate sectors of the slant tracks, and so one of the

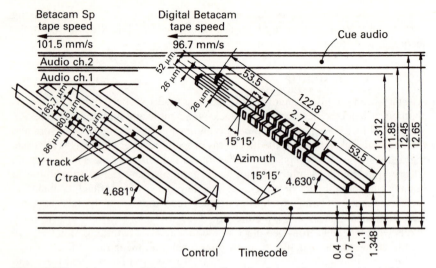

Figure 8.8 The track pattern of Digital Betacam. Control and timecode tracks are identical in location to the analog format, as is the single analog audio cue track. Note the use of a small guard band between segments.

linear audio channels of the analog format is dispensed with, leaving one linear audio track for cueing.

Azimuth recording is employed, with two tracks being recorded simultaneously by adjacent heads. Electronic delays are used to align the position of the edit gaps in the two tracks of a segment, allowing double-width flying-erase heads to be used. Three segments are needed to record one field, requiring one and a half drum revolutions. Thus the drum speed is three times that of the analog format. However, the track pitch is less than one third that of the analog format so the linear speed of the digital tape is actually slower. Track width is 24 micrometres, with a 4 micrometre guard band between segments making the effective track pitch 26 micrometres, compared with 18 for D-3/D-5 and 39 for D-2/DCT.

There is a linear timecode track whose structure is identical to the analog Betacam timecode, and a control track shown in Figure 8.9 having a fundamental frequency of 50 Hz. Ordinarily the duty cycle is 50 per cent, but this changes to 65/35 in field 1 and 35/65 in field 5. The rising edge of the CTL signal coincide with the first segment of a field, and the duty cycle variations allow four- or eight-field colour framing if decoded composite sources are used. As the drum speed is 75 Hz, CTL and drum phase coincide every three revolutions.

Figure 8.10 shows the track layout in more detail. Unlike earlier digital formats, DB incorporates tracking pilot tones recorded between the audio and video sectors. The first tone has a frequency of approximately 4 MHz and appears once per drum revolution. The second is recorded at approximately 400 kHz and appears twice per drum revolution. The pilot tones are recorded when a recording is first made on a blank tape, and will be rerecorded following an assemble edit, but during an insert edit the tracking pilots are not rerecorded, but used as a guide to the insertion of the new tracks.

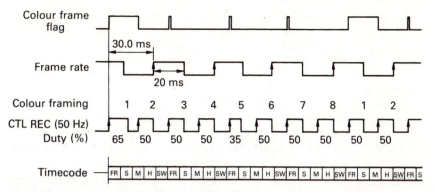

Figure 8.9 The control track of Digital Betacam uses duty cycle modulation for colour framing purposes.

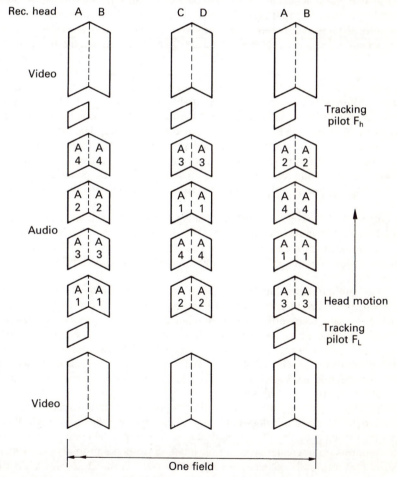

Figure 8.10 The sector structure of Digital Betacam. Note the tracking tones between the audio and video sectors which are played back for alignment purposes during insert edits.

The amplitude of the pilot signal is a function of the head tracking. The replay heads are somewhat wider than the tracks, and so a considerable tracking error will have to be present before a loss of amplitude is noted. This is partly offset by the use of a very long wavelength pilot tone in which fringeing fields increase the effective track width. The low-frequency tones are used for automatic playback tracking.

With the tape moving at normal speed, the capstan phase is changed by steps in one direction then the other as the pilot tone amplitude is monitored. The phase which results in the largest amplitude will be retained. During the edit preroll the record heads play back the high-frequency pilot tone and capstan phase is set for largest amplitude. The record heads are the same width as the tracks and a short-wavelength pilot tone is used such that any mistracking will cause immediate amplitude loss. This is an edit optimize process and results in new tracks being inserted in the same location as the originals. As the pilot tones are played back in insert editing, there will be no tolerance build-up in the case of multiple inserts.

The compression technique of DB^2 works on an intra-field basis to allow complete editing freedom and uses processes described in Chapter 5.

Figure 8.11 shows a block diagram of the record section of DB. Analog component inputs are sampled at 13.5 and 6.75 MHz. Alternatively the input may be SDI at 270 Mbits/s which is deserialized and demultiplexed to separate components. The raster scan input is first converted to blocks which are 8 pixels wide by 4 pixels high in the luminance channel and 4 pixels by 4 in the two colour difference channels. When two fields are combined on the screen, the result is effectively an interlaced 8×8 luminance block with colour difference pixels having twice the horizontal luminance pixel spacing. The pixel blocks are then subject to a field shuffle. A shuffle based on individual pixels is impossible because it would raise the high-frequency content of the image and destroy the power of the compression process. Instead the block shuffle helps the compression process by making the average entropy of the image more constant.

This works because the shuffle exchanges blocks from flat areas of the image with blocks from highly detailed areas. The shuffle algorithm also has to consider the requirements of picture in shuttle. The blocking and shuffle take place when the read addresses of the input memory are permuted with respect to the write addresses.

Following the input shuffle the blocks are associated into sets of ten in each component and are then subject to the discrete cosine transform. The resulting coefficients are then subject to an iterative requantizing process followed by variable-length coding. The iteration adjusts the size of the quantizing step until the overall length of the ten coefficient sets is equal to the constant capacity of an entropy block which is 364 bytes. Within that entropy block the amount of data representing each individual DCT blocks may vary considerably, but the overall block size stays the same.

The DCT process results in coefficients whose wordlength exceeds the input wordlength. As a result it does not matter if the input wordlength is eight bits or ten bits; the requantizer simply adapts to make the output data rate constant. Thus the compression is greater with ten-bit input, corresponding to about 2.4 to 1.

The next step is the generation of product codes as shown in Figure 8.12(a). Each entropy block is divided into two halves of 162 bytes each and loaded into the rows of the outer code RAM which holds 114 such rows, corresponding to

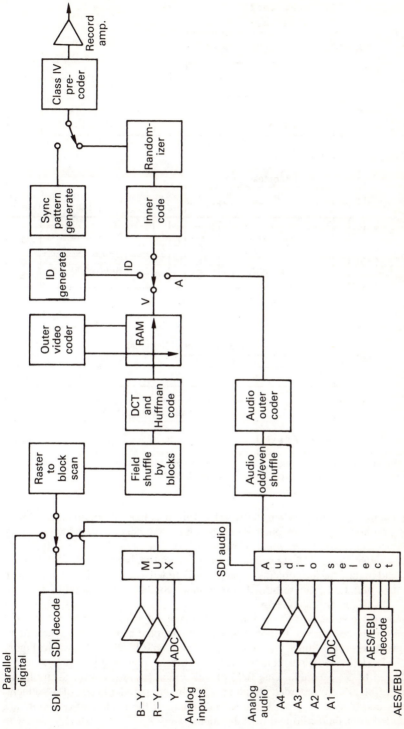

Figure 8.11 Block diagram of Digital Betacam record channel. Note that the use of data reduction makes this rather different to the block layout of full-bit formats.

(a) Video . . . 12 ECC blocks/field (2 ECC blocks/track)

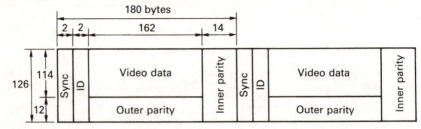

(b) Audio . . . 2 ECC blocks/ (CH. × field)

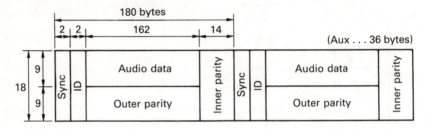

(Aux . . . 36 bytes)

(c) Sync block

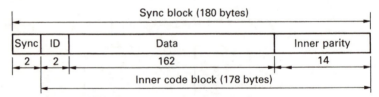

Figure 8.12 Video product codes of Digital Betacam are shown in (a); twelve of these are needed to record one field. Audio product codes are shown in (b); two of these record samples corresponding to one field period. Sync blocks are common to audio and video as shown in (c). The ID code discriminates between video and audio channels.

one twelfth of a field. When the RAM is full, it is read in columns by address mapping and 12 bytes of outer Reed–Solomon redundancy are added to every column, increasing the number of rows to 126.

The outer code RAM is read out in rows once more, but this time all 126 rows are read in turn. To the contents of each row is added a two-byte ID code and then the data plus ID bytes are turned into an inner code by the addition of 14 bytes of Reed–Solomon redundancy.

Inner codewords pass through the randomizer and are then converted to serial form for class-IV partial response precoding. With the addition of a sync pattern of two bytes, each inner codeword becomes a sync block as shown in Figure 8.12(c). Each video block contains 126 sync blocks, preceded by a preamble and followed by a postamble. One field of video data requires twelve such blocks. Pairs of blocks are recorded simultaneously by the parallel heads of a segment. Two video blocks are recorded at the beginning of the track, and two more are recorded after the audio and tracking tones.

The audio data for each channel are separated into odd and even samples for concealment purposes and assembled in RAM into two blocks corresponding to one field period. Two twenty-bit samples are stored in 5 bytes. Figure 8.12(b) shows that each block consists of 1458 bytes including auxiliary data from the AES/EBU interface, arranged as a block of 162×9 bytes. One hundred per cent outer code redundancy is obtained by adding 9 bytes of Reed–Solomon check bytes to each column of the blocks.

The inner codes for the audio blocks are produced by the same circuitry as the video inner codes on a time-shared basis. The resulting sync blocks are identical in size and differ only in the provision of different ID codes. The randomizer and precoder are also shared. It will be seen from Figure 8.10 that there are three segments in a field and that the position of an audio sector corresponding to a particular audio channel is different in each segment. This means that damage due to a linear tape scratch is distributed over three audio channels instead of being concentrated in one.

Each audio product block results in 18 sync blocks. These are accommodated in audio sectors of six sync blocks each in three segments. The audio sectors are preceded by preambles and followed by postambles. Between these are edit gaps which allow each audio channel to be independently edited.

By spreading the outer codes over three different audio sectors the correction power is much improved because data from two sectors can be used to correct errors in the third.

Figure 8.13 shows the replay channel of DB. The RF signal picked up by the replay head passes first to the Class IV partial response playback circuit in which it becomes a three-level signal as was shown in Chapter 6. The three-level signal is passed to an ADC which converts it into a digitally represented form so that the Viterbi detection can be carried out in logic circuitry. The sync detector identifies the synchronizing pattern at the beginning of each sync block and resets the block bit count. This allows the entire inner codeword of the sync block to be deserialized into bytes and passed to the inner error checker. Random errors will be corrected here, whereas burst errors will result in the block being flagged as in error.

Sync blocks are written into the de-interleave RAM with error flags where appropriate. At the end of each video sector the product code RAM will be filled, and outer code correction can be performed by reading the RAM at right-angles and using the error flags to initiate correction by erasure. Following outer code correction the RAM will contain corrected data or uncorrectable error flags which will later be used to initiate concealment.

The sync blocks can now be read from memory and assembled in pairs into entropy blocks. The entropy block is of fixed size, but contains coefficient blocks of variable length. The next step is to identify the individual coefficients and separate the luminance and colour difference coefficients by decoding the Huffman-coded sequence. Following the assembly of coefficient sets the inverse DCT will result in pixel blocks in three components once more. The pixel blocks are de-shuffled by mapping the write address of a field memory. When all the tracks of a field have been decoded, the memory will contain a de-shuffled field containing either correct sample data or correction flags. By reading the memory without address mapping the de-shuffled data are then passed through the concealment circuit where flagged data are concealed by data from nearby in the same field or from a previous field. The memory readout process is buffered from

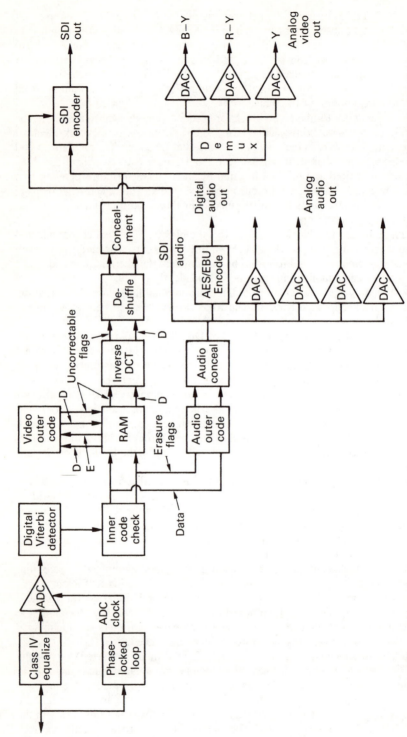

Figure 8.13 The replay channel of Digital Betacam. This differs from a full-bit system primarily in the requirement to deserialize variable-length coefficient blocks prior to the inverse DCT.

the offtape timing by the RAM and as a result the timebase correction stage is inherent in the replay process. Following concealment the data can be output as conventional raster-scan video either formatted to parallel or serial digital standards or converted to analog components.

8.11 DVC and DVCPRO

This component format uses quarter-inch wide metal evaporated (ME) tape which is only 7 micrometres thick in conjunction with compression to allow realistic playing times in miniaturized equipment. The format has jointly been developed by all the leading VCR manufacturers. Whilst intended as a consumer format it was clear that such a format is ideal for professional applications such as news gathering and simple production because of the low cost and small size. This led to the development of the DVCPRO format.

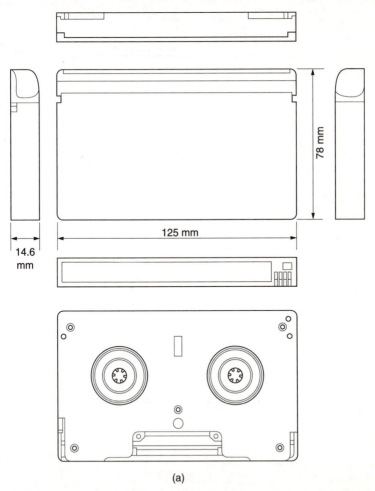

(a)

Figure 8.14 The cassettes developed for the $\frac{1}{4}$-inch DVC format. At (a) the standard cassette which holds 4.5 hours of program material.

In addition to component video there are also two channels of sixteen-bit uniformly quantized digital audio at 32, 44.1 or 48 kHz, with an option of four audio channels using twelve-bit non-uniform quantizing at 32 kHz.

Figure 8.14 shows that two cassette sizes are supported. The standard size cassette offers $4\frac{1}{2}$ hours of recording time and yet is only a little larger than an audio Compact Cassette. The small cassette is even smaller than a DAT cassette yet plays for one hour. Machines designed to play both tape sizes will be equipped with moving-reel motors. Both cassettes are equipped with fixed identification tabs and a moveable write-protect tab. These tabs are sensed by switches in the transport.

DVC (Digital Video Cassette) has adopted many of the features first seen in small formats such as the DAT digital audio recorder and the 8 mm analog video tape format. Of these the most significant is the elimination of the control track permitted by recording tracking signals in the slant tracks themselves. The adoption of metal evaporated tape and embedded tracking allows extremely high

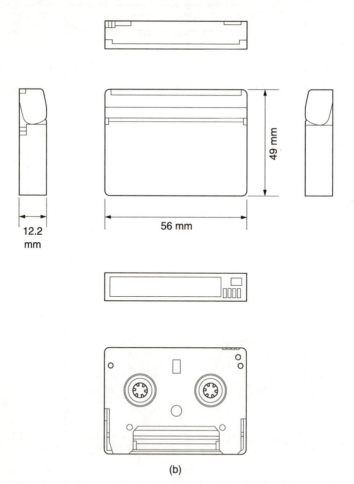

(b)

Figure 8.14 (*Continued*) The small cassette, shown at (b) is intended for miniature equipment and plays for 1 hour.

recording density. Tracks recorded with slant azimuth are only $10\,\mu m$ wide and the minimum wavelength is only $0.49\,\mu m$ resulting in a superficial density of over 0.4 Megabits per square millimetre.

Segmentation is used in DVC in such a way that as much commonality as possible exists between 50 and 60 Hz versions. The transport runs at 300 tape tracks per second; Figure 8.15 shows that 50 Hz frames contain 12 tracks and 60 Hz frames contain ten tracks.

The tracking mechanism relies upon buried tones in the slant tracks. From a tracking standpoint there are three types of track shown in Figure 8.16; F_0, F_1 and F_2. F_1 contains a low-frequency pilot and F_2 contains a high-frequency pilot. F_0 contains no pilot tone, but the recorded data spectrum contains notches at the frequencies of the two tones. Figure 8.16 also shows that every other track will contain F_0 following a four-track sequence.

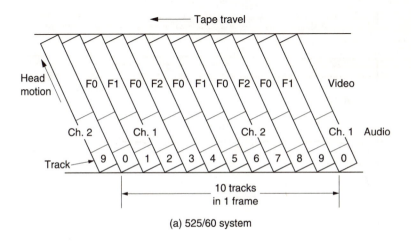

(a) 525/60 system

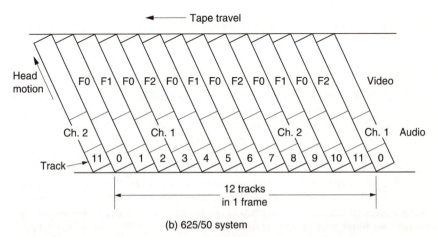

(b) 625/50 system

Figure 8.15 In order to use a common transport for 50 and 60 Hz standards the segmentation shown here is used. The segment rate is constant but 10 or 12 segments can be used in a frame.

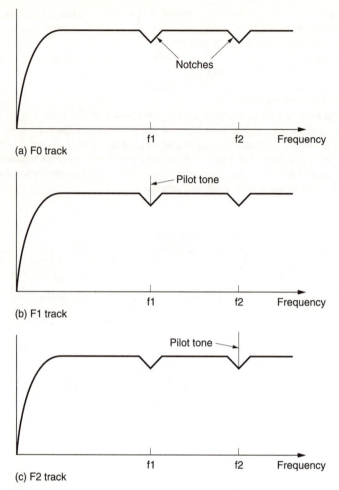

(a) F0 track

(b) F1 track

(c) F2 track

Figure 8.16 The tracks are of three types shown here. The F_0 track (a) contains spectral notches at two selected frequencies. The other two track types (b), (c) place a pilot tone in one or other of the notches.

The embedded tracking tones are recorded throughout the track by inserting a low frequency into the channel-coded data. Every twenty-four data bits an extra bit is added whose value has no data meaning but whose polarity affects the average voltage of the waveform. By controlling the average voltage with this bit, low frequencies can be introduced into the channel-coded spectrum to act as tracking tones. The tracking tones have sufficiently long wavelength that they are not affected by head azimuth and can be picked up by the 'wrong' head. When a head is following an F_0 type track, one edge of the head will detect F_1 and the other edge will detect F_2. If the head is centralized on the track, the amplitudes of the two tones will be identical. Any tracking error will result in the relative amplitudes of the F_1 F_2 tones changing. This can be used to modify the capstan phase in order to correct the tracking error. As azimuth recording is used requiring a minimum of two heads, one head of the pair will always be able to play a type F_0 track.

In simple machines only one set of heads will be fitted and these will record or play as required. In more advanced machines, separate record and replay heads will be fitted. In this case the replay head will read the tracking tones during normal replay, but in editing modes, the record head would read the tracking tones during the preroll in order to align itself with the existing track structure.

Figure 8.17 shows the track dimensions. The tracks are approximately 33 mm long and lie at approximately 9° to the tape edge. A transport with a 180° wrap would need a drum of only 21 mm diameter. For camcorder applications with the small cassette this allows a transport no larger than an audio 'Walkman'. With the larger cassette it would be advantageous to use time compression to allow a larger drum with partial wrap to be used. This would simplify threading and make room for additional heads in the drum for editing functions.

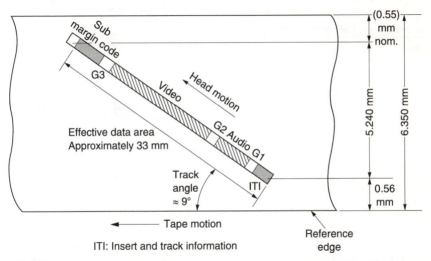

Figure 8.17 The dimensions of the DVC track. Audio, video and subcode can independently be edited. Insert and Track Information block aligns heads during insert.

The audio, video and subcode data are recorded in three separate sectors with edit gaps between so that they can be independently edited in insert mode. In the case where all three data areas are being recorded in insert mode, there must be some mechanism to keep the new tracks synchronous with those which are being overwritten. In a conventional VTR this would be the job of the control track.

In DVC there is no control track and the job of tracking during insert is undertaken by part of each slant track. Figure 8.17 shows that the track begins with the insert and track information (ITI) block. During an insert edit the ITI block in each track is always read by the record head. This identifies the position of the track in the segmentation sequence and in the tracking tone sequence and allows the head to identify its physical position both along and across the track prior to an insert edit. The remainder of the track can then be recorded as required.

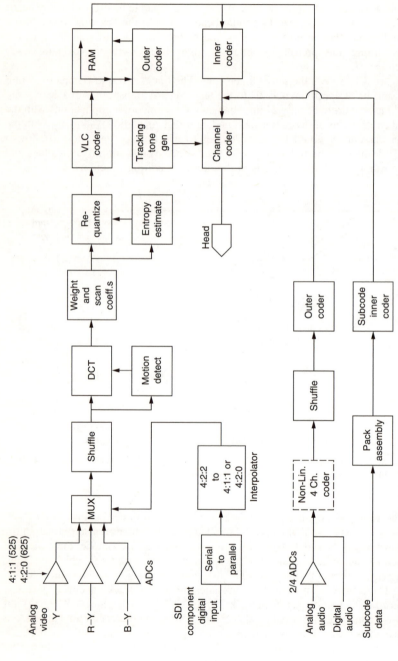

Figure 8.18 Block diagram of DVC signal system. This is similar to larger formats except that a high compression factor allows use of a single channel with no distribution.

As there are no linear tracks, the subcode is designed to be read in shuttle for access control purposes. It will contain timecodes and flags.

Figure 8.18 shows a block diagram of the DVC signal system. The input video is eight-bit component digital according to CCIR-601, but compression of about 5:1 is used. The colour difference signals are subsampled prior to compression. In 60 Hz machines, 4:1:1 sampling is used, allowing a colour difference bandwidth in excess of that possible with NTSC. In 50 Hz machines, 4:2:0 sampling is used. The colour difference sampling rate is still 6.75 MHz, but the two colour difference signals are sent on sequential lines instead of simultaneously. The result is that the vertical colour difference resolution matches the horizontal resolution. A similar approach is used in SECAM and MAC video. Studio standard 4:2:2 parallel or serial inputs and outputs can be handled using simple interpolators in the colour difference signal paths.

A 16:9 aspect ratio can be supported in standard definition by increasing the horizontal pixel spacing as is done in Digital Betacam. High-definition signals can be supported using a higher compression factor.

As in other DVTRs, the error-correction strategy relies upon a combination of shuffle and product codes. Frames are assembled in RAM, and partitioned into blocks of 8×8 pixels. In the luminance channel, four of these blocks cover the same screen area as one block in each colour difference signal as Figure 8.19 shows. The four luminance blocks and the two colour difference blocks are together known as a macroblock. The shuffle is based upon reordering of macroblocks. Following the shuffle compression takes place. The compression system is DCT based and uses techniques described in Chapter 5. Compression acts within frame boundaries so as to permit frame-accurate editing. This contrasts with the intra-field compression used in DB.

Intra-frame compression uses 8×8 pixel DCT blocks and allows a higher compression factor because advantage can be taken of redundancy between the two fields when there is no motion. If motion is detected, then moving areas of the two fields will be independently coded in 8×4 pixel blocks to prevent motion blur. Following the motion compensation the DCT coefficients are weighted, zig-

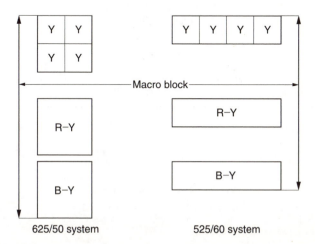

625/50 system 525/60 system

Figure 8.19 In DVC a macroblock contains information from a fixed screen area. As the colour resolution is reduced, there are twice as many luminance pixels.

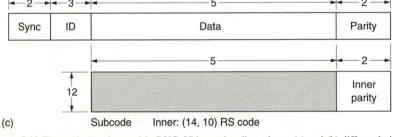

Figure 8.20 The product codes used in DVC. Video and audio codes at (a) and (b) differ only in size and use the same inner code structure. Subcode at (c) is designed to be read in shuttle and uses short sync blocks to improve chances of recovery.

zag scanned and requantized prior to variable-length coding. As in other compressed VTR formats, the requantizing is adaptive so that the same amount of data is output irrespective of the input picture content. The entropy block occupies one sync block and contains data compressed from five macroblocks.

The DVC product codes are shown in Figure 8.20. The video product block is shown at (a). This block is common to both 525- and 625-line formats. Ten such blocks record one 525-line frame whereas 12 blocks are required for a 625-line frame.

The audio channels are shuffled over a frame period and assembled into the product codes shown in Figure 8.20(b). Video and audio sync blocks are identical except for the ID numbering. The subcode structure is different. Figure 8.20(c) shows the structure of the subcode block. The subcode is not a product block because these can only be used for error correction when the entire block is recovered. The subcode is intended to be read in shuttle where only parts of the track are recovered. Accordingly only inner codes are used and these are much shorter than the video/audio codes, containing only 5 data bytes, known as a pack. The structure of a pack is shown In Figure 8.21. The subcode block in each track can accommodate 12 packs. Packs are repeated throughout the frame so that they have a high probability of recovery in shuttle. The pack header identifies the type of pack, leaving four bytes for pack data, e.g. timecode.

Following the assembly of product codes, the data are then channel coded for recording on tape. A scrambled NRZI channel code is used which is similar to the system used in D-1 except that the tracking tones are also inserted by the modulation process.

In the DVCPRO format the extremely high recording density and long playing time of the consumer DVC was not a requirement. Instead ruggedness and reliable editing were needed. In developing DVCPRO, Panasonic chose to revert to metal particle tape as used in most other DVTRs. This requires wider tracks, and this was achieved by increasing the tape linear speed. The wider tracks also reduce the mechanical precision needed for interchange and editing. However, the DVCPRO transport can still play regular DVC tapes.

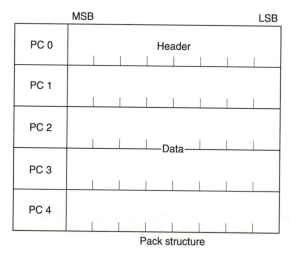

Pack structure

Figure 8.21 Structure of a pack.

The DVCPRO format has proved to be extremely popular and a number of hard disk editors are now designed to import native DVCPRO data to cut down on generation loss. With a suitable tape drive this can be done at 4 × normal speed. The SDTI interface (see Chapter 9) can also carry native DVC data.

8.12 The D-9 format

The D-9 format (formerly known as Digital-S) was developed by JVC and is based on half-inch metal particle tape. D-9 is intended as a fully specified 4:2:2 production format and so has four uncompressed 48 kHz digital audio tracks and a frame-based compression scheme with a mild compression factor.

In the same way that the Sony format used technology from the Betamax, the cassette and transport design of D-9 are refinements of the analog VHS format. Whereas the compression algorithm of Digital Betacam is unique, in D-9 the same algorithm as in DVC has been adopted. This gives an economic advantage as the DVC chip set is produced in large quantities. An increasing number of manufacturers are producing equipment which can accept native DVC-coded data to avoid generation loss and D-9 fits well with that concept.

In D-9 the bit rate on tape is 50 Mbits/s; twice that of the DVC format. As a result, the DCT coefficients will suffer less requantizing allowing a lower noise floor and genuine multi-generation performance. Tests performed by the EBU/ SMPTE task force found that D-9 and Digital Betacam had almost identical performance.

8.13 Digital audio in VTRs

Digital audio recording with video is rather more difficult than in an audio-only environment. The audio samples are carried by the same channel as the video samples. The audio could have used separate stationary heads, but this would have increased tape consumption and machine complexity. In order to permit independent audio and video editing, the tape tracks are given a block structure. Editing will require the heads momentarily to go into record as the appropriate audio block is reached. Accurate synchronization is necessary if the other parts of the recording are to remain uncorrupted. The sync block structure of the video sector continues in the audio sectors because the same read/write circuitry is used for audio and video data. Clearly, the ID code structure must also continue through the audio. In order to prevent audio samples from arriving in the framestore in shuttle, the audio addresses are different from the video addresses.

Despite the additional complexity of sharing the medium with video, the professional DVTR must support such functions as track bouncing, synchronous recording and split audio edits with variable crossfade times.

The audio samples in a DVTR are binary numbers just like the video samples, and although there is an obvious difference in sampling rate and wordlength, the use of time compression means that this affects only the relative areas of tape devoted to the audio and video samples. The most important difference between audio and video samples is the tolerance to errors. The acuity of the ear means that uncorrected audio samples must not occur more than once every few hours. There is little redundancy in sound when compared to video, and concealment of errors is not desirable on a routine basis. In video, the samples are highly redundant, and concealment can be effected using samples from previous or subsequent lines or, with care, from the previous frame.

Whilst subjective considerations require greater data reliability in the audio samples, audio data form a small fraction of the overall data and it is difficult to protect them with an extensive interleave whilst still permitting independent editing. For these reasons major differences can be expected between the ways that audio and video samples are handled in a digital video recorder. One such difference is that the error-correction strategy for audio samples uses a greater amount of redundancy. Whilst this would cause a serious playing-time penalty in an audio recorder, even doubling the audio data rate in a video recorder only raises the overall data rate by a few per cent. The arrangement of the audio blocks is also designed to maximize data integrity in the presence of tape defects and head clogs. The audio blocks are at the ends of the head sweeps in D-2 and D-3, but are placed in the middle of the segment in D-1, D-5, Digital Betacam and DCT.

The audio sample interleave varies in complexity between the various formats. It will be seen from Figure 8.22 that the physical location of a given audio channel rotates from segment to segment. In this way a tape scratch will cause slight damage to all channels instead of serious damage to one. Data are also distributed between the heads, and so if one (or sometimes two) of the heads clogs, the audio is still fully recovered.

In each sector, the track commences with a preamble to synchronize the phase-locked loop in the data separator on replay. Each of the sync blocks begins, as the name suggests, with a synchronizing pattern which allows the read sequencer to deserialize the block correctly. At the end of a sector, it is not possible simply to turn off the write current after the last bit, as the turnoff transient would cause data corruption. It is necessary to provide a postamble such that current can be turned off away from the data. It should now be evident that any editing has to take place a sector at a time. Any attempt to rewrite one sync block would result in damage to the previous block owing to the physical inaccuracy of replacement, damage to the next block due to the turnoff transient, and inability to synchronize to the replaced block because of the random phase jump at the point where it began. The sector in a DVTR is analogous to the cluster in a disk drive. Owing to the difficulty of writing in exactly the same place as a previous recording, it is necessary to leave tolerance gaps between sectors where the write current can turn on and off to edit individual write blocks. For convenience, the tolerance gaps are made the same length as a whole number of sync blocks. Figure 8.23 shows that in D-1 the edit gap is two sync blocks long, as it is in D-3, whereas in D-2 it is only one sync block long. The first half of the tolerance gap is the postamble of the previous block, and the second half of the tolerance gap acts as the preamble for the next block. The tolerance gap following editing will contain, somewhere in the centre, an arbitrary jump in bit phase, and a certain amount of corruption due to turnoff transients. Provided that the postamble and preamble remain intact, this is of no consequence.

The number of audio sync blocks in a given time is determined by the number of video fields in that time. It is only possible to have a fixed tape structure if the audio sampling rate is locked to video. With 625/50 machines, the sampling rate of 48 kHz results in exactly 960 audio samples in every field.

For use on 525/60, it must be recalled that the 60 Hz is actually 59.94 Hz. As this is slightly slow, it will be found that in sixty fields, exactly 48 048 audio

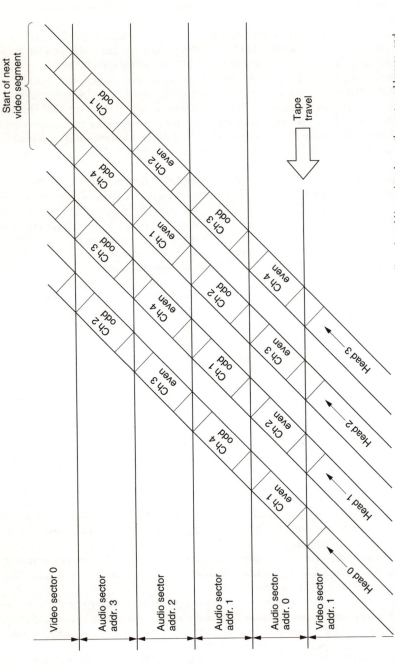

Figure 8.22 The structure of the audio blocks in D-1 format showing the double recording, the odd/even interleave, the sector addresses, and the distribution of audio channels over all heads. The audio samples recorded in this area represent a 6.666 ms time slot in the audio recording.

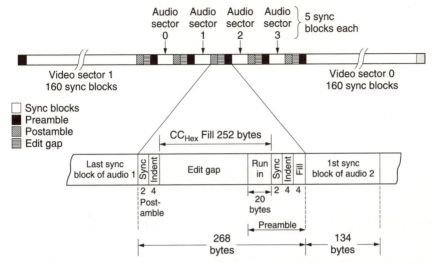

Figure 8.23 The position of preambles and postambles with respect to each sector is shown along with the gaps necessary to allow individual sectors to be written without corrupting others. When the whole track is recorded, 252 bytes of CC hex fill are recorded after the postamble before the next sync pattern. If a subsequent recording commences in this gap, it must do so at least 20 bytes before the end in order to write a new run-in pattern for the new recording.

samples will be necessary. Unfortunately 60 will not divide into 48 048 without a remainder. The largest number which will divide 60 and 48 048 is 12; thus in 60/12 = 5 fields there will be 48 048/12 = 4004 samples. Over a five-field sequence the fields contain 801, 801, 801, 801 and 800 samples respectively, adding up to 4004 samples.

8.14 AES/EBU compatibility

In order to comply with the AES/EBU digital audio interconnect, wordlengths between sixteen and twenty bits can be supported, but it is necessary to record a code in the sync block to specify the wordlength in use. Pre-emphasis may have been used prior to conversion, and this status is also to be conveyed, along with the four channel-use bits. The AES/EBU digital interconnect uses a block-sync pattern which repeats after 192 sample periods corresponding to 4 ms at 48 kHz. Since the block size is different to that of the DVTR interleave block, there can be any phase relationship between interleave-block boundaries and the AES/EBU block-sync pattern. In order to re-create the same phase relationship between block sync and sample data on replay, it is necessary to record the position of block sync within the interleave block. It is the function of the interface control word in the audio data to convey these parameters.

There is no guarantee that the 192-sample block-sync sequence will remain intact after audio editing; most likely there will be an arbitrary jump in block-sync phase. Strictly speaking, a DVTR, playing back an edited tape would have to ignore the block-sync positions on the tape, and create new block sync at the standard 192-sample spacing. Unfortunately the DVTR formats are not totally transparent to the whole of the AES/EBU data stream, as certain information is not recorded.

References

1. Huckfield, D., Sato, N. and Sato, I. Digital Betacam – The application of state of the art technology to the development of an affordable component DVTR. *Record of 18th ITS*, 180–199 (Montreux, 1993)
2. Creed, D. and Kaminaga, K. Digital compression strategies for video tape recorders in studio applications. *Record of 18th ITS*, 291–301 (Montreux, 1993)

Chapter 9

Digital communication

9.1 Introduction

Digital communication includes any system which can deliver data over distance. Figure 9.1 shows some of the ways in which the subject can be classified. The simplest is a unidirectional point-to-point signal path shown at (a). This is common in digital production equipment and includes the AES/EBU digital audio interface and the serial digital interface (SDI) for digital video. Bidirectional point-to-point signals include the RS-232 and RS-422 duplex systems. Bidirectional signal paths may be symmetrical, i.e. have the same capacity in both directions (b), or asymmetrical, having more capacity in one direction than the other (c). In this case the low-capacity direction may be known as a *back channel*.

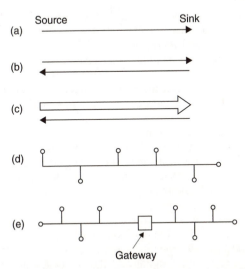

Figure 9.1 Some ways of classifying communications systems. At (a) the unidirectional point-to-point connection used in many digital audio and video interconnects. (b) Symmetrical bidirectional point-to-point system. (c) Asymmetrical point-to-point system. (d) A network must have some switching or addressing ability in addition to delivering data. (e) Networks can be connected by gateways.

Back channels are useful in a number of applications. Video-on demand and interactive video are both systems in which the inputs from the viewer are relatively small, but result in extensive data delivery to the viewer. Archives and databases have similar characteristics.

When more than two devices can be interconnected in such a way that any one can communicate at will with any other, the result is a network as in Figure 9.1(d). The traditional telephone system is a network, and although the original infrastructure assumed analog speech transmission, subsequent developments in modems have allowed data transmission.

The computer industry has developed its own network technology, a long-serving example being Ethernet. Computer networks can work over various distances, giving rise to LANs (local area networks), MANs (metropolitan area networks) and WANs (wide area networks). Such networks can be connected together to form *internetworks* or internets for short, including *the* Internet. A private network, linking all employees of a given company, for example, may be referred to as an intranet.

Figure 9.1(e) shows that networks are connected together by *gateways*. In this example a private network (typically a local area network (LAN) within an office block) is interfaced to an *access network* (typically a metropolitan area network (MAN) with a radius of the order of a few kilometres) which in turn connects to the *transport network*. The access networks and the transport network together form a public network.

The different requirements of networks of different sizes have led to different protocols being developed. Where a gateway exists between two such networks, the gateway will often be required to perform protocol conversion. Protocol conversion represents unnecessary cost and delay and recent protocols such as ATM are sufficiently flexible that they can be adopted in any type of network to avoid conversion.

Networks also exist which are optimized for storage devices. These range from the standard buses linking hard drives with their controllers to SANs (storage area networks) in which distributed storage devices behave as one large store.

Communication must also include broadcasting, which initially was analog, but has also adopted digital techniques so that transmitters effectively radiate data. Traditional analog broadcasting was unidirectional, but with the advent of digital techniques, various means for providing a back channel have been developed.

To have an understanding of communications it is important to appreciate the concept of layers shown in Figure 9.2(a). The lowest layer is the *physical medium dependent* layer. In the case of a cabled interface, this layer would specify the dimensions of the plugs and sockets so that a connection could be made, and the use of a particular type of conductor such as co-axial, STP (screened twisted pair) or UTP (unscreened twisted pair). The impedance of the cable may also be specified. The medium may also be optical fibre which will need standardization of the terminations and the wavelength(s) in use.

Once a connection is made, the physical medium dependent layer standardizes the voltage of the transmitted signal and the frequency at which the voltage changes (the channel bit rate). This may be fixed at a single value, chosen from a set of fixed values, or, rarely, variable. Practical interfaces need some form of channel coding (see Chapter 6) in order to embed a bit clock in the data transmission.

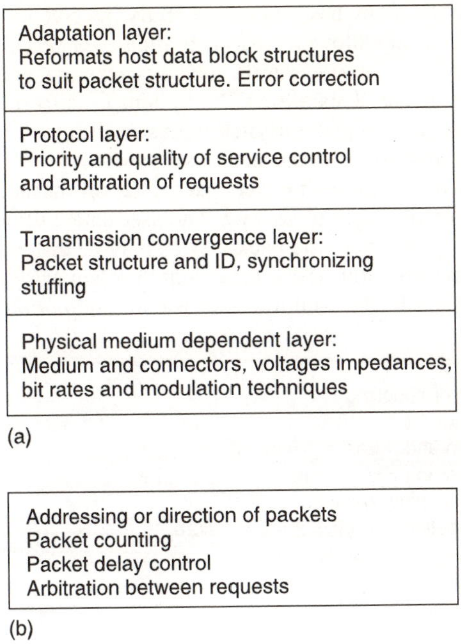

Adaptation layer:
Reformats host data block structures
to suit packet structure. Error correction

Protocol layer:
Priority and quality of service control
and arbitration of requests

Transmission convergence layer:
Packet structure and ID, synchronizing
stuffing

Physical medium dependent layer:
Medium and connectors, voltages impedances,
bit rates and modulation techniques

(a)

Addressing or direction of packets
Packet counting
Packet delay control
Arbitration between requests

(b)

Figure 9.2 (a) Layers are important in communications because they have a degree of independence such that one can be replaced by another leaving the remainder undisturbed. (b) The functions of a network protocol. See text.

The physical medium dependent layer allows binary transmission, but this needs to be structured or formatted. The *transmission convergence* layer takes the binary signalling of the physical medium dependent layer and builds a packet or cell structure. This consists at least of some form of synchronization system so that the start and end of serialized messages can be recognized and an addressing or labelling scheme so that packets can reliably be routed and recognized. Real cables and optical fibres run at fixed bit rates and a further function of the transmission convergence layer is the insertion of null or stuffing packets where insufficient user data exist.

In broadcasting, the physical medium dependent layer may be one which contains some form of radio signal and a modulation scheme. The modulation scheme will be a function of the kind of service, for example a satellite modulation scheme would be quite different from one used in a terrestrial service.

In all real networks requests for transmission will arise randomly. Network resources need to be applied to these requests in a structured way to prevent chaos, data loss or lack of throughput. This raises the requirement for a protocol layer. TCP (transmission control protocol) and ATM (asynchronous transfer mode) are protocols. A protocol is an agreed set of actions in given circumstances. In a point-to-point interface the protocol is trivial, but in a network it is complex. Figure 9.2(b) shows some of the functions of a network protocol. There must be an addressing mechanism so that the sender can direct the data to the desired location, and a mechanism by which the receiving device confirms that all the data have been correctly received. In more advanced systems the protocol may allow variations in *quality of service* whereby the user can select (and pay for) various criteria such as packet delay and delay variation and the packet error rate. This allows the system to deliver isochronous (near-real-time) MPEG data alongside asynchronous (non-time-critical) data such as e-mail by appropriately prioritizing packets.

The protocol layer arbitrates between demands on the network and delivers packets at the required quality of service. The user data will not necessarily have been packeted, or if they were the packet sizes may be different from those used in the network. This situation arises, for example, when MPEG transport packets are to be sent via ATM. The solution is to use an *adaptation layer*.

```
1  No correction or checking
2  Detection only
3  Error detection and retransmit request
4  Error detection and FEC to handle random errors
5  FEC and interleaving to handle packet loss
6  Automatic rerouting following channel failure
```

Figure 9.3 Different approaches to error checking used in various communications systems.

Adaptation layers reformat the original data into the packet structure needed by the network at the sending device, and reverse the process at the destination device. Practical networks must have error checking/correction. Figure 9.3 shows some of the possibilities. In short interfaces, no errors are expected and a simple parity check or checksum with an error indication is adequate. In bidirectional applications a checksum failure would result in a retransmission request or cause the receiver to fail to acknowledge the transmission so that the sender would try again. In real-time systems, there may not be time for a retransmission, and an FEC (forward error correction) system will be needed in which enough redundancy is included with every data block to permit on-the-fly correction at the receiver. The sensitivity to error is a function of the type of data, and so it is a further function of the adaptation layer to take steps such as interleaving and the addition of FEC codes.

9.2 Production-related interfaces

As audio and video production equipment made the transition from analog to digital technology, computers and networks were still another world and the potential of the digital domain was largely neglected because the digital interfaces which were developed simply copied analog practice but transmitted binary numbers instead of the original signal waveform. These interfaces are simple and have no addressing or handshaking ability. Creating a network requires switching devices called routers which are controlled independently of the signals themselves. Although obsolescent, there are substantial amounts of equipment in service adhering to these standards which will remain in use for some time.

The AES/EBU (Audio Engineering Society/European Broadcast Union) interface was developed to provide a short-distance point-to-point connection for PCM digital audio and subsequently evolved to handle compressed audio data.

The serial digital interface (SDI) was developed to allow up to ten-bit samples of standard definition interlaced component or composite digital video to be communicated serially.[1] 16:9 format component signals with 18 MHz sampling rate can also be handled. SDI as first standardized had no error-detection ability at all. This was remedied by a later option known as EDH (error detection and handling). The interface allows ancillary data including transparent conveyance of embedded AES/EBU digital audio channels during video blanking periods.

SDI is highly specific to two broadcast television formats. Subsequently the electrical and channel coding layer of SDI was used to create SDTI (serial data transport interface) which is used for transmitting, among other things, elementary streams from video compressors. ASI (asynchronous serial interface) uses only the electrical interface of SDI but with a different channel code and protocol and is used for transmitting MPEG transport streams through SDI-based equipment.

9.3 SDI

The serial digital interface was designed to allow easy conversion to and from traditional analog component video for production purposes. Only 525/59.94/2:1 and 625/50/2:1 formats are supported with 4:2:2 sampling. The sampling structure of SDI was detailed in section 7.14 and only the transmission technique will be considered here.

Chapter 6 introduced the concepts of DC components and uncontrolled clock content in serial data for recording and the same issues are important in interfacing, leading to a coding requirement. SDI uses convolutional randomizing, as shown in section 6.13, in which the signal sent down the channel is the serial data waveform which has been convolved with the impulse response of a digital filter. On reception the signal is deconvolved to restore the original data.

The components necessary for an SDI link are shown in Figure 9.4. Parallel component or composite data having a wordlength of up to ten bits form the input. These are fed to a ten-bit shift register which is clocked at ten times the input rate, which will be 270 MHz or $40 \times F_{sc}$. If there are only eight bits in the input words, the missing bits are forced to zero for transmission except for the all ones condition which will be forced to ten ones. The serial data from the shift register are then passed through the scrambler, in which a given bit is converted to the exclusive-OR of itself and two bits which are five and nine clocks ahead. This is followed by another stage, which converts channel ones into transitions. The resulting signal is fed to a line driver which converts the logic level into an alternating waveform of 800 mV peak-to-peak. The driver output impedance is carefully matched so that the signal can be fed down 75 Ohm co-axial cable using BNC connectors.

The scrambling process at the encoder spreads the signal spectrum and makes that spectrum reasonably constant and independent of the picture content. It is possible to assess the degree of equalization necessary by comparing the energy in a low-frequency band with that in higher frequencies. The greater the disparity, the more equalization is needed. Thus fully automatic cable equalization is easily achieved. The receiver must generate a bit clock at 270 MHz or $40 \times F_{sc}$ from the input signal, and this clock drives the input sampler and slicer which converts the cable waveform back to serial binary. The local bit clock also drives a circuit which simply reverses the scrambling at the transmitter. The first stage returns transitions to ones, and the second stage is a mirror image of the encoder which reverses the exclusive-OR calculation to output the original data. Since transmission is serial, it

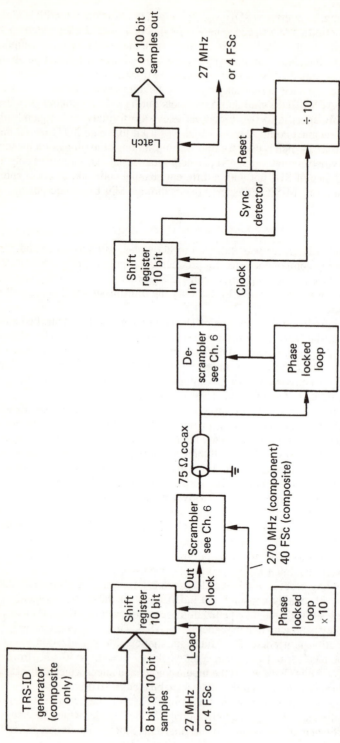

Figure 9.4 Major components of a serial scrambled link. Input samples are converted to serial form in a shift register clocked at ten times the sample rate. The serial data are then scrambled for transmission. On reception, a phase-locked loop recreates the bit rate clock and drives the de-scrambler and serial-to-parallel conversion. On detection of the sync pattern, the divide-by-ten counter is rephased to load parallel samples correctly into the latch. For composite working the bit rate will be 40 times subcarrier, and a sync pattern generator (top left) is needed to inject TRS-ID into the composite data stream.

is necessary to obtain word synchronization, so that correct deserialization can take place.

In the component parallel input, the *SAV* and *EAV* sync patterns are present and the all-ones and all-zeros bit patterns these contain can be detected in the thirty-bit shift register and used to reset the deserializer.

On detection of the synchronizing symbols, a divide-by-ten circuit is reset, and the output of this will clock words out of the shift register at the correct times. This output will also become the output word clock.

It is a characteristic of all randomizing techniques that certain data patterns will interact badly with the randomizing algorithm to produce a channel waveform which is low in clock content. These so-called pathological data patterns[2] are extremely rare in real program material, but can be specially generated for testing purposes.

9.4 SDTI

SDI is closely specified and is only suitable for transmitting 2:1 interlaced 4:2:2 digital video in 525/60 or 625/50 systems. Since the development of SDI, it has become possible economically to compress digital video and the SDI standard cannot handle this. SDTI (serial data transport interface) is designed to overcome that problem by converting SDI into an interface which can carry a variety of data types whilst retaining compatibility with existing SDI router infrastructures.

SDTI[3] sources produce a signal which is electrically identical to an SDI signal and which has the same timing structure. However, the digital active line of SDI becomes a data packet or *item* in SDTI. Figure 9.5 shows how SDTI fits into the existing SDI timing. Between EAV and SAV (horizontal blanking in SDI) an

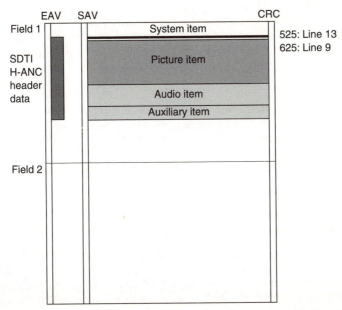

Figure 9.5 SDTI is a variation of SDI which allows transmission of generic data. This can include compressed video and non-real-time transfer.

ancillary data block is incorporated. The structure of this meets the SDI standard, and the data within describe the contents of the following digital active line.

The data capacity of SDTI is about 200 Mbits/s because some of the 270 Mbits/s is lost due to the retention of the SDI timing structure. Each digital active line finishes with a CRCC (cyclic redundancy check character) to check for correct transmission.

SDTI raises a number of opportunities, including the transmission of compressed data at faster than real time. If a video signal is compressed at 4:1, then one quarter as much data would result. If sent in real time the bandwidth required would be one quarter of that needed by uncompressed video. However, if the same bandwidth is available, the compressed data could be sent in 1/4 of the usual time. This is particularly advantageous for data transfer between compressed camcorders and non-linear editing workstations. Alternatively, four different 50 Megabit/s signals could be conveyed simultaneously.

Thus an SDTI transmitter takes the form of a multiplexer which assembles packets for transmission from input buffers. The transmitted data can be encoded according to MPEG, MotionJPEG, Digital Betacam or DVC formats and all that is necessary is that compatible devices exist at each end of the interface. In this case the data are transferred with bit accuracy and so there is no generation loss associated with the transfer. If the source and destination are different, i.e. having different formats or, in MPEG, different group structures, then a conversion process with attendant generation loss would be needed.

9.5 ASI

The asynchronous serial interface is designed to allow MPEG transport streams to be transmitted over standard SDI cabling and routers. ASI offers higher performance than SDTI because it does not adhere to the SDI timing structure. Transport stream data do not have the same statistics as PCM video and so the scrambling technique of SDI cannot be used. Instead ASI uses an 8/10 group code (see section 6.12) to eliminate DC components and ensure adequate clock content).

SDI equipment is designed to run at a closely defined bit rate of 270 Mbits/s and has phase-locked loops in receiving and repeating devices which are intended to remove jitter. These will lose lock if the channel bit rate changes. Transport streams are fundamentally variable in bit rate and to retain compatibility with SDI routing equipment ASI uses stuffing bits to keep the transmitted bit rate constant.

The use of an 8/10 code means that although the channel bit rate is 270 Mbits/s, the data bit rate is only 80 per cent of that, i.e 216 Mbits/s. A small amount of this is lost to overheads.

9.6 AES/EBU

The AES/EBU digital audio interface, originally published in 1985[4] was proposed to embrace all the functions of existing formats in one standard. The goal was to ensure interconnection of professional digital audio equipment irrespective of origin. The EBU ratified the AES proposal with the proviso that the optional transformer coupling was made mandatory and led to the term AES/EBU interface, also called EBU/AES by some Europeans and standardized as IEC 958.

The interface has to be self-clocking and self-synchronizing, i.e. the single signal must carry enough information to allow the boundaries between individual bits, words and blocks to be detected reliably. To fulfil these requirements, the FM channel code is used (see Chapter 6) which is DC-free, strongly self-clocking and capable of working with a changing sampling rate. Synchronization of deserialization is achieved by violating the usual encoding rules.

The use of FM means that the channel frequency is the same as the bit rate when sending data ones. Tests showed that in typical analog audio-cabling installations, sufficient bandwidth was available to convey two digital audio channels in one twisted pair. The standard driver and receiver chips for RS–422A[5] data communication (or the equivalent CCITT-V.11) are employed for professional use, but work by the BBC[6] suggested that equalization and transformer coupling were desirable for longer cable runs, particularly if several twisted pairs occupy a common shield. Successful transmission up to 350 m has been achieved with these techniques.[7] Figure 9.6 shows the standard configuration. The output impedance of the drivers will be about 110 Ohms, and the impedance of the cable and receiver should be similar at the frequencies of interest. The driver was specified in AES-3-1985 to produce between 3 and 10 V peak-to-peak into such an impedance but this was changed to between 2 and 7 V in AES-3-1992 to better reflect the characteristics of actual RS-422 driver chips.

In Figure 9.7, the specification of the receiver is shown in terms of the minimum eye pattern (see section 6.9) which can be detected without error. It

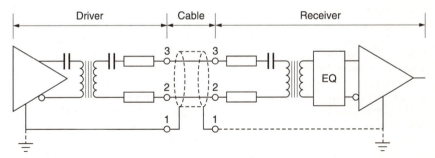

Figure 9.6 Recommended electrical circuit for use with the standard two-channel interface.

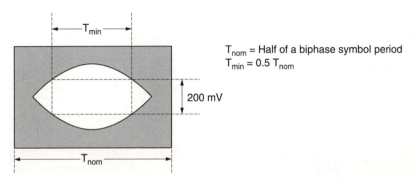

T_{nom} = Half of a biphase symbol period
$T_{min} = 0.5\,T_{nom}$

200 mV

Figure 9.7 The minimum eye pattern acceptable for correct decoding of standard two-channel data.

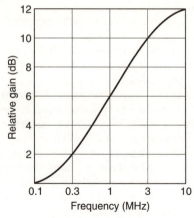

Figure 9.8 EQ characteristic recommended by the AES to improve reception in the case of long lines.

will be noted that the voltage of 200 mV specifies the height of the eye opening at a width of half a channel bit period. The actual signal amplitude will need to be larger than this, and even larger if the signal contains noise. Figure 9.8 shows the recommended equalization characteristic which can be applied to signals received over long lines.

The purpose of the standard is to allow the use of existing analog cabling, and as an adequate connector in the shape of the XLR is already in wide service, the connector made to IEC 268 Part 12 has been adopted for digital audio use. Effectively, existing analog audio cables having XLR connectors can be used without alteration for digital connections.

There is a separate standard[8] for a professional interface using coaxial cable for distances of around 1000 m. This is simply the AES/EBU protocol but with a 75 Ohm coaxial cable carrying a one volt signal so that it can be handled by analog video distribution amplifiers. Impedance converting transformers allow balanced 110 Ohm to unbalanced 75 Ohm matching.

In Figure 9.9 the basic structure of the professional and consumer formats can be seen. One subframe consists of 32 bit-cells, of which four will be used by a synchronizing pattern. Subframes from the two audio channels, A and B,

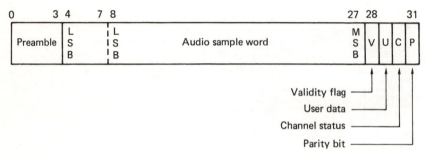

Figure 9.9 The basic subframe structure of the AES/EBU format. Sample can be twenty bits with four auxiliary bits, or twenty-four bits. LSB is transmitted first.

alternate on a time-division basis, with the least significant bit sent first. Up to twenty-four-bit sample wordlength can be used, which should cater for all conceivable future developments, but normally twenty-bit maximum length samples will be available with four auxiliary data bits, which can be used for a voice-grade channel in a professional application.

The format specifies that audio data must be in two's complement coding. If different wordlengths are used, the MSBs must always be in the same bit position otherwise the polarity will be misinterpreted. Thus the MSB has to be in bit 27 irrespective of wordlength. Shorter words are leading-zero filled up to the twenty-bit capacity. The channel status data included from AES-3-1992 signalling of the actual audio wordlength used so that receiving devices could adjust the digital dithering level needed to shorten a received word which is too long or pack samples onto a storage device more efficiently.

Four status bits accompany each subframe. The validity flag will be reset if the associated sample is reliable. Whilst there have been many aspirations regarding what the V bit could be used for, in practice a single bit cannot specify much, and if combined with other V bits to make a word, the time resolution is lost. AES-3-1992 described the V bit as indicating that the information in the associated subframe is 'suitable for conversion to an analog signal'. Thus it might be reset if the interface was being used for non-PCM audio data such as the output of an audio compressor.

The parity bit produces even parity over the subframe, such that the total number of ones in the subframe is even. This allows for simple detection of an odd number of bits in error, but its main purpose is that it makes successive sync patterns have the same polarity, which can be used to improve the probability of detection of sync. The user and channel-status bits are discussed later.

Two of the subframes described above make one frame, which repeats at the sampling rate in use. The first subframe will contain the sample from channel A, or from the left channel in stereo working. The second subframe will contain the sample from channel B, or the right channel in stereo. At 48 kHz, the bit rate will be 3.072 MHz, but as the sampling rate can vary, the clock rate will vary in proportion.

In order to separate the audio channels on receipt the synchronizing patterns for the two subframes are different as Figure 9.10 shows. These sync patterns begin with a run length of 1.5 bits which violates the FM channel coding rules and so cannot occur due to any data combination. The type of sync pattern is denoted by the position of the second transition which can be 0.5, 1.0 or 1.5 bits away from the first. The third transition is designed to make the sync patterns DC-free.

The channel status and user bits in each subframe form serial data streams with one bit of each per audio channel per frame. The channel status bits are given a block structure and synchronized every 192 frames, which at 48 kHz gives a block rate of 250 Hz, corresponding to a period of 4 ms. In order to synchronize the channel-status blocks, the channel A sync pattern is replaced for one frame only by a third sync pattern which is also shown in Figure 9.10. The AES standard refers to these as X, Y and Z whereas IEC 958 calls them M, W and B. As stated, there is a parity bit in each subframe, which means that the binary level at the end of a subframe will always be the same as at the beginning. Since the sync patterns have the same characteristic, the effect is that sync patterns always have the same polarity and the receiver can use that information to reject noise.

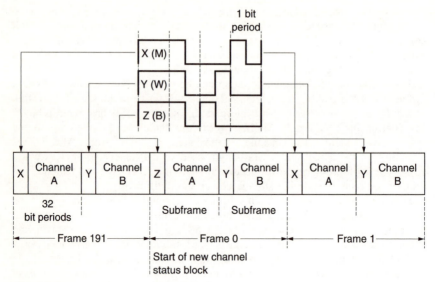

Figure 9.10 Three different preambles (X, Y and Z) are used to synchronize a receiver at the start of subframes.

The polarity of transmission is not specified, and indeed an accidental inversion in a twisted pair is of no consequence, since it is only the transition that is of importance, not the direction.

In both the professional and consumer formats, the sequence of channel-status bits over 192 subframes builds up a 24-byte channel-status block. However, the contents of the channel status data is completely different between the two applications. The professional channel status structure is shown in Figure 9.11. Byte 0 determines the use of emphasis and the sampling rate. Byte 1 determines the channel usage mode, i.e. whether the data transmitted are a stereo pair, two unrelated mono signals or a single mono signal, and details the user bit handling and byte 2 determines wordlength. Byte 3 is applicable only to multichannel applications. Byte 4 indicates the suitability of the signal as a sampling rate reference. There are two slots of four bytes each which are used for alphanumeric source and destination codes. These can be used for routing. The bytes contain seven-bit ASCII characters (printable characters only) sent LSB first with the eighth bit set to zero acording to AES-3-1992. The destination code can be used to operate an automatic router, and the source code will allow the origin of the audio and other remarks to be displayed at the destination.

Bytes 14–17 convey a thirty-two-bit sample address which increments every channel status frame. It effectively numbers the samples in a relative manner from an arbitrary starting point. Bytes 18–21 convey a similar number, but this is a time-of-day count, which starts from zero at midnight. As many digital audio devices do not have real-time clocks built in, this cannot be relied upon. AES-3-92 specified that the time-of-day bytes should convey the real time at which a recording was made, making it rather like timecode. There are enough combinations in thirty-two bits to allow a sample count over 24 hours at

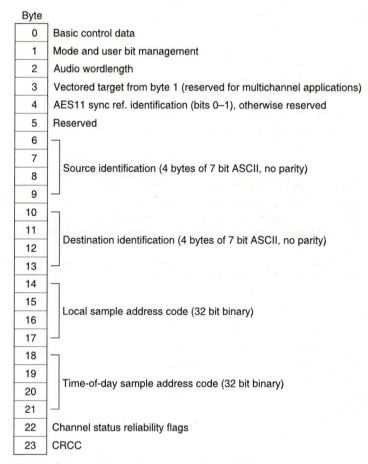

Byte

0	Basic control data
1	Mode and user bit management
2	Audio wordlength
3	Vectored target from byte 1 (reserved for multichannel applications)
4	AES11 sync ref. identification (bits 0–1), otherwise reserved
5	Reserved
6	
7	Source identification (4 bytes of 7 bit ASCII, no parity)
8	
9	
10	
11	Destination identification (4 bytes of 7 bit ASCII, no parity)
12	
13	
14	
15	Local sample address code (32 bit binary)
16	
17	
18	
19	Time-of-day sample address code (32 bit binary)
20	
21	
22	Channel status reliability flags
23	CRCC

Figure 9.11 Overall format of the professional channel-status block.

48 kHz. The sample count has the advantage that it is universal and independent of local supply frequency. In theory if the sampling rate is known, conventional hours, minutes, seconds, frames timecode can be calculated from the sample count, but in practice it is a lengthy computation and users have proposed alternative formats in which the data from EBU or SMPTE timecode are transmitted directly in these bytes. Some of these proposals are in service as *de facto* standards.

The penultimate byte contains four flags which indicate that certain sections of the channel-status information are unreliable. This allows the transmission of an incomplete channel-status block where the entire structure is not needed or where the information is not available. The final byte in the message is a CRCC which converts the entire channel-status block into a codeword (see Chapter 6). The channel status message takes 4 ms at 48 kHz and in this time a router could have switched to another signal source. This would damage the transmission, but will also result in a CRCC failure so the corrupt block is not used.

9.7 Telephone-based systems

The success of the telephone has led to vast number of subscribers being connected with copper wires and this is a valuable network infrastructure. As technology has developed, the telephone has become part of a global telecommunications industry. Simple economics suggests that in many cases improving the existing telephone cabling with modern modulation schemes is a good way of providing new communications services.

For economic reasons, there are fewer paths through the telephone system than there are subscribers. This is because telephones were not used continuously until teenagers discovered them. Before a call can be made, the exchange has to find a free path and assign it to the calling telephone. Traditionally this was done electromechanically. A path which was already in use would be carrying loop current. When the exchange sensed that a handset was off-hook, a rotary switch would advance and sample all the paths until it found one without loop current where it would stop. This was signalled to the calling telephone by sending a dial tone.

The development of electronics revolutionized telephone exchanges. Whilst the loop current, AC ringing and hook switch sensing remained for compatibility, the electromechanical exchange gave way to electronic exchanges where the dial pulses were interpreted by digital counters which then drove crosspoint switches to route the call. The communication remained analog.

The next advance permitted by electronic exchanges was touch-tone dialling, also called DTMF. Touch-tone dialling is based on seven discrete frequencies shown in Figure 9.12. The telephone contains tone generators and tuned filters in the exchange can detect each frequency individually. The numbers 0 through 9 and two non-numerical symbols, asterisk and hash, can be transmitted using twelve unique tone pairs. A tone pair can reliably be detected in about 100 ms and this makes dialling much faster than the pulse system.

Tone	1	2	3	4	5	6	7	8
Frequency Hz	697	770	852	941	1209	1336	1477	1633

Key	0	1	2	3	4	5	6	7	8	9	*	#
Tone pair	4,6	1,5	1,6	1,7	2,5	2,6	2,7	3,5	3,6	3,7	4,5	4,7

Figure 9.12 DTMF dialling works on tone pairs.

The frequencies chosen for DTMF are logarithmically spaced so that the filters can have constant bandwidth and response time, but they do not correspond to the conventional musical scale. In addition to dialling speed, because the DTMF tones are within the telephone audio bandwidth, they can also be used for signalling during a call.

The first electronic exchanges simply used digital logic to perform the routing function. The next step was to use a fully digital system where the copper wires from each subscriber terminate in an interface or *line card* containing ADCs and DACs. The sampling rate of 8 kHz retains the traditional analog bandwidth, and eight-bit quantizing is used. This is not linear, but uses logarithmically sized quantizing steps so that the quantizing error is greater on larger signals. The result is a 64 kbit/s data rate in each direction.

Packets of data can be time-division multiplexed into high bit-rate data buses which can carry many calls simultaneously. The routing function becomes simply one of watching the bus until the right packet comes along for the selected destination. 64 kbit/s data switching came to be known as IDN (Integrated Digital Network). As a data bus doesn't care whether it carries 64 kbit/s of speech or 64 kbit/s of something else, communications systems based on IDN tend to be based on multiples of that rate.

Such a system is called ISDN (integrated services digital network) which is basically a use of the telephone system that allows dial-up data transfer between subscribers in much the same way as a conventional phone call is made.

With the subsequent development of broadband networks (B-ISDN) the original ISDN is now known as N-ISDN where the N stands for narrow-band. B-ISDN is the ultimate convergent network able to carry any type of data and uses the well-known ATM (asynchronous transfer mode) protocol. Broadband and ATM are considered in a later section.

One of the difficulties of the AMI coding used in N-ISDN are that the data rate is limited and new cabling is needed to the exchange. ADSL (asymmetric digital subscriber line) is an advanced coding scheme which obtains high bit rate delivery and a back channel down existing subscriber telephone wiring.

ADSL works on frequency-division multiplexing using 4 kHz wide channels, 249 of these provide the delivery or downstream channel and 25 provide the back channel. Figure 9.13(a) shows that the existing bandwidth used by the traditional analog telephone is retained. The back channel occupies the lowest-frequency channels, with the downstream channels above. Figure 9.13(b) shows that at each end of the existing telephone wiring a device called a

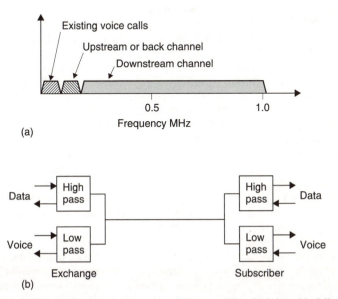

Figure 9.13 (a) ADSL allows the existing analog telephone to be retained, but adds delivery and back channels at higher frequencies. (b) A splitter is needed at each end of the subscriber's line.

splitter is needed. This is basically a high-pass/low-pass filter which directs audio frequency signals to the telephones and high-frequency signals to the modems.

Telephone wiring was never designed to support high-frequency signalling and is non-ideal. There will be reflections due to impedance mismatches which will cause an irregular frequency response in addition to high-frequency losses and noise which will all vary with cable length. ADSL can operate under these circumstances because it constantly monitors the conditions in each channel. If a given channel has adequate signal level and low noise, the full bit rate can be used, but in another channel there may be attenuation and the bit rate will have to be reduced. By independently coding the channels, the optimum data throughput for a given cable is obtained.

Each channel is modulated using DMT (discrete multitone technique) in which combinations of discrete frequencies are used. Within one channel symbol, there are 15 combinations of tones and so the coding achieves 15 bits/ s/Hz. With a symbol rate of 4 kHz, each channel can deliver 60 kbits/s, making 14.9 Mbits/s for the downstream channel and 1.5 Mbits/s for the back channel. It should be stressed that these figures are theoretical maxima which are not reached in real cables. Practical ADSL systems deliver multiples of the ISDN channel rate up to about 6 Mbits/s, enough to deliver MPEG-2 coded video.

Over shorter distances, VDSL can reach up to 50 Mbits/s. Where ADSL and VDSL are being referred to as a common technology, the term xDSL will be found.

9.8 Digital television broadcasting

Digital television broadcasting relies on the combination of a number of fundamental technologies. These are: MPEG-2 compression to reduce the bit rate, multiplexing to combine picture and sound data into a common bitstream, digital modulation schemes to reduce the RF bandwidth needed by a given bit rate and error correction to reduce the error statistics of the channel down to a value acceptable to MPEG data.

MPEG compressed video is highly sensitive to bit errors, primarily because they confuse the recognition of variable length codes so that the decoder loses synchronization. However, MPEG is a compression and multiplexing standard and does not specify how error correction should be performed. Consequently a transmission standard must define a system which has to correct essentially all errors such that the delivery mechanism is transparent.

Essentially a transmission standard specifies all the additional steps needed to deliver an MPEG transport stream from one place to another. This transport stream will consist of a number of elementary streams of video and audio, where the audio may be coded according to MPEG audio standard or AC-3. In a system working within its capabilities, the picture and sound quality will be determined only by the performance of the compression system and not by the RF transmission channel.

Whilst in one sense an MPEG transport stream is only data, it differs from generic data in that it must be presented to the viewer at a particular rate. Generic data are usually asynchronous, whereas baseband video and audio are synchronous. However, after compression and multiplexing audio and video are no longer precisely synchronous and so the term *isochronous* is used. This

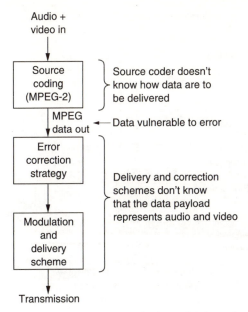

Figure 9.14 Source coder doesn't know delivery mechanism and delivery doesn't need to know what the data mean.

means a signal that was at one time synchronous and will be displayed synchronously, but which uses buffering at transmitter and receiver to accommodate moderate timing errors in the transmission.

Clearly another mechanism is needed so that the time axis of the original signal can be re-created on reception. The time stamp and program clock reference system of MPEG does this.

Figure 9.14 shows that the concepts involved in digital television broadcasting exist at various levels which have an independence not found in analog technology. In a given configuration a transmitter can radiate a given payload data bit rate. This represents the useful bit rate and does not include the necessary overheads needed by error correction, multiplexing or synchronizing. It is fundamental that the transmission system does not care what this payload bit rate is used for. The entire capacity may be used up by one high-definition channel, or a large number of heavily compressed channels may be carried. The details of this data usage are the domain of the *transport stream*. The multiplexing of transport streams is defined by the MPEG standards, but these do not define any error-correction or transmission technique.

At the lowest level in Figure 9.15 the source coding scheme, in this case MPEG compression, results in one or more elementary streams, each of which carries a video or audio channel. Elementary streams are multiplexed into a transport stream. The viewer then selects the desired elementary stream from the transport stream. Metadata in the transport stream ensure that when a video elementary stream is chosen, the appropriate audio elementary stream will automatically be selected.

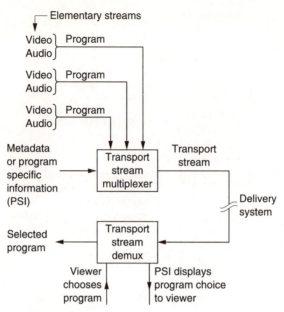

Figure 9.15 Program Specific Information helps the demultiplexer to select the required program.

9.9 MPEG packets and time stamps

The video elementary stream is an endless bitstream representing pictures which take a variable length of time to transmit. Bidirection coding means that pictures are not necessarily in the correct order. Storage and transmission systems prefer discrete blocks of data and so elementary streams are packetized to form a PES (packetized elementary stream). Audio elementary streams are also packetized. A packet is shown in Figure 9.16. It begins with a header containing an unique packet start code and a code which identifies the type of data stream. Optionally the packet header also may contain one or more *time stamps* which are used for synchronizing the video decoder to real time and for obtaining lip-sync.

Figure 9.17 shows that a time stamp is a sample of the state of a counter which is driven by a 90 kHz clock. This is obtained by dividing down the master 27 MHz clock of MPEG-2. This 27 MHz clock must be locked to the video frame

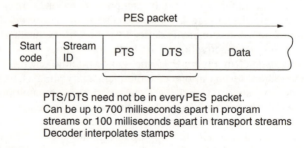

Figure 9.16 A PES packet structure is used to break up the continuous elementary stream.

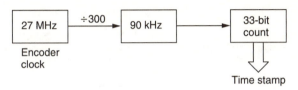

Figure 9.17 Time stamps are the result of sampling a counter driven by the encoder clock.

rate and the audio sampling rate of the program concerned. There are two types of time stamp: PTS and DTS. These are abbreviations for presentation time stamp and decode time stamp. A presentation time stamp determines when the associated picture should be displayed on the screen, whereas a decode time stamp determines when it should be decoded. In bidirectional coding these times can be quite different.

Audio packets have only presentation time stamps. Clearly if lip-sync is to be obtained, the audio sampling rate of a given program must have been locked to the same master 27 MHz clock as the video and the time stamps must have come from the same counter driven by that clock.

In practice, the time between input pictures is constant and so there is a certain amount of redundancy in the time stamps. Consequently PTS/DTS need not appear in every PES packet. Time stamps can be up to 100 ms apart in transport streams. As each picture type (*I, P* or *B*) is flagged in the bitstream, the decoder can infer the PTS/DTS for every picture from the ones actually transmitted.

The MPEG-2 transport stream is intended to be a multiplex of many TV programs with their associated sound and data channels, although a single program transport stream (SPTS) is possible. The transport stream is based upon packets of constant size so that multiplexing, adding error-correction codes and interleaving in a higher layer is eased. Figure 9.18 shows that these are always 188 bytes long.

Transport stream packets always begin with a header. The remainder of the packet carries data known as the payload. For efficiency, the normal header is relatively small, but for special purposes the header may be extended. In this case the payload gets smaller so that the overall size of the packet is unchanged. Transport stream packets should not be confused with PES packets which are larger and vary in size. PES packets are broken up to form the payload of the transport stream packets.

The header begins with a sync byte which is an unique pattern detected by a demultiplexer. A transport stream may contain many different elementary streams and these are identified by giving each an unique thirteen-bit Packet Identification Code or PID which is included in the header. A multiplexer seeking a particular elementary stream simply checks the PID of every packet and accepts only those which match.

In a multiplex there may be many packets from other programs in between packets of a given PID. To help the demultiplexer, the packet header contains a continuity count. This is a four-bit value which increments at each new packet having a given PID.

This approach allows statistical multiplexing as it does matter how many or how few packets have a given PID; the demux will still find them. Statistical multiplexing has the problem that it is virtually impossible to make the sum of

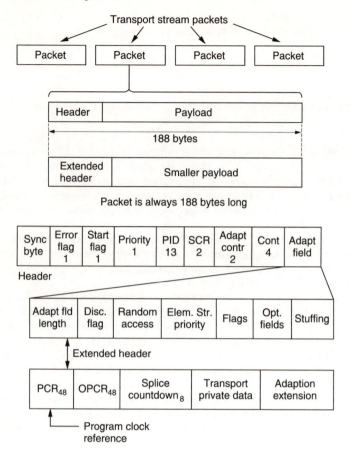

Figure 9.18 Transport stream packets are always 188 bytes long to facilitate multiplexing and error correction.

the input bit rates constant. Instead the multiplexer aims to make the average data bit rate slightly less than the maximum and the overall bit rate is kept constant by adding 'stuffing' or null packets. These packets have no meaning, but simply keep the bit rate constant. Null packets always have a PID of 8191 (all ones) and the demultiplexer discards them.

9.10 Program clock reference

A transport stream is a multiplex of several TV programs and these may have originated from widely different locations. It is impractical to expect all the programs in a transport stream to be genlocked and so the stream is designed from the outset to allow unlocked programs. A decoder running from a transport stream has to genlock to the encoder and the transport stream has to have a mechanism to allow this to be done independently for each program. The synchronizing mechanism is called Program Clock Reference (PCR).

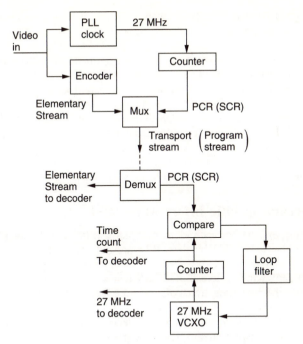

Figure 9.19 Program or System Clock Reference codes regenerate a clock at the decoder. See text for details.

Figure 9.19 shows how the PCR system works. The goal is to re-create at the decoder a 27 MHz clock which is synchronous with that at the encoder. The encoder clock drives a forty-eight-bit counter which continuously counts up to the maximum value before overflowing and beginning again.

A transport stream multiplexer will periodically sample the counter and place the state of the count in an extended packet header as a PCR (see Figure 9.18). The demultiplexer selects only the PIDs of the required program, and it will extract the PCRs from the packets in which they were inserted.

The PCR codes are used to control a numerically locked loop (NLL) described in section 2.9. The NLL contains a 27 MHz VCXO (voltage controlled crystal oscillator), a variable-frequency oscillator based on a crystal which has a relatively small frequency range.

The VCXO drives a forty-eight-bit counter in the same way as in the encoder. The state of the counter is compared with the contents of the PCR and the difference is used to modify the VCXO frequency. When the loop reaches lock, the decoder counter would arrive at the same value as is contained in the PCR and no change in the VCXO would then occur. In practice the transport stream packets will suffer from transmission jitter and this will create phase noise in the loop. This is removed by the loop filter so that the VCXO effectively averages a large number of phase errors.

A heavily damped loop will reject jitter well, but will take a long time to lock. Lockup time can be reduced when switching to a new program if the decoder counter is jammed to the value of the first PCR received in the new program. The loop filter may also have its time constants shortened during lockup.

Once a synchronous 27 MHz clock is available at the decoder, this can be divided down to provide the 90 kHz clock which drives the time stamp mechanism.

The entire timebase stability of the decoder is no better than the stability of the clock derived from PCR. MPEG-2 sets standards for the maximum amount of jitter which can be present in PCRs in a real transport stream.

Clearly, if the 27 MHz clock in the receiver is locked to one encoder it can only receive elementary streams encoded with that clock. If it is attempted to decode, for example, an audio stream generated from a different clock, the result will be periodic buffer overflows or underflows in the decoder. Thus MPEG defines a program in a manner which relates to timing. A program is a set of elementary streams which have been encoded with the same master clock.

9.11 Program Specific Information (PSI)

In a real transport stream, each elementary stream has a different PID, but the demultiplexer has to be told what these PIDs are and what audio belongs with what video before it can operate. This is the function of PSI which is a form of metadata. Figure 9.20 shows the structure of PSI. When a decoder powers up, it knows nothing about the incoming transport stream except that it must search for all packets with a PID of zero. PID zero is reserved for the Program Association Table (PAT). The PAT is transmitted at regular intervals and contains a list of all the programs in this transport stream. Each program is further described by its own Program Map Table (PMT) and the PIDs of of the PMTs are contained in the PAT.

Figure 9.20 also shows that the PMTs fully describe each program. The PID of the video elementary stream is defined, along with the PID(s) of the associated

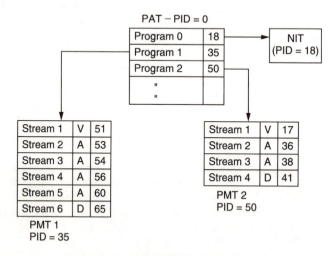

Figure 9.20 MPEG-2 Program Specific Information (PSI) is used to tell a demultiplexer what the transport stream contains.

audio and data streams. Consequently when the viewer selects a particular program, the demultiplexer looks up the program number in the PAT, finds the right PMT and reads the audio, video and data PIDs. It then selects elementary streams having these PIDs from the transport stream and routes them to the decoders.

Program 0 of the PAT contains the PID of the Network Information Table (NIT). This contains information about what other transport streams are available. For example, in the case of a satellite broadcast, the NIT would detail the orbital position, the polarization, carrier frequency and modulation scheme. Using the NIT a set-top box could automatically switch between transport streams.

Apart from 0 and 8191, a PID of 1 is also reserved for the Conditional Access Table (CAT). This is part of the access control mechanism needed to support pay per view or subscription viewing.

9.12 Transport stream multiplexing

A transport stream multiplexer is a complex device because of the number of functions it must perform. A fixed multiplexer will be considered first. In a fixed multiplexer, the bit rate of each of the programs must be specified so that the sum does not exceed the payload bit rate of the transport stream. The payload bit rate is the overall bit rate less the packet headers and PSI rate.

In practice, the programs will not be synchronous to one another, but the transport stream must produce a constant packet rate given by the bit rate divided by 188 bytes, the packet length. Figure 9.21 shows how this is handled. Each elementary stream entering the multiplexer passes through a buffer which is divided into payload-sized areas. Note that periodically the payload area is made smaller because of the requirement to insert PCR.

MPEG-2 decoders also have a quantity of buffer memory. The challenge to the multiplexer is to take packets from each program in such a way that neither its

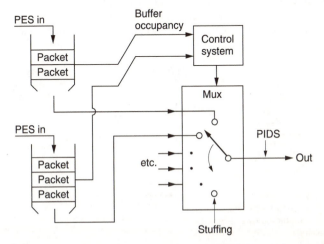

Figure 9.21 A transport stream multiplexer can handle several programs which are asynchronous to one another and to the transport stream clock. See text for details.

own buffers nor the buffers in any decoder either overflow or underflow. This requirement is met by sending packets from all programs as evenly as possible rather than bunching together a lot of packets from one program. When the bit rates of the programs are different, the only way this can be handled is to use the buffer contents indicators. The more full a buffer is, the more likely it should be that a packet will be read from it. This a buffer content arbitrator can decide which program should have a packet allocated next.

If the sum of the input bit rates is correct, the buffers should all slowly empty because the overall input bit rate has to be less than the payload bit rate. This allows for the insertion of Program Specific Information. Whilst PATs and PMTs are being transmitted, the program buffers will fill up again. The multiplexer can also fill the buffers by sending more PCRs as this reduces the payload of each packet. In the event that the multiplexer has sent enough of everything but still can't fill a packet then it will send a null packet with a PID of 8191. Decoders will discard null packets and as they convey no useful data, the multiplexer buffers will all fill whilst null packets are being transmitted.

The use of null packets means that the bit rates of the elementary streams do not need to be synchronous with one another or with the transport stream bit rate. As each elementary stream can have its own PCR, it is not necessary for the different programs in a transport stream to be genlocked to one another; in fact they don't even need to have the same frame rate.

This approach allows the transport stream bit rate to be accurately defined and independent of the timing of the data carried. This is important because the transport stream bit rate determines the spectrum of the transmitter and this must not vary.

In a statistical multiplexer or statmux, the bit rate allocated to each program can vary dynamically. Figure 9.22 shows that there must be a tight connection between the statmux and the associated compressors. Each compressor has a buffer memory which is emptied by a demand clock from the statmux. In a normal, fixed bit rate, coder the buffer content feeds back and controls the requantizer. In statmuxing this process is less severe and only takes place if the buffer is very close to full, because the degree of coding difficulty is also fed to the statmux.

The statmux contains an arbitrator which allocates more packets to the program with the greatest coding difficulty. Thus if a particular program encounters difficult

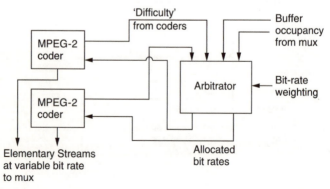

Figure 9.22 A statistical multiplexer contains an arbitrator which allocates bit rate to each program as a function of program difficulty.

matérial it will produce large prediction errors and begin to fill its output buffer. As the statmux has allocated more packets to that program, more data will be read out of that buffer, preventing overflow. Of course this is only possible if the other programs in the transport stream are handling typical video.

In the event that several programs encounter difficult material at once, clearly the buffer contents will rise and the requantizing mechanism will have to operate.

9.13 Broadcast modulation techniques

A key difference between analog and digital transmission is that the transmitter output is switched between a number of discrete states rather than continuously varying. A good code minimizes the channel bandwidth needed for a given bit rate. This quality of the code is measured in bits/s/Hz and is the equivalent of the density ratio in recording. Figure 9.23 shows, not surprisingly, that the less bandwidth required, the better the signal-to-noise ratio has to be. The figure shows the theoretical limit as well as the performance of a number of codes which offer different balances of bandwidth/noise performance.

Where the SNR is poor, as in satellite broadcasting, the amplitude of the signal will be unstable, and phase modulation is used. Figure 9.24 shows that phase-shift keying (PSK) can use two or more phases. When four phases in quadrature are used, the result is Quadrature Phase Shift Keying or QPSK. Each period of

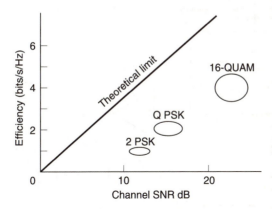

Figure 9.23 Where a better SNR exists, more data can be sent down a given bandwidth channel.

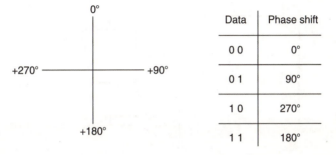

Data	Phase shift
0 0	0°
0 1	90°
1 0	270°
1 1	180°

Figure 9.24 Differential quadrature phase shift keying (DQPSK).

the transmitted waveform can have one of four phases and therefore conveys the value of two data bits. 8-PSK uses eight phases and can carry three bits per symbol where the SNR is adequate. PSK is generally encoded in such a way that a knowledge of absolute phase is not needed at the receiver. Instead of encoding the signal phase directly, the data determine the magnitude of the phase shift between symbols. A QPSK coder is shown in Figure 9.25.

In terrestrial transmission more power is available than, for example from a satellite and so a stronger signal can be delivered to the receiver. Where a better SNR exists, an increase in data rate can be had using multi-level signalling or *m*-ary coding instead of binary. Figure 9.26 shows that the ATSC system uses an eight-level signal (8-VSB) allowing three bits to be sent per symbol. Four of the levels exist with normal carrier phase and four with inverted phase so that a phase-sensitive rectifier is needed in the receiver. Clearly, the data separator must have a three-bit ADC which can resolve the eight signal levels. The gain and offset of the signal must be precisely set so that the quantizing levels register precisely with the centres of the eyes. The transmitted signal contains sync pulses

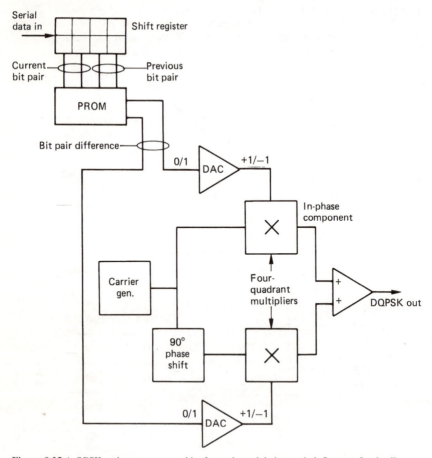

Figure 9.25 A QPSK coder conveys two bits for each modulation period. See text for details.

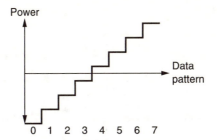

Figure 9.26 In 8-VSB the transmitter operates in eight different states enabling three bits to be sent per symbol.

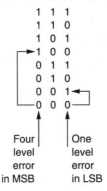

Figure 9.27 In multi-level signalling the error probability is not the same for each bit.

which are encoded using specified code levels so that the data separator can set its gain and offset.

Multi-level signalling systems have the characteristic that the bits in the symbol have different error probability. Figure 9.27 shows that a small noise level will corrupt the low-order bit, whereas twice as much noise will be needed to corrupt the middle bit and four times as much will be needed to corrupt the high-order bit. In ATSC the solution is that the lower two bits are encoded together in an inner error-correcting scheme so that they represent only one bit with similar reliability to the top bit. As a result the 8-VSB system actually delivers two data bits per symbol even though eight-level signalling is used.

The modulation of the carrier results in a double-sideband spectrum, but following analog TV practice most of the lower sideband is filtered off leaving a vestigial sideband only, hence the term 8-VSB. A small DC offset is injected into the modulator signal so that the four in-phase levels are slightly higher than the four out-of-phase levels. This has the effect of creating a small pilot at the carrier frequency to help receiver locking.

Multi-level signalling can be combined with PSK to obtain multi-level Quadrature Amplitude Modulation (QUAM). Figure 9.28 shows an example of 64-QUAM. Incoming six-bit data words are split into two three-bit words and each is used to amplitude modulate a pair of sinusoidal carriers which are generated in quadrature. The modulators are four-quadrant devices such that 2^3 amplitudes are available, four which are in phase with the carrier and four which

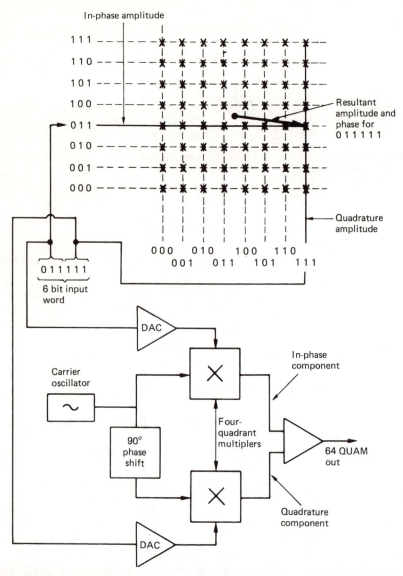

Figure 9.28 In 64-QUAM, two carriers are generated with a quadrature relationship. These are independently amplitude modulated to eight discrete levels in four quadrant multipliers. Adding the signals produces a QUAM signal having 64 unique combinations of amplitude and phase. Decoding requires the waveform to be sampled in quadrature like a colour TV subcarrier.

are antiphase. The two AM carriers are linearly added and the result is a signal which has 2^6 or 64 combinations of amplitude and phase. There is a great deal of similarity between QUAM and the colour subcarrier used in analog television in which the two colour difference signals are encoded into one amplitude- and phase-modulated waveform. On reception, the waveform is sampled twice per cycle in phase with the two original carriers and the result is a pair of eight-level

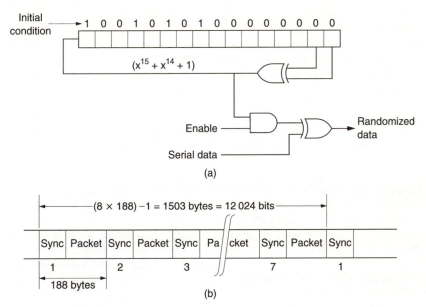

Figure 9.29 (a) The randomizer of DVB is preset to the initial condition once every 8 transport stream packets. The maximum length of the sequence is 65 535 bits, but only the first 12 024 bits are used before resetting again (b).

signals. 16-QUAM is also possible, delivering only four bits per symbol but requiring a lower SNR.

The data bit patterns to be transmitted can have any combinations whatsoever, and if nothing were done, the transmitted spectrum would be non-uniform. This is undesirable because peaks cause interference with other services, whereas energy troughs allow external interference in. The randomizing technique of section 6.13 is used to overcome the problem. The process is known as energy dispersal. The signal energy is spread uniformly throughout the allowable channel bandwidth so that it has less energy at a given frequency.

A pseudo-random sequence generator is used to generate the randomizing sequence. Figure 9.29 shows the randomizer used in DVB. This sixteen-bit device has a maximum sequence length of 65 535 bits, and is preset to a standard value at the beginning of each set of eight transport stream packets. The serialized data are XORed with the LSB of the Galois field, which randomizes the output which then goes to the modulator. The spectrum of the transmission is now determined by the spectrum of the prs.

9.14 OFDM

The way that radio signals interact with obstacles is a function of the relative magnitude of the wavelength and the size of the object. AM sound radio transmissions with a wavelength of several hundred metres can easily diffract around large objects. The shorter the wavelength of a transmission, the larger objects in the environment appear to it, and these objects can then become reflectors. Reflecting objects produce a delayed signal at the receiver in addition to the direct signal. In analog television transmissions this causes the familiar ghosting. In digital transmissions, the symbol rate may be so high that the reflected signal may be one or more symbols behind the direct signal, causing intersymbol interference. As the reflection may be continuous, the result may be that almost every symbol is corrupted. No error-correction system can handle this. Raising the transmitter power is no help at all as it simply raises the power of the reflection in proportion.

The only solution is to change the characteristics of the RF channel in some way to either prevent the multipath reception or stop it being a problem. The RF channel includes the modulator, transmitter, antennae, receiver and demodulator.

As with analog UHF TV transmissions, a directional antenna is useful with digital transmission as it can reject reflections. However, directional antennae tend to be large and they require a skilled permanent installation. Mobile use on a vehicle or vessel is simply impractical.

Another possibility is to incorporate a ghost canceller in the receiver. The transmitter periodically sends a standardized waveform known as a training sequence. The receiver knows what this waveform looks like and compares it with the received signal. In theory it is possible for the receiver to compute the delay and relative level of a reflection and so insert an opposing one. In practice if the reflection is strong it may prevent the receiver finding the training sequence.

The most elegant approach is to use a system in which multipath reception conditions cause only a small increase in error rate which the error-correction system can manage. This approach is used in DVB. Figure 9.30(a) shows that when using one carrier with a high bit rate, reflections can easily be delayed by one or more bit periods, causing interference between the bits. Figure 9.30(b) shows that instead, OFDM sends many carriers each having a low bit rate. When a low bit rate is used, the energy in the reflection will arrive during the same bit period as the direct signal. Not only is the system immune to multipath reflections, but the energy in the reflections can actually be used. This characteristic can be enhanced by using guard intervals shown in Figure 9.30(c). These reduce multipath bit overlap even more.

Note that OFDM is not a modulation scheme, and each of the carriers used in a OFDM system still needs to be modulated using any of the digital coding schemes described above. What OFDM does is to provide an efficient way of packing many carriers close together without mutual interference.

A serial data waveform basically contains a train of rectangular pulses. The transform of a rectangle is the function $\sin x/x$ and so the baseband pulse train has a $\sin x/x$ spectrum. When this waveform is used to modulate a carrier the result is a symmetrical $\sin x/x$ spectrum centred on the carrier frequency. Figure 9.31(a) shows that nulls in the spectrum appear spaced at multiples of the bit rate away from the carrier.

Further carriers can be placed at spacings such that each is centred at the nulls of the others as is shown in Figure 9.31(b). The distance between the carriers is equal to 90° or one quadrant of $\sin x$. Owing to the quadrant spacing, these carriers are mutually orthogonal, hence the term 'orthogonal frequency division'. A large number of such carriers (in practice, several thousand) will be interleaved to produce an overall spectrum which is almost rectangular and which fills the available transmission channel.

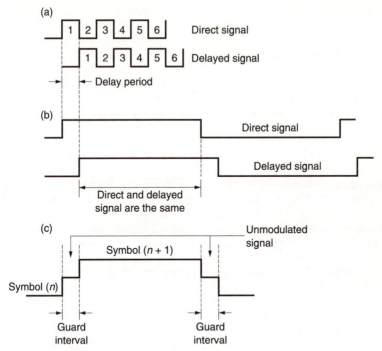

Figure 9.30 (a) High bit rate transmissions are prone to corruption due to reflections. (b) If the bit rate is reduced the effect of reflections is eliminated, in fact reflected energy can be used. (c) Guard intervals may be inserted between symbols.

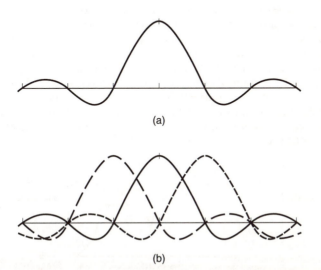

Figure 9.31 In OFDM the carrier spacing is critical, but when correct the carriers become independent and most efficient use is made of the spectrum. (a) Spectrum of bitstream has regular nulls. (b) Peak of one carrier occurs at null of another.

When guard intervals are used, the carrier returns to an unmodulated state between bits for a period which is greater than the period of the reflections. Then the reflections from one transmitted bit decay during the guard interval before the next bit is transmitted. The use of guard intervals reduces the bit rate of the carrier because for some of the time it is radiating carrier, not data. A typical reduction is to around 80 per cent of the capacity without guard intervals.

This capacity reduction does, however, improve the error statistics dramatically, such that much less redundancy is required in the error-correction system. Thus the effective transmission rate is improved. The use of guard intervals also moves more energy from the sidebands back to the carrier. The frequency spectrum of a set of carriers is no longer perfectly flat but contains a small peak at the centre of each carrier.

The ability to work in the presence of multipath cancellation is one of the great strengths of OFDM. In DVB, more than 2000 carriers are used in single-transmitter systems. Provided there is exact synchronism, several transmitters can radiate exactly the same signal so that a single-frequency network can be created throughout a whole country. SFNs require a variation on OFDM which uses over 8000 carriers.

With OFDM, directional antennae are not needed and, given sufficient field strength, mobile reception is perfectly feasible. Of course, directional antennae may still be used to boost the received signal outside of normal service areas or to enable the use of low-powered transmitters.

An OFDM receiver must perform fast Fourier transforms (FFTs) on the whole band at the symbol rate of one of the carriers. The amplitude and/or phase of the carrier at a given frequency effectively reflects the state of the transmitted symbol at that time slot and so the FFT partially demodulates as well.

In order to assist with tuning in, the OFDM spectrum contains pilot signals. These are individual carriers which are transmitted with slightly more power than the remainder. The pilot carriers are spaced apart through the whole channel at agreed frequencies which form part of the transmission standard.

Practical reception conditions, including multipath reception, will cause a significant variation in the received spectrum and some equalization will be needed. Figure 9.32 shows what the possible spectrum looks like in the presence of a powerful reflection. The signal has almost been cancelled at certain frequencies. However, the FFT performed in the receiver is effectively a spectral analysis of the signal and so the receiver computes for free the received spectrum. As in a flat spectrum the peak magnitude of all the coefficients would be the same (apart from the pilots), equalization is easily performed by multiplying the coefficients by suitable constants until this characteristic is obtained.

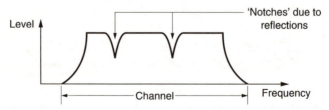

Figure 9.32 Multipath reception can place notches in the channel spectrum. This will require equalization at the receiver.

Although the use of transform-based receivers appears complex, when it is considered that such an approach simultaneously allows effective equalization the complexity is not significantly higher than that of a conventional receiver which needs a separate spectral analysis system just for equalization purposes.

The only drawback of OFDM is that the transmitter must be highly linear to prevent intermodulation between the carriers. This is readily achieved in terrestrial transmitters by derating the transmitter so that it runs at a lower power than it would in analog service. This is not practicable in satellite transmitters which are optimized for efficiency so OFDM is not really suitable for satellite use.

9.15 Error correction in digital television broadcasting

As in recording, broadcast data suffer from both random and burst errors and the error-correction strategies of digital television broadcasting have to reflect that. Figure 9.33 shows a typical system in which inner and outer codes are employed.

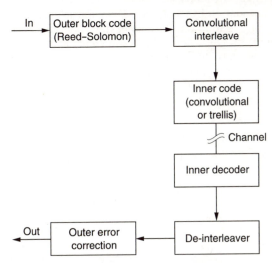

Figure 9.33 Error-correcting strategy of digital television broadcasting systems.

The Reed–Solomon codes are universally used for burst-correcting outer codes, along with an interleave which will be convolutional rather than the block-based interleave used in recording media. The inner codes will not be R–S, as more suitable codes exist for the statistical conditions prevalent in broadcasting. DVB uses a parity-based variable-rate system in which the amount of redundancy can be adjusted according to reception conditions. ATSC uses a fixed-rate parity-based system along with trellis coding to overcome co-channel interference from analog NTSC transmitters.

9.16 DVB

The DVB system is subdivided into systems optimized for satellite, cable and terrestrial delivery. This section concentrates on the terrestrial delivery system. Figure 9.34 shows a block diagram of a DVB-T transmitter.

Transport
stream
in

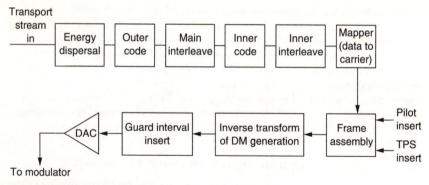

To modulator

Figure 9.34 DVB-T transmitter block diagram. See text for details.

Incoming transport stream packets of 188 bytes each are first subject to R–S outer coding. This adds 16 bytes of redundancy to each packet, resulting in 204 bytes. Outer coding is followed by interleaving. The interleave mechanism is shown in Figure 9.35. Outer code blocks are commutated on a byte basis into twelve parallel channels. Each channel contains a different amount of delay, typically achieved by a ring-buffer RAM. The delays are integer multiples of 17 bytes, designed to skew the data by one outer block ($12 \times 17 = 204$). Following the delays, a commutator reassembles interleaved outer blocks. These have 204

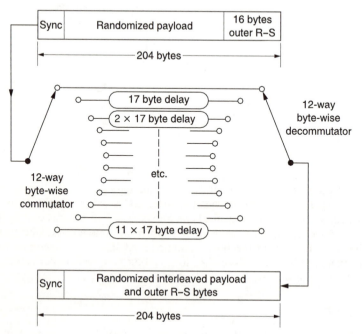

Figure 9.35 The interleaver of DVB uses 12 incrementing delay channels to reorder the data. The sync byte passes through the undelayed channel and so is still at the head of the packet after interleave. However, the packet now contains non-adjacent bytes from 12 different packets.

bytes as before, but the effect of the interleave is that adjacent bytes in the input are 17 bytes apart in the output. Each output block contains data from twelve input blocks making the data resistant to burst errors.

Following the interleave, the energy-dispersal process takes place. The pseudo-random sequence runs over eight outer blocks and is synchronized by inverting the transport stream packet sync symbol in every eighth block. The packet sync symbols are not randomized.

The inner coding process of DVB is shown in Figure 9.36. Input data are serialized and pass down a shift register. Exclusive-OR gates produce convolutional parity symbols X and Y, such that the output bit rate is twice the input bit rate. Under the worst reception conditions, this 100 per cent redundancy offers the most powerful correction with the penalty that a low data rate is delivered. However, Figure 9.36 also shows that a variety of inner redundancy factors can be used from 1/2 down to 1/8 of the transmitted bit rate. The X, Y data from the inner coder are subsampled, such that the coding is punctured.

The DVB standard allows the use of QPSK, 16-QUAM or 64-QUAM coding in an OFDM system. There are five possible inner code rates, and four different guard intervals which can be used with each modulation scheme, Thus for each modulation scheme there are twenty possible transport stream bit rates in the standard DVB channel, each of which requires a different receiver SNR. The broadcaster can select any suitable balance between transport stream bit rate and

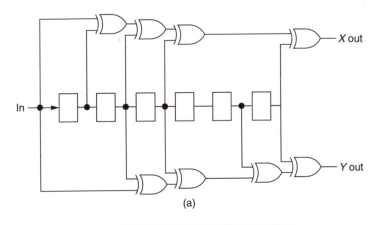

(a)

Rate	Transmitted sequence
1/2	X_1 Y_1
2/3	X_1 Y_1 Y_2
3/4	X_1 Y_1 Y_2 X_3
5/6	X_1 Y_1 Y_2 X_3 Y_4 X_6
7/8	X_1 Y_1 Y_2 Y_3 Y_4 X_5 X_6 X_7

(b)

Figure 9.36 (a) The mother inner coder of DVB produces 100 per cent redundancy, but this can be punctured by subsampling the X and Y data to give five different code rates as (b) shows.

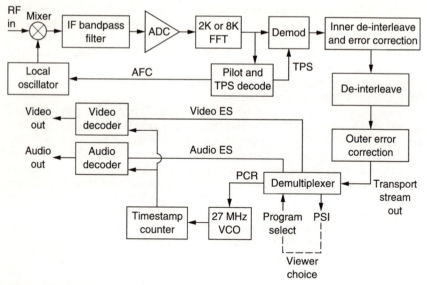

Figure 9.37 DVB receiver block diagram. See text for details.

coverage area. For a given transmitter location and power, reception over a larger area may require a channel code with a smaller number of bits/s/Hz and this reduces the bit rate which can be delivered in a standard channel. Alternatively a higher amount of inner redundancy means that the proportion of the transmitted bit rate which is data goes down. Thus for wider coverage the broadcaster will have to send fewer programs in the multiplex or use higher compression factors.

Figure 9.37 shows a block diagram of a DVB receiver. The off-air RF signal is fed to a mixer driven by the local oscillator. The IF output of the mixer is bandpass filtered and supplied to the ADC which outputs a digital IF signal for the FFT stage. The FFT is analysed initially to find the higher-level pilot signals. If these are not in the correct channels the local oscillator frequency is incorrect and it will be changed until the pilots emerge from the FFT in the right channels. The data in the pilots will be decoded in order to tell the receiver how many carriers, what inner redundancy rate, guard band rate and modulation scheme are in use in the remaining carriers. The FFT magnitude information is also a measure of the equalization required.

The FFT outputs are demodulated into 2K or 8K bitstreams and these are multiplexed to produce a serial signal. This is subject to inner error correction which corrects random errors. The data are then de-interleaved to break up burst errors and then the outer R–S error-correction operates. The output of the R–S correction will then be derandomized to become an MPEG transport stream once more. The derandomizing is synchronized by the transmission of inverted sync patterns.

The receiver must select a PID of 0 and wait until a Program Association Table (PAT) is transmitted. This will describe the available programs by listing the PIDs of the Program Map Tables (PMT). By looking for these packets the receiver can determine what PIDs to select to receive any video and audio elementary streams.

When an elementary stream is selected, some of the packets will have extended headers containing program clock reference (PCR). These codes are used to synchronize the 27 MHz clock in the receiver to the one in the MPEG encoder of the desired program. The 27 MHz clock is divided down to drive the time stamp counter so that audio and video emerge from the decoder at the correct rate and with lip sync.

It should be appreciated that time stamps are relative, not absolute. The time stamp count advances by a fixed amount each picture, but the exact count is meaningless. Thus the decoder can only establish the frame rate of the video from time stamps, but not the precise timing. In practice the receiver has finite buffering memory between the demultiplexer and the MPEG decoder. If the displayed video timing is too late, the buffer will tend to overflow whereas if the displayed video timing is too early the decoding may not be completed. The receiver can advance or retard the time stamp counter during lockup so that it places the output timing mid-way between these extremes.

9.17 ATSC

The ATSC system is an alternative way of delivering a transport stream, but it is considerably less sophisticated than DVB, and supports only one transport stream bit rate of 19.28 Mbits/s. If any change in the service area is needed, this will require a change in transmitter power.

Figure 9.38 shows a block diagram of an ATSC transmitter. Incoming transport stream packets are randomized, except for the sync pattern, for energy dispersal. Figure 9.39 shows the randomizer.

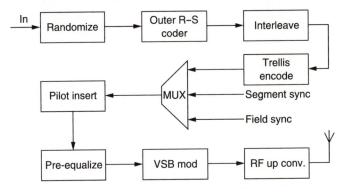

Figure 9.38 Block diagram of ATSC transmitter. See text for details.

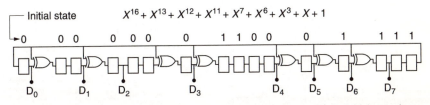

Figure 9.39 The randomizer of ATSC. The twisted ring counter is preset to the initial state shown each data field. It is then clocked once per byte and the eight outputs D_0-D_7 are X-ORed with the data byte.

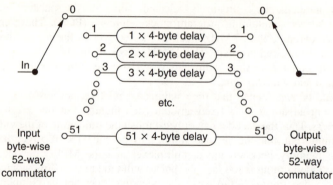

Figure 9.40 The ATSC convolutional interleaver spreads adjacent bytes over a period of about 4 ms.

The outer correction code includes the whole packet except for the sync byte. Thus there are 187 bytes of data in each codeword. Twenty bytes of R–S redundancy are added to make a 207-byte codeword. After outer coding, a convolutional interleaver shown in Figure 9.40 is used. This reorders data over a time span of about 4 ms. Interleave simply exchanges content between packets, but without changing the packet structure.

Figure 9.41 shows that the result of outer coding and interleave is a data frame which is divided into two fields of 313 segments each. The frame is tranmitted by scanning it horizontally a segment at a time. There is some similarity with a traditional analog video signal here, because there is a sync pulse at the beginning of each segment and a field sync which occupies two segments of the

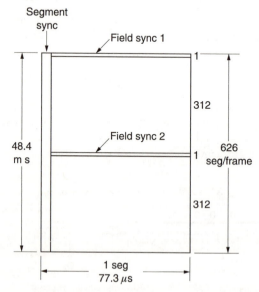

Figure 9.41 The ATSC data frame is transmitted one segment at a time. Segment sync denotes the beginning of each segment and the segments are counted from the field sync signals.

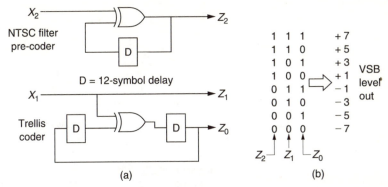

Figure 9.42 (a) The precoder and trellis coder of ATSC converts two data bits X_1, X_2 to three output bits Z_0, Z_1, Z_2. (b) The Z_0, Z_1, Z_2 output bits map to the eight-level code as shown.

frame. *Data segment sync* repeats every 77.3 ms, a segment rate of 12 933 Hz, whereas a frame has a period of 48.4 ms. The field sync segments contain a training sequnce to drive the adaptive equalizer in the receiver.

The data content of the frame is subject to trellis coding which converts each pair of data bits into three channel bits inside an inner interleave. The trellis coder is shown in Figure 9.42 and the interleave in Figure 9.43. Figure 9.42 also shows how the three channel bits map to the eight signal levels in the 8-VSB modulator.

Figure 9.44 shows the data segment after eight-level coding. The sync pattern of the transport stream packet, which was not included in the error-correction code, has been replaced by a segment sync waveform. This acts as a timing reference to allow deserializing of the segment, but as the two levels of the sync pulse are standardized, it also acts as an amplitude reference for the eight-level slicer in the receiver.

The eight-level signal is subject to a DC offset so that some transmitter energy appears at the carrier frequency to act as a pilot. Each eight-level symbol carries two data bits and so there are 832 symbols in each segment. As the segment rate is 12 933 Hz, the symbol rate is 10.76 MHz and so this will require 5.38 MHz of bandwidth in a single sideband,

Figure 9.45 shows the transmitter spectrum. The lower sideband is vestigial and an overall channel width of 6 MHz results.

Figure 9.46 shows an ATSC receiver. The first stages of the receiver are designed to lock to the pilot in the transmitted signal. This then allows the eight-level signal to be sampled at the right times. This process will allow location of the segment sync and then the field sync signals. Once the receiver is synchronized, the symbols in each segment can be decoded. The inner or trellis coder corrects for random errors, then following de-interleave the R–S coder corrects burst errors, After derandomizing, standard transport stream sync patterns are added to the output data.

In practice, ATSC transmissions will experience co-channel interference from NTSC transmitters and the ATSC scheme allows the use of an NTSC rejection filter. Figure 9.47 shows that most of the energy of NTSC is at the carrier, subcarrier and sound carrier frequencies. A comb filter with a suitable delay can produce nulls or notches at these frequencies. However, the delay-and-add

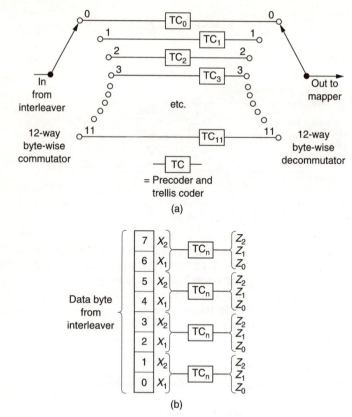

Figure 9.43 The inner interleave (a) of ATSC makes the trellis coding operate as twelve parallel channels working on every twelfth byte to improve error resistance. The interleave is byte-wise, and, as (b) shows, each byte is divided into four di-bits for coding into the tri-bits Z_0, Z_1, Z_2.

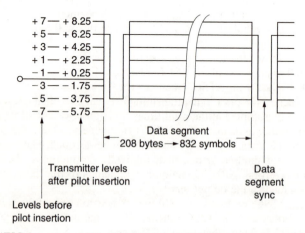

Figure 9.44 ATSC data segment. Note the sync pattern which acts as a timing and amplitude reference. The eight levels are shifted up by 1.25 to create a DC component resulting in a pilot at the carrier frequency.

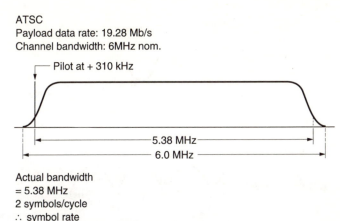

ATSC
Payload data rate: 19.28 Mb/s
Channel bandwidth: 6MHz nom.

Pilot at + 310 kHz

5.38 MHz
6.0 MHz

Actual bandwidth
= 5.38 MHz
2 symbols/cycle
∴ symbol rate
= 2 × 5.38 MHz = 10.76 MHz

Figure 9.45 The spectrum of ATSC and its associated bit and symbol rates. Note pilot at carrier frequency created by DC offset in multi-level coder.

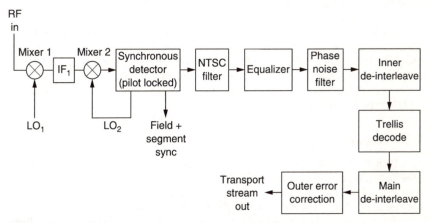

RF
in

Mixer 1   Mixer 2   Synchronous detector (pilot locked)   NTSC filter   Equalizer   Phase noise filter   Inner de-interleave

IF_1

LO_1 LO_2 Field + segment sync

Trellis decode

Transport stream out   Outer error correction   Main de-interleave

Figure 9.46 An ATSC receiver. Double conversion can be used so that the second conversion stage can be arranged to lock to the transmitted pilot.

process in the comb filter also causes another effect. When two eight-level signals are added together, the result is a sixteen-level signal. This will be corrupted by noise of half the level that would corrupt an eight-level signal. However, the sixteen-level signal contains redundancy because it corresponds to the combinations of four bits whereas only two bits are being transmitted. This allows a form of error correction to be used.

The ATSC inner precoder results in a known relationship existing between symbols independent of the data. The time delays in the inner interleave are designed to be compatible with the delay in the NTSC rejection comb filter. This limits the number of paths the received waveform can take through a time/voltage graph called a trellis. Where a signal is in error it takes a path sufficiently near to the correct one that the correct one can be implied.

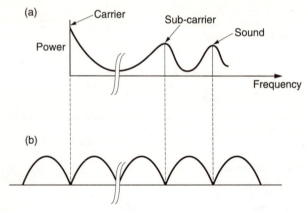

Figure 9.47 Spectrum of typical analog transmitter showing (a) maximum power at carrier, subcarrier and audio carrier. A comb filter (b) with a suitable delay can notch out NTSC interference. The precoding of ATSC is designed to work with the necessary receiver delay.

ATSC uses a training sequence sent once every data field, but is otherwise helpless against multipath reception as tests have shown. In urban areas, ATSC must have a correctly oriented directional antenna to reject reflections. Unfortunately the American viewer has been brought up to believe that television reception is possible with a pair of 'rabbit's ears' on top of the TV set and ATSC will not work like this. Mobile reception is not practicable.

As a result, the majority of the world's broadcasters appear to be favouring an OFDM-based system.

9.18 Networks

A network is basically a communication resource which is shared for economic reasons. Like any shared resource, decisions have to be made somewhere and somehow about how the resource is to be used. In the absence of such decisions the resultant chaos will be such that the resource might as well not exist.

In communications networks the resource is the ability to convey data from any node or port to any other. On a particular cable, clearly only one transaction of this kind can take place at any one instant even though in practice many nodes will simultaneously be wanting to transmit data. Arbitration is needed to determine which node is allowed to transmit.

There are a number of different arbitration protocols and these have evolved to support the needs of different types of network. In small networks, such as LANs, a single point failure which halts the entire network may be acceptable, whereas in a public transport network owned by a telecommunications company, the network will be redundant so that if a particular link fails data may be sent via an alternative route. A link which has reached its maximum capacity may also be supplanted by transmission over alternative routes.

In physically small networks, arbitration may be carried out in a single location. This is fast and efficient, but if the arbitrator fails it leaves the system completely crippled. The processor buses in computers work in this way. In centrally arbitrated systems the arbitrator needs to know the structure of the system and the status of all the nodes. Following a configuration change, due perhaps to the installation of new equipment, the arbitrator needs to be told what the new configuration is, or have a mechanism which allows it to explore the network and learn the configuration. Central arbitration is only suitable for small networks which change their configuration infrequently.

In other networks the arbitration is distributed so that some decision-making ability exists in every node. This is less efficient but is does allow at least some of the network to continue operating after a component failure. Distributed arbitration also means that each node is self-sufficient and so no changes need to be made if the network is reconfigured by adding or deleting a node. This is the only possible approach in wide area networks where the structure may be very complex and change dynamically in the event of failures or overload.

Ethernet uses distributed arbitration. FireWire is capable of using both types of arbitration. A small amount of decision-making ability is built into every node so that distributed arbitration is possible. However, if one of the nodes happens to be a computer, it can run a centralized arbitration algorithm.

The physical structure of a network is subject to some variation as Figure 9.48 shows. In radial networks, (a), each port has a unique cable connection to a device called a *hub*. The hub must have one connection for every port and this limits the number of ports. However, a cable failure will only result in the loss of one port. In a ring system (b) the nodes are connected like a daisy chain with each node acting as a feedthrough. In this case the arbitration requirement must be distributed. With some protocols, a single cable break doesn't stop the network operating. Depending on the protocol, simultaneous transactions may be possible provided they don't require the same cable. For example, in a storage network a disk drive may be outputting data to an editor while another drive is backing up data to a tape streamer. For the lowest cost, all nodes are physically

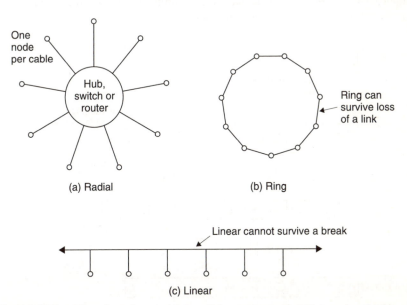

(a) Radial   (b) Ring

(c) Linear

Figure 9.48 Network configurations. At (a) the radial system uses one cable to each node. (b) Ring system uses less cable than radial. (c) Linear system is simple but has no redundancy.

connected in parallel to the same cable. Figure 9.48(c) shows that a cable break would divide the network into two halves, but it is possible that the impedance mismatch at the break could stop both halves working.

One of the concepts involved in arbitration is priority, which is fundamental to providing an appropriate quality of service. If two processes both want to use a network, the one with the highest priority would normally go first. Attributing priority must be done carefully because some of the results are non-intuitive. For example, it may be beneficial to give a high priority to a humble device which has a low data rate for the simple reason that if it is given use of the network it won't need it for long. In a television environment transactions concerned with on-air processes would have priority over file transfers concerning production and editing.

When a device gains access to the network to perform a transaction, generally no other transaction can take place until it has finished. Consequently it is important to limit the amount of time that a given port can stay on the bus. In this way when the time limit expires, a further arbitration must take place. The result is that the network resource rotates between transactions rather than one transfer hogging the resource and shutting everyone else out.

It follows from the presence of a time (or data quantity) limit that ports must have the means to break large files up into frames or cells and reassemble them on reception. This process is sometimes called *adaptation*. If the data to be sent originally exist at a fixed bit rate, some buffering will be needed so that the data can be time-compressed into the available frames. Each frame must be contiguously numbered and the system must transmit a file size or word count so that the receiving node knows when it has received every frame in the file.

The error-detection system interacts with this process because if any frame is in error on reception, the receiving node can ask for a retransmission of the frame. This is more efficient than retransmitting the whole file. Figure 9.49 shows the flow chart for a receiving node.

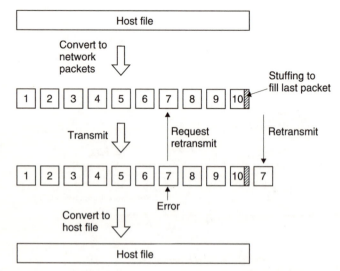

Figure 9.49 Receiving a file which has been divided into packets allows for the retransmission of just the packet in error.

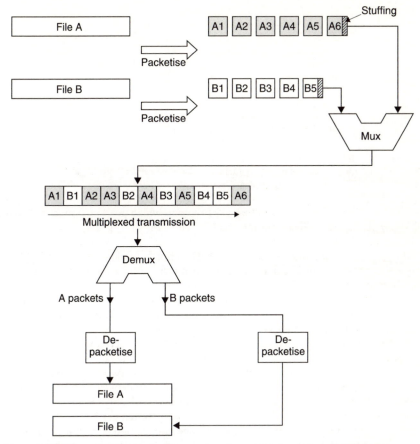

Figure 9.50 Files are broken into frames or packets for multiplexing with packets from other users. Short packets minimize the time between the arrival of successive packets. The priority of the multiplexing must favour isochronous data over asynchronous data.

Breaking files into frames helps to keep down the delay experienced by each process using the network. Figure 9.50 shows that each frame may be stored ready for transmission in a silo memory. It is possible to make the priority a function of the number of frames in the silo, as this is a direct measure of how long a process has been kept waiting. Isochronous systems must do this in order to meet maximum delay specifications. In Figure 9.50 once frame transmission has completed, the arbitrator will determine which process sends a frame next by examining the depth of all the frame buffers. MPEG transport stream multiplexers and networks delivering MPEG data must work in this way because the transfer is isochronous and the amount of buffering in a decoder is limited for economic reasons.

A central arbitrator is relatively simple to implement because when all decisions are taken centrally there can be no timing difficulty (assuming a well-engineered system). In a distributed system, there is an extra difficulty due to the finite time taken for signals to travel down the data paths between nodes.

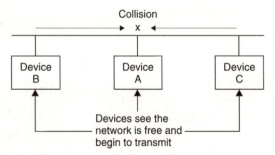

Figure 9.51 In Ethernet collisions can occur because of the finite speed of the signals. A 'back-off' algorithm handles collisions, but they do reduce the network throughput.

Figure 9.51 shows the structure of Ethernet which uses a protocol called CSMA/CD (carrier sense multiple access with collision detect) developed by DEC and Xerox. This is a distributed arbitration network where each node follows some simple rules. The first of these is not to transmit if an existing bus signal is detected. The second is not to transmit more than a certain quantity of data before releasing the bus. Devices wanting to use the bus will see bus signals and so will wait until the present bus transaction finishes. This must happen at some point because of the frame size limit. When the frame is completed, signalling on the bus should cease. The first device to sense the bus becoming free and to assert its own signal will prevent any other nodes transmitting according to the first rule. Where numerous devices are present it is possible to give them a priority structure by providing a delay between sensing the bus coming free and beginning a transaction. High-priority devices will have a short delay so they get in first. Lower-priority devices will only be able to start a transaction if the high-priority devices don't need to transfer.

It might be thought that these rules would be enough and everything would be fine. Unfortunately the finite signal speed means that there is a flaw in the system. Figure 9.51 shows why. Device A is transmitting and devices B and C both want to transmit and have equal priority. At the end of A's transaction, devices B and C see the bus become free at the same instant and start a transaction. With two devices driving the bus, the resultant waveform is meaningless. This is known as a collision and all nodes must have means to recover from it. First, each node will read the bus signal at all times. When a node drives the bus, it will also read back the bus signal and compare it with what was sent. Clearly if the two are the same all is well, but if there is a difference, this must be because a collision has occurred and two devices are trying to determine the bus voltage at once.

If a collision is detected, both colliding devices will sense the disparity between the transmitted and readback signals, and both will release the bus to terminate the collision. However, there is no point is adhering to the simple protocol to reconnect because this will simply result in another collision. Instead each device has a built-in delay which must expire before another attempt is made to transmit. This delay is not fixed, but is controlled by a random number generator and so changes from transaction to transaction.

The probability of two node devices arriving at the same delay is infinitesimally small. Consequently if a collision does occur, both devices will drop the bus, and they will start their back-off timers. When the first timer expires, that device will transmit and the other will see the transmission and remain silent. In this way the collision is not only handled, but prevented from happening again.

The performance of Ethernet is usually specified in terms of the bit rate at which the cabling runs. However, this rate is academic because it is not available all the time. In a real network bit rate is lost by the need to send headers and error-correction codes and by the loss of time due to interframe spaces and collision handling. As the demand goes up, the number of collisions increases and throughput goes down. Collision-based arbitrators do not handle congestion well.

An alternative method of arbitration developed by IBM is show in Figure 9.52. This is known as a *token ring* system. All the nodes have an input and an output and are connected in a ring which must be complete for the system to work. Data circulate in one direction only. If data are not addressed to a node which receives them, the data will be passed on. When the data arrive at the addressed node, that node will capture the data as well as passing them on with an acknowledge added. Thus the data packet travels right around the ring back to the sending node. When the sending node receives the acknowledge, it will transmit a token packet. This token packet passes to the next node, which will pass it on if it does not wish to transmit. If no device wishes to transmit, the token will circulate endlessly. However, if a device has data to send, it simply waits until the token arrives again and captures it. This node can now transmit data in the knowledge that there cannot be a collision because no other node has the token.

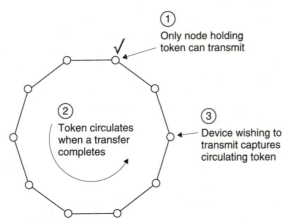

Figure 9.52 In a token ring system only the node in possession of the token can transmit so collisions are impossible. In very large rings the token circulation time causes loss of throughput.

In simple token ring systems, the transmitting node transmits idle characters after the data packet has been sent in order to maintain synchronization. The idle character transmission will continue until the acknowledge arrives. In the case of long packets the acknowledge will arrive before the packet has all been sent and no idle characters are necessary. However, with short packets idle characters will be generated. These idle characters use up ring bandwidth.

Later token ring systems use *early token release* (ETR). After the packet has been transmitted, the sending node sends a token straight away. Another node wishing to transmit can do so as soon as the current packet has passed.

It might be thought that the nodes on the ring would transmit in their physical order, but this is not the case because a priority system exists. Each node can have a different priority if necessary. If a high-priority node wishes to transmit, as a packet from elsewhere passes through that node, the node will set *reservation bits* with its own priority level. When the sending node finishes and tranmits a token, it will copy that priority level into the token. In this way nodes with a lower priority level will pass the token on instead of capturing it. The token will ultimately arrive at the high-priority node.

The token ring system has the advantage that it does not waste throughput with collisions and so the full capacity is always available. However, if the ring is broken the entire network fails.

In Ethernet the performance is degraded by the number of transactions, not the number of nodes, whereas in token ring the performance is degraded by the number of nodes.

9.19 FireWire

FireWire[9] is actually an Apple Computers Inc. trade name for the interface which is formally known as IEEE 1394-1995. It was originally intended as a digital audio network, but grew out of recognition. FireWire is more than just an interface as it can be used to form networks and if used with a computer effectively extends the computer's data bus. Figure 9.53 shows that devices are simply connected together as any combination of daisy-chain or star network.

Any pair of devices can communicate in either direction, and arbitration ensures that only one device transmits at once. Intermediate devices simply pass on transmissions. This can continue even if the intermediate device is powered down as the FireWire carries power to keep repeater functions active.

Communications are divided into *cycles* which have a period of 125 μs. During a cycle, there are 64 time slots. During each time slot, any one node can communicate with any other, but in the next slot, a different pair of nodes may communicate. Thus FireWire is best described as a time-division multiplexed

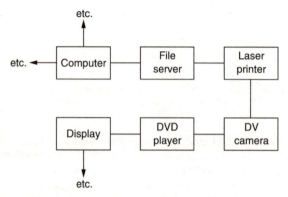

Figure 9.53 FireWire supports radial (star) or daisy-chain connection. Two-port devices pass on signals destined for a more distant device – they can do this even when powered down.

(TDM) system. There will be a new arbitration between the nodes for each cycle.

FireWire is eminently suitable for video/computer convergent applications because it can simultaneously support asynchronous transfers of non-real-time computer data and isochronous transfers of real-time audio/video data. It can do this because the arbitration process allocates a fixed proportion of slots for isochronous data (about 80 per cent) and these have a higher priority in the arbitration than the asynchronous data. The higher the data rate a given node needs, the more time slots it will be allocated. Thus a given bit rate can be guaranteed throughout a transaction; an prerequisite of real-time A/V data transfer.

It is the sophistication of the arbitration system which makes FireWire remarkable. Some of the arbitration is in hardware at each node, but some is in software which only needs to be at one node. The full functionality requires a computer somewhere in the system which runs the isochronous bus management arbitration. Without this only asynchronous transfers are possible. It is possible to add or remove devices whilst the system is working. When a device is added the system will recognize it through a periodic learning process. Essentially every node on the system transmits in turn so that the structure becomes clear.

The electrical interface of FireWire is shown in Figure 9.54. It consists of two twisted pairs for signalling and a pair of power conductors. The twisted pairs carry differential signals of about 220 mV swinging around a common mode voltage of about 1.9 V with an impedance of 112 W. Figure 9.55 shows how the

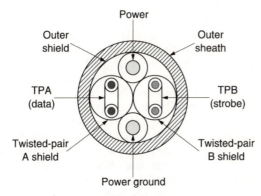

Figure 9.54 FireWire uses twin twisted pairs and a power pair.

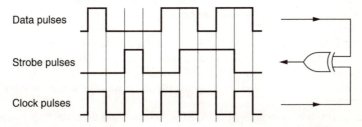

Figure 9.55 The strobe signal is the X-OR of the data and the bit clock. The data and strobe signals together form a self-clocking system.

data are transmitted. The host data are simply serialized and used to modulate twisted pair A. The other twisted pair (B) carries a signal called *strobe*, which is the exclusive-OR of the data and the clock. Thus whenever a run of identical bits results in no transitions in the data, the strobe signal will carry transitions. At the receiver another exclusive-OR gate adds data and strobe to recreate the clock.

This signalling technique is subject to skew between the two twisted pairs and this limits cable lengths to about 10 metres between nodes. Thus FireWire is not a long-distance interface technique, instead it is very useful for interconnecting a large number of devices in close proximity. Using a copper interconnect, FireWire can run at 100, 200 or 400 Mbits/s, depending on the specific hardware. It is proposed to create an optical fibre version which would run at gigabit speeds.

9.20 Broadband networks and ATM

Broadband ISDN (B-ISDN) is the successor to N-ISDN and in addition to offering more bandwidth, offers practical solutions to the delivery of any conceivable type of data. The flexibility with which ATM operates means that intermittent or one-off data transactions which only require asynchronous delivery can take place alongside isochronous MPEG video delivery. This is known as *application independence* whereby the sophistication of isochronous delivery does not raise the cost of asynchronous data. In this way, generic data, video, speech and combinations of the above can co-exist.

ATM is multiplexed, but it is not time-division multiplexed. TDM is inefficient because if a transaction does not fill its allotted bandwidth, the capacity is wasted. ATM does not offer fixed blocks of bandwidth, but allows infinitely variable bandwidth to each transaction. This is done by converting all host data into small fixed-size cells at the adaptation layer. The greater the bandwidth needed by a transaction, the more cells per second are allocated to that transaction. This approach is superior to the fixed bandwidth approach, because if the bit rate of a particular transaction falls, the cells released can be used for other transactions so that the full bandwidth is always available.

As all cells are identical in size, a multiplexer can assemble cells from many transactions in an arbitrary order. The exact order is determined by the quality of service required, where the time positioning of isochronous data would be determined first, with asynchronous data filling the gaps.

Figure 9.56 shows how a broadband system might be implemented. The transport network would typically be optical fibre based, using SONET (synchronous optical network) or SDH (synchronous digital hierarchy). These standards differ in minor respects. Figure 9.57 shows the bit rates available in each. Lower bit rates will be used in the access networks which will use different technology such as xDSL.

SONET and SDH assemble ATM cells into a structure known as a *container* in the interests of efficiency. Containers are passed intact between exchanges in the transport network. The cells in a container need not belong to the same transaction, they simply need to be going the same way for at least one transport network leg.

The cell-routing mechanism of ATM is unusual and deserves explanation. In conventional networks, a packet must carry the complete destination address so that at every exchange it can be routed closer to its destination. The exact route

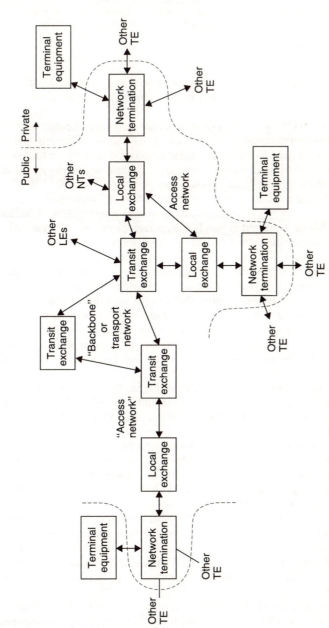

Figure 9.56 Structure and terminology of a broadband network. See text.

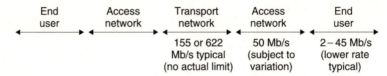

Figure 9.57 Bit rates available in SONET and SDH.

by which the packet travels cannot be anticipated and successive packets in the same transaction may take different routes. This is known as a *connectionless* protocol.

In contrast, ATM is a *connection oriented* protocol. Before data can be transferred, the network must set up an end-to-end route. Once this is done, the ATM cells do not need to carry a complete destination address. Instead they only need to carry enough addressing so that an exchange or switch can distinguish between all the expected transactions.

The end-to-end route is known as a *virtual channel* which consists of a series of *virtual links* between switches. The term 'virtual channel' is used because the system acts like a dedicated channel even though physically it is not. When the transaction is completed the route can be dismantled so that the bandwidth is freed for other users. In some cases, such as delivery of a TV station's output to a transmitter, or as a replacement for analog cable TV the route can be set up continuously to form what is known as a *permanent virtual channel*.

The addressing in the cells ensures that all cells with the same address take the same path, but owing to the multiplexed nature of ATM, at other times and with other cells a completely different routing scheme may exist. Thus the routing structure for a particular transaction always passes cells by the same route, but the next cell may belong to another transaction and will have a different address causing it to routed in another way.

The addressing structure is hierarchical. Figure 9.58(a) shows the ATM cell and its header. The cell address is divided into two fields, the virtual channel identifier and the virtual path identifier. Virtual paths are logical groups of virtual channels which happen to be going the same way. An example would be the output of a video-on-demand server travelling to the first switch. The virtual path concept is useful because all cells in the same virtual path can share the same container in a transport network. A virtual path switch shown in Figure 9.58(b) can operate at the container level whereas a virtual channel switch (c) would need to dismantle and reassemble containers.

When a route is set up, at each switch a table is created. When a cell is received at a switch the VPI and/or VCI code is looked up in the table and used for two purposes. First, the configuration of the switch is obtained, so that this switch will correctly route the cell, second, the VPI and/or VCI codes may be updated so that they correctly control the next switch. This process repeats until the cell arrives at its destination.

In order to set up a path, the initiating device will initially send cells containing an ATM destination address, the bandwidth and quality of service required. The first switch will reply with a message containing the VPI/VCI codes which are to be used for this channel. The message from the initiator will propagate to the destination, creating look-up tables in each switch. At each switch the logic will add the requested bandwidth to the existing bandwidth in use to check that the

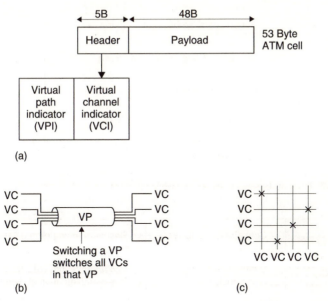

Figure 9.58 The ATM cell (a) carries routing information in the header. ATM paths carrying a group of channels can be switched in a virtual path switch (b). Individual channel switching requires a virtual channel switch which is more complex and causes more delay.

requested quality of service can be met. If this succeeds for the whole channel, the destination will reply with a connect message which propagates back to the initiating device as confirmation that the channel has been set up. The connect message contains an unique call reference value which identifies this transaction. This is necessary because an initiator such a file server may be initiating many channels and the connect messages will not necessarily return in the same order as the set-up messages were sent.

The last switch will confirm receipt of the connect message to the destination and the initiating device will confirm receipt of the connect message to the first switch.

ATM works by dividing all real data messages into cells of 48 bytes each. At the receiving end, the original message must be recreated. This can take many forms. Figure 9.59 shows some possibilities. The message may be a generic data file having no implied timing structure. The message may be a serial bitstream with a fixed clock frequency, known as UDT (unstructured data transfer). It may be a burst of data bytes from a TDM system.

The application layer in ATM has two sub-layers shown in Figure 9.60. The first is the segmentation and reassembly (SAR) sublayer which must divide the message into cells and rebuild it to get the binary data right. The second is the convergence sublayer (CS) which recovers the timing structure of the original message. It is this feature which makes ATM so appropriate for delivery of audio/ visual material. Conventional networks such as the Internet don't have this ability.

In order to deliver a particular quality of service, the adaptation layer and the ATM layer work together. Effectively the adaptation layer will place constraints

Generic data file having no timebase
Constant bit rate serial data stream
Audio/video data requiring a timebase
Compressed A/V data with fixed bit rate
Compressed A/V data with variable bit rate

Figure 9.59 Types of data which may need adapting to ATM.

ATM application layer	Convergence sublayer	Recovers timing of original data
	Segmentation and reassembly	Divides data into cells for transport Reassembles original data format

Figure 9.60 ATM adaption layer has two sublayers, segmentation and convergence.

on the ATM layer, such as cell delay, and the ATM layer will meet those constraints without needing to know why. Provided the constraints are met, the adaptation layer can rebuild the message. The variety of message types and timing constraints leads to the adaptation layer having a variety of forms.

The adaptation layers which are most relevant to MPEG applications are AAL-1 and AAL-5. AAL-1 is suitable for transmitting MPEG-2 multi-program transport streams at constant bit rate and is standardized for this purpose in ETS 300814 for DVB application. AAL-1 has an integral forward error correction (FEC) scheme. AAL-5 is optimized for single-program transport streams (SPTS) at a variable bit rate and has no FEC.

AAL-1 takes as an input the 188-byte transport stream packets which are created by a standard MPEG-2 multiplexer. The transport stream bit rate must be constant but it does not matter if statistical multiplexing has been used within the transport stream.

The Reed–Solomon FEC of AAL-1 uses a codeword of size 128 so that the codewords consist of 124 bytes of data and 4 bytes of redundancy, making 128 bytes in all. Thirty-one 188-byte TS packets are restructured into this format. The 256-byte codewords are then subject to a block interleave. Figure 9.61 shows that 47 such codewords are assembled in rows in RAM and then columns are read out. These columns are 47 bytes long and, with the addition of an AAL header byte make up a 48-byte ATM packet payload. In this way the interleave block is transmitted in 128 ATM cells.

The result of the FEC and interleave is that the loss of up to four cells in 128 can be corrected, or a random error of up to two bytes can be corrected in each cell. This FEC system allows most errors in the ATM layer to be corrected so that no retransmissions are needed. This is important for isochronous operation.

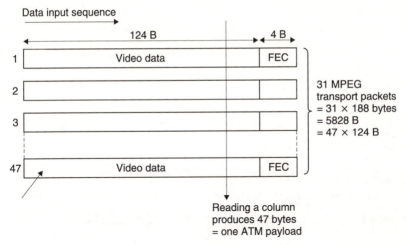

Figure 9.61 The interleave structure used in AAL-1.

The AAL header has a number of functions. One of these is to identify the first ATM cell in the interleave block of 128 cells. Another function is to run a modulo-8 cell counter to detect missing or out-of sequence ATM cells. If a cell simply fails to arrive, the sequence jump can be detected and used to flag the FEC system so that it can correct the missing cell by erasure (see section 6.22). In a manner similar to the use of program clock reference (PCR) in MPEG, AAL-1 embeds a timing code in ATM cell headers. This is called the synchronous residual time stamp (SRTS) and in conjunction with the ATM network clock allows the receiving AAL device to reconstruct the original data bit rate. This is important because in MPEG applications it prevents the PCR jitter specification being exceeded.

In AAL-5 there is no error correction and the adaptation layer simply reformats MPEG TS blocks into ATM cells. Figure 9.62 shows one way in which this can be done. Two TS blocks of 188 bytes are associated with an 8-byte trailer known as CPCS (common part convergence sublayer). The presence of the trailer makes a total of 384 bytes which can be carried in eight ATM cells. AAL-5 does not offer constant delay and external buffering will be required, controlled by reading the MPEG PCRs in order to reconstruct the original time axis.

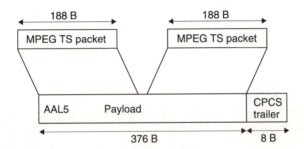

Figure 9.62 The AAL-5 adaptation layer can pack MPEG transport packets in this way.

References

1. SMPTE 259M – 10-bit 4:2:2 Component and 4FSc NTSC Composite Digital Signals – Serial Digital Interface
2. Eguchi, T., Pathological check codes for serial digital interface systems. Presented at SMPTE Conference, Los Angeles, October 1991
3. SMPTE 305M – Serial Data Transport Interface
4. Audio Engineering Society, AES recommended practice for digital audio engineering – serial transmission format for linearly represented digital audio data. *J. Audio Eng. Soc.*, **33**, 975–984 (1985)
5. EIA RS-422A. Electronic Industries Association, 2001 Eye Street NW, Washington, DC 20006, USA
6. Smart, D.L., Transmission performance of digital audio serial interface on audio tie lines. *BBC Designs Dept Technical Memorandum*, 3.296/84
7. European Broadcasting Union, Specification of the digital audio interface. *EBU Doc. Tech.*, 3250
8. Rorden, B. and Graham, M., A proposal for integrating digital audio distribution into TV production. *J. SMPTE*, 606–608 (September, 1992)
9. Wicklegren, I.J., The facts about FireWire. *IEEE Spectrum*, 19–25 (1997)

Glossary

3:2 Pulldown Process used in telecine to produce 60 Hz video from 24fps film.

AAU Audio access unit (*see* Access unit).

AC-3 Audio coding technique developed by Dolby used with ATSC (q.v.) and DVD (q.v.); aka Dolby Digital.

ACATS Advisory Committee on Advanced Television Service.

AES/EBU interface Standardized interface for transmitting digital audio information between two devices (*see* Channel status).

ATM Asynchronous transfer mode; a protocol for transport in broadband digital networks.

ATSC Advanced Television Systems Committee.

ATV Advanced Television; television transmission systems using modern techniques such as compression and error correction.

Access unit The coded data for a picture or block of sound and any stuffing which follows it.

Active line Part of television line on which visible part of image is portrayed. Distinguishes the line blanking area which is not visible.

Aliasing Generation of incorrect samples due to input frequencies exceeding one-half the sampling rate.

Anchor picture Picture used as basis for predictive coding.

Anti-aliasing filter Filter which restricts the frequency range of an analog signal to less than one half the sampling rate.

Aperture effect A loss of resolution caused by the image being scanned or sampled by a non-ideal system in which the scanning beam has finite area.

Artifact Undesirable erroneous effect visible on-screen.

Aspect ratio Ratio of image width to height; SDTV is 4:3, widescreen TV is 16:9. May also refer to spacing of pixels.

Azimuth recording Twisting alternate magnetic heads left and right to record adjacent tracks which can be read without crosstalk. Also called guard-bandless recording.

B Video signal representing blue component of image.

BER Bit error rate (*see* Bit); rate at which incorrect bits are received.

BNC Bayonet Neill-Concelman; coaxial connector used for video signals.

Back porch Blanked area between sync pulse and beginning of video.

Bit Abbreviation for binary digit

Blanking   Voltage corresponding to black on-screen. Below blanking = invisible.

Block matching   Simple technique used to estimate motion between images in compression and standards conversion.

Bottles   Colour ident signals inserted in vertical blanking to synchronize *Dr,Db* sequence in SECAM. No longer mandatory but still in use.

Bouquet   Group of transport streams in which programs are identified by combination of network ID and PID. Part of DVB-SI.

Bowtie   Test signal used to check relative timing of analog component video signals.

Breezeway   Space between end of burst and active line.

Burst   Signal used to synchronize local subcarrier oscillator in composite decoder.

Buzzword   A specialist term which performs two functions. (a) To those who understand its meaning it makes technical conversations briefer. (b) To those who do not understand its meaning it is a way of preventing communication.

Byte   Set of bits, generally eight.

CAT   Conditional Access Table. Packets having PID (q.v.) code of 1 which contain information about the scrambling system (*see* ECM and EMM).

CCIR   International Radio Consultative Committee, now known as ITU-RB (radio branch of International Telecommunications Union).

CCTV   Closed circuit television; cameras connected to monitors by private cabling (as opposed to broadcast). Often used for security.

CIE   Commission Internationale d'Eclairage; standards body for colorimetry. Branch of ISO (q.v.).

CIF   Common Intermediate Format; 352×240 pixel format for 30fps videoconferencing.

CRC   Cyclic redundancy check; error-detection technique.

Channel code   Modulation technique which converts raw data into a signal which can be recorded, or transmitted by radio or cable.

Clamp   Circuit which forces video signal to correct level during blanking.

Closed GOP   Group of pictures in which the last pictures do not need data from the next GOP for bidirectional coding. Used to make a splice point in a bitstream.

Codeword   Entity used in error correction which has constant testable characteristic.

Coefficient   Number specifying the amplitude of a particular frequency in a transform or binary number used to control a multiplier.

Coercivity   Measure of the erasure difficulty, hence replay energy, of a magnetic recording.

Colour framing   A process which synchronizes the subcarrier and sync timing in composite VTRs to a reference. This allows edits without jumps or break-up on replay.

Colour under   Recording technique which downconverts chroma signal to lower frequency to save tape.

Companding   Abbreviation of compressing and expanding; means for increasing dynamic range.

Composite video   (a) (old usage) Video signal carrying sync pulses as well as picture. (b) (modern usage) Video signal carrying subcarrier-based chroma.

Compression Also called bit rate reduction or data reduction. Process which allows pictures to be represented with fewer data. Picture quality may suffer if used to excess.

Concatenation Connecting more than one codec in tandem.

Concealment Method of making uncorrectable errors less visible, e.g. interpolation.

Constellation Pattern seen on test instrument when viewing QUAM (q.v.) signal.

Contouring Video artifact caused by insufficient sample wordlength.

Contribution quality Describes a signal which is going to be further post-produced and so must have high quality.

Convergence Causing all three electron beams in a shadow mask tube to fall in the same place.

Crosstalk Unwanted signal breaking through from adjacent wiring or recording track.

Crushing Some of contrast range is lost typically because of wrongly adjusted brightness control.

Curie temperature Temperature at which magnetic materials demagnetize.

Cutting list Data sheet containing edit frame numbers which is used by film laboratory to conform negative. Equivalent of EDL.

Cylinder In disks, set of tracks having the same radius.

Cylinder address Code sent to disk drive positioner to control radial head movement.

Db Colour difference signal used in SECAM; $Db = 1.505(B-Y)$

Dr Colour difference signal used in SECAM; $Dr = -1.902(R-Y)$

DCT Discrete cosine transform; mathematical technique which analyses blocks of an image to determine which spatial frequencies are present. Used in compression.

DSB Direct satellite broadcasting; system where a consumer antenna can receive a broadcast.

DSP Digital signal processor; computer optimized for waveform processing.

DTS Decoding time stamp; part of PES header indicating when an access unit is to be decoded.

DVB Digital video broadcasting; the broadcasting of television programs using digital modulation of a radio frequency carrier. This can be terrestrial or from a satellite.

DVB-IRD *See* IRD.

DVB-SI DVB service information; information carried in a DVB multiplex describing the contents of different multiplexes. Includes NIT, SDT, EIT, TDT, BAT, RST, ST (q.v.).

DVC Digital video cassette.

DVD Digital video disk, aka digital versatile disk; optical disk for consumer video or data applications.

Data integrity General term for any action or strategy which minimizes the proportion of data bits in a system which are corrupted.

Decimation Reduction of sampling rate by omitting samples.

Dielectric Insulating material used between the conductors in transmission lines.

Differential coding Sending a value not in an absolute sense but as the difference between the current and previous values.

Digitizing   In non-linear editors the process of transferring input images and audio onto the system disks. May include a compression step.

Distribution   (a) In DVTRs, sharing data between two or more heads in a DVTR allows concealment if one head clogs. (b) In statistics, the shape of the probability curve.

Dither   Random signal added to analog input to linearize subsequent quantizing step.

Downconvertor   Unit which allows HDTV signals to be converted to standard definition.

Drift   Error build-up when bit rate is further reduced by additional truncation of coefficients.

EBU   European Broadcasting Union.

ECM   Entitlement control message; conditional access information specifying control words or other stream-specific scrambling parameters.

EDH   Error detection and handling; a data integrity option for SDI digital interface (q.v.).

EDL   Edit decision list; used to control editing process with timecode.

EDTV   Extended definition television; television systems which enhance the quality of existing TV standards.

EFM   Eight to fourteen modulation; the channel code (q.v.) used in D-3 and D-5 DVTRs.

EIT   Event information table; part of DVB-SI.

EMC   Electromagnetic compatibility; legislation controlling sensitivity of equipment to external interference and limiting radiation of interference.

EMM   Entitlement management message; conditional access information specifying authorization level or services of specific decoders. An individual decoder or a group of decoders may be addressed.

ENG   Electronic news gathering; term used to describe use of videorecording instead of film in news coverage.

EOB   End of block; code sent when all remaining coefficients in a DCT block are zero.

EPG   Electronic program guide; program guide delivered by data transfer rather than printed paper.

E–E   Electronics to electronics; a mode in a VTR where the tape and heads are bypassed but the signal passes through everything else.

ESP   Extended studio PAL; BBC-developed wide bandwidth composite signal compatible with composite digital VTRs.

Edit gap   In a DVTR, a space left on the recorded track allowing data on one side of the space to be edited without corrupting data on the other side.

Elementary stream   Raw output of a compressor carrying a single video or audio signal.

Embedded audio   Digital audio signals multiplexed into ancillary data capacity of SDI (q.v.).

Entropy   The unpredictable part of a signal which has to transmitted by a compression system if quality is not to be lost.

Entropy coding   Coding system which achieves compression of signals having non-uniform statistics.

Equalizer   Circuit needed to compensate for a reduction of certain frequencies in recording or transmission.

Error propagation   When a small error in a critical bit results in protracted errors in a decoder.

Event A set of elementary streams, typically audio and video, having common clocks, start times and end times.

Eye pattern Characteristic pattern seen on oscilloscope when viewing channel-coded (q.v.) signals.

FEC Forward error correction; system in which redundancy is added to the message so that errors can be corrected dynamically at the receiver.

FM Frequency modulation; used in analog VTRs.

FSc Frequency of subcarrier.

Faraday effect Rotation of plane of polarization of light by magnetic field.

Ferrite Hard non-conductive magnetic material used for heads and transformers.

Film-for-video System in which material intended for television broadcast or videocassette release is shot on film.

Flash convertor High-speed ADC used for video conversion.

Flatbed A traditional film editing machine having a viewer and means to drive the film to and fro.

Flyback Retrace prior to a further scan.

Flywheel sync System used in TV sets to resist sync loss due to interference. Effectively a phase-locked loop.

Four-field sequence A repetition rate which occurs in NTSC due to the subcarrier frequency having a half-line offset. The subcarrier can only return to a given phase after an even number of lines and this requires two frames or four fields.

Fourier transform Frequency domain or spectral representation of a signal.

Front porch Blanked area after video but before H sync pulse.

Fukinuki hole Space theoretically clear in ideal three-dimensional PAL spectrum.

G Video signal representing green component of image.

Gs Video signal representing green component of image with synchronizing pulses added.

GOP Group of pictures; starts with an *I* picture and ends with last picture before next *I* picture.

Galois field Mathematical entity on which Reed–Solomon (q.v.) error correction is based.

Gamma Non-linear relationship between video signal voltage and screen brightness.

Gamut Allowable range of signal voltage.

Gigabyte Measure of data storage; equal to 1024 megabytes.

Guard band In older VTRs a space left between the tracks to reduce crosstalk.

H Horizontal.

H-coherent Having a frequency which is an integer multiple of H.

HDTV High-definition television; television systems which produce a better quality image than SDTV by use of e.g. progressive scanning.

Hamming distance Number of bits different between two words.

Helix angle In VTRs the mechanical inclination of the tape path.

Helper Additional signal which when optionally decoded in a more complex receiver enhances picture.

Horizontal editing Edit process which primarily involves the time axis. (e.g. cuts). Contrasts with vertical editing.

Hue   The dominant wavelength in a colour (*see* Saturation).

Huffman coding   Form of entropy coding (q.v.) using variable-length parameters.

I   Signal used in NTSC chroma; $I = -0.27(B-Y) + 0.74(R-Y)$.

IEC   International Electrotechnical Commission; branch of ISO (q.v.).

IRD   Integrated receiver decoder; combined RF receiver and MPEG decoder used to adapt TV set to digital transmissions, aka STB (set-top box).

IRE **(unit)**   Institute of Radio Engineers unit which is 1 per cent of black to white voltage swing.

ISO   International Standards Organization.

IT   Information technology; computers, software, data storage and communications.

Illuminant   Light source having standardized position on chromaticity diagram.

Insertion loss   Reduction in signal level due to incorporating a device into a series circuit.

Inter-coding   Compression which uses redundancy between successive pictures, aka temporal coding.

Interlaced scan   The frame is scanned by two or more fields where not all lines are included in each scan.

Interleaving   Reordering of data on medium to reduce effect of defects.

Interpolation   Replacing missing sample with average of those either side.

Intra-coding   Compression which works entirely within one picture, aka spatial coding.

JPEG (Joint Photographic Experts Group)   A compression standard for still images or individual frames.

Jitter   Statistical distribution of events on time axis which ideally are equally spaced.

Judder   Artifact occurring when motion is incorrectly portrayed.

K-factor   Figure of merit for linear distortions in a transmission system. Tested with pulse and bar signal.

Kell factor   Degree by which a display approaches ideal resolution set by line spacing.

Kerr effect   *See* Faraday effect.

Keycode   Signal recorded on the edge of film which allows individual frames to be located according to the time at which they were shot (*see* Timecode).

Keying   Electronic equivalent of matte in which part of one image is replaced with part of another.

Kilobyte   Measure of data storage equal to 1024 bytes.

Latency   Access delay in disk drive due to mechanical motion.

Level   In MPEG (q.v.), the size of the input picture in use with a given profile (q.v.).

Lightning   Screen display of all three colour difference signals allowing rapid assessment of correct adjustment.

Loop-through   Connecting the same signal to several destinations in a daisy chain.

MAC   Multiplexed analog components; an alternative to composite video for colour broadcast from satellites.

MPEG (Moving Picture Experts Group)   A compression technique using inter-coding and intra-coding for moving pictures.

MTF Modulation transfer function; the ratio of output to input contrast index in an optical system.

Macroblock Screen area represented by several luminance and colour difference DCT blocks which are all steered by one motion vector.

Margining Checking performance of system by measuring the amount of deliberate degradation needed to cause a failure.

Masking In human hearing the reduced sensitivity to one sound in the presence of another.

Matrix (a) Circuit for converting between component and colour difference signals. (b) Switching element of a router.

Megabyte Measure of data storage; equal to 1024 kilobytes.

Mezzanine level System using lower compression factor to permit concatenation.

Moiré Artifact in composite analog VTRs caused by high-order sidebands folding into baseband.

Motion compensation Technique used to eliminate judder in standards convertor (*see* Phase correlation; Block matching; Motion vector).

Motion vector Parameter in a compression system or standards convertor which tells the decoder how to shift pixels from a previous picture so it more nearly resembles the current picture.

NIT Network information table; information in one transport stream which describes many transport streams.

NTSC Never Twice the Same Colour; a television system in which the colours are a function of where the hue control was left by an unskilled viewer.

Non-linear An editing system in which random access storage is used so that the time axis of access to the material can be non-linear.

Null packets Packets of 'stuffing' which carry no data but which are necessary to maintain a constant bit rate with a variable payload. Null packets always have a PID (q.v.) of 8191 (all 1s).

OIRT East European equivalent of EBU, merged with EBU in 1992.

Off-line System where low-quality images are used for decision-making purposes. Low quality is not seen by end viewer.

On-line System where the quality seen by the operator is the same as that seen by the end viewer.

Optic flow axis Axis passing through space-time relative to which part of a moving image appears stationary.

Orthogonal Signals or processes which are on independent axes. e.g. U and V in PAL chroma.

Overflow When a buffer memory consistently receives more data than is being removed or when a counter exceeds its maximum value.

Oversampling Temporary use of a higher than necessary sampling rate in convertors in order to simplify analog filters.

PAL Phase alternating line; composite video colour system.

PALPlus 16:9 version of PAL.

PAT Program Association Table. Data appearing in packets having PID (q.v.) code of zero which the MPEG decoder uses to determine which programs exist in a transport stream. PAT points to PMT (q.v.) which in turn points to the video, audio and data content of each program.

Pb Standard colour difference signal which is $0.56433(B-Y)$.

Pr Standard colour difference signal which is $0.71327(R-Y)$.

PCM   Pulse code modulation; technical term for analog source waveform e.g. audio or video, expressed as periodic numerical samples. PCM is an uncompressed digital signal.

PCR   Program clock reference; sample of encoder clock count sent in program header to synchronize decoder clock.

PCRI   Interpolated program clock reference; PCR estimated from previous PCR to measure jitter.

PID   Packet identifier; thirteen-bit code in transport packet header. PID 0 indicates packet contains PAT (q.v.). PID 1 indicates packet contains CAT (q.v.). PID 8191 (all 1s) indicates null (stuffing) packets. All packets belonging to the same elementary stream have the same PID.

PLUGE   Picture line-up generator; a test signal for adjusting monitors.

PMT   Program Map Tables; tables in PAT (q.v.) which point to video, audio and data content of a transport stream.

PSI   Program Specific Information; information which keeps track of the different programs in an MPEG transport stream and the elementary streams in each program. PSI includes PAT, PMT, NIT, CAT, ECM and EMM.

PSI/SI   General term for MPEG PSI and DVB-SI combined.

PTS   Presentation time stamp; time at which a presentation unit is to be available to the viewer.

PU   Presentation unit; one compressed picture or block of audio.

Pack   A set of PES packets; the pack header contains a SCR code.

Packets   Beware! the term is used in two contexts and these are not the same. In program streams, a packet contains one or more presentation units. In transport streams a packet is a small fixed-size data quantum.

Padding   *See* stuffing.

Partial response   Channel coding (q.v.) technique used in Digital Betacam.

Patch   Tube face area illuminated by the flying spot in telecine.

Payload   Content of a packet other than the header.

Pedestal   In NTSC black level may be raised to 7.5 *IRE* units instead of zero to guarantee retrace is invisible.

Pel   *See* Pixel.

Phase correlation   Technique used for motion estimation in compression and standards conversion.

Phase-linear   Describes a circuit which has constant delay at all frequencies.

Phase-locked loop   An electronic circuit which extracts the average phase from a jittery signal in a manner analogous to a flywheel. *See* Reclocker.

Phosphor   Substance which emits light when struck by electron beam.

Pixel   Short for picture cell; a point sample of a picture. Also called a pel (*see* also Square pixel).

Power factor   In electrical supplies, power factor measures the efficiency of transmission. If current is out of phase with voltage, the power factor is poor and transmission losses increase.

Preprocessing   Noise reduction, downsampling, cut edit identification and 3:2 pulldown identification are all preprocessing steps needed before compression.

Primary colour   Colour emitted by one phosphor in a colour display.

Product code   Combination of two one-dimensional error-correcting codes in an array.

Profile   A subset of the entire coding repertoire of MPEG.

Program stream   Bitstream containing compressed video and audio and timing information.

Progressive scan   Scanning proceeds from top to bottom of frame taking in every line in sequence.

Pseudo-random code   Number sequence which is sufficiently random for practical purposes but which is repeatable.

Purity   Degree to which colour monitor can produce a primary colour.

Q   Signal used in NTSC chroma; $Q = 0.41(B-Y) + 0.48(R-Y)$.

QCIF   One-quarter resolution (176×144 pixels) common intermediate format (*see* CIF).

QSIF   One-quarter resolution source input format. (*see* SIF).

Quadrature   When two signals have 90° relative phase.

Quantizer   A device which breaks an analog signal's voltage range into even intervals and outputs the number of the interval in which the analog input lies.

Quincunx   Sampling grid where pixels on alternate rows are shifted to create a pattern resembling the five of dice.

R   Video signal representing red component of image.

RLC   Run-length coding; coding scheme which counts number of similar bits instead of sending them individually.

RFI   Radio frequency interference; interference which is radiated rather than conducted.

Random access   Storage device like a disk where contents can be output in any order. Contrasts with serial access.

Randomizing   *See* Scrambling.

Reclocker   A combination of a slicer and a phase-locked loop which can remove noise and jitter from a digital signal.

Reconstruction   Filtering process which converts a series of samples back to a continuous waveform.

Redundancy   (a) In error correction, extra check bits appended to the wanted data. (b) In compression, that part of a signal which can be predicted and so need not be sent.

Reed–Solomon code   Error-correction code which is popular because it is as powerful as theory allows.

Requantizing   Shortening of sample wordlength.

Router   Equivalent of a telephone exchange for video, audio and control signals.

Rushes   Film developed urgently after a day's shooting for confirmation purposes.

Sc-H   Subcarrier to horizontal sync phase.

SAV   Start of Active Video; *see* TRS.

SCR   System clock reference; clock data carried in program stream pack header.

SDI   Serial digital interface; standardized coaxial cable digital video interface

SECAM   System essentially contrary to the American method. French colour system.

STC   System time clock; common clock used to encode video and audio in the same program.

SDT   Service description table; table listing the providers of each service in a transport stream.

SI   *See* DVB-SI.

SIF   Source input format; half-resolution input signal used by MPEG-1.

SMPTE   Society of Motion Picture and Television Engineers (USA).

ST   Stuffing table.

STB   Set-top box (*see* IRD).

Sampling   A process in which some continous variable is measured at discrete (usually uniform) intervals.

Saturation   Lack of dilution of a colour by white; e.g. red is saturated, pink is desaturated.

Scalability   System where more complex decoder produces better picture but simple decoder only uses part of data.

Scrambling   Process in digital transmission which spreads signal spectrum and increases clock content.

Segmentation   In VTRs the use of several parallel tracks to record one field.

Seek   Process of moving disk drive heads from one track to another.

Serial access   Storage system such as tape where data come out in a fixed sequence. Contrasts with random access.

Set-up   *See* Pedestal.

Shuffling   Random pixel reordering process used in DVTRs which spreads uncorrected pixels over a large area to aid concealment (q.v) of uncorrectable errors.

Signature analysis   Test technique for verifying quality of digital transmission system.

Skew   In analog VTRs a timing error caused by incorrect tape tension.

Slicer   Electronic circuit which judges an input to be above or below a threshold. Used to clean up binary signals (*see* Reclocker).

Spline   Motion interpolating algorithm used in DVEs.

Split edit   Edit in which the image and soundtrack transition at different times.

Square pixel   An image-sampling process in which the vertical and horizontal sampling spacing is the same.

Stripview   Graphical representation of the time axis through an edit sequence.

Stuffing   Meaningless data added to maintain constant bit rate.

Syndrome   Initial result of an error-checking calculation. Generally if zero there is assumed to be no error.

TDAC   Time domain aliasing cancellation; coding technique used in AC-3 audio compression.

TDT   Time and date table; used in DVB-SI.

TRS   Timing reference signal; equivalent of sync pulses in digital interfaces.

T-STD   Transport stream system target decoder; decoder having a certain amount of buffer memory assumed present by an encoder.

TSO   Tape speed override; means of changing speed of VTR to stretch or squeeze duration of a program.

Tally   Signal which retraces signal routing path to operate on-air light.

Telecine   Machine which drives film past an optical scanning system in order to output a video signal.

Television   Literally 'seeing at a distance'. This implies some long cable or radio transmission between camera and display. If there is no such distance television becomes video.

Terminator   Device fitted at the ends of a transmission line to match its characteristic impedance and prevent signal reflections.

Timecode   A signal recorded down the side of a video tape allowing individual frames to be located according to the time at which they were shot (*see* Keycode).

Time compression   Process used to squeeze the time taken to send a given quantity of data by raising the data rate.

Track angle   In VTRs angle between track and edge of tape.

Transmission line   Cable which is long compared to wavelength of signals carried.

Transport stream   Multiplex of several program streams which are carried in packets. Demultiplexing is achieved by different packet IDs (PIDs) (*see* PSI; PAT; PMT; PCR).

Triad   Set of three primary phosphor dots in CRT.

Truncation   Shortening wordlength of sample or coefficient by removing low-order bits.

Twitter   Artifact due to interlace where horizontal picture transitions flicker.

U   Scaled component signal used in PAL; $U = 0.493(B-Y)$.

Underflow   When a buffer memory consistently receives less data than is being removed from it.

Upconvertor   Device which allows standard definition signals to be displayed on HDTV equipment

V   Scaled component signal used in PAL; $V = 0.877(R-Y)$.

VAU   Video access unit; one compressed picture in program stream.

VBV   Video buffer verifier; parameter specifying amount of buffer memory needed to decode a given elementary stream.

VCR   Video cassette recorder; generally implying a consumer device.

VLC   Variable-length coding; compression technique which allocates short codes to frequent values and long codes to infrequent values.

VOD   Video-on-demand; system in which television programs or movies are transmitted to a single consumer only when required.

VSB   Vestigial sideband; system used for terrestrial TV transmission in which lower sideband is heavily curtailed.

VTR   Video tape recorder.

V-switch   Vertical axis switch reversing phase of $R-Y$ component on alternate lines in PAL.

Vector   Multi-dimensional quantity. In TV generally a two-dimensional representation of a pair of colour difference signals.

Vertical editing   Editing where manipulations take place within the picture. e.g. layering, chroma key. Contrasts with horizontal editing.

Video   Literally 'I see'; an electrical signal which represents a picture.

Wavelet transform   Technique for analysing spatial frequencies of an image which does not use blocks. Used in compression.

Weighting   Modifying measurements to give better correspondence to human perception.

Winchester disk   Disk drive having heads and disks sealed in one unit allowing higher speed and capacity.

Wordlength   Number of bits in a sample; typically eight or ten in video.

Working positive   A viewable print made from master negative used for making edit decisions. Damage to the working positive is irrelevant as it is only used to create a cutting list (q.v.).

Wrap angle   Number of degrees of circumference of VTR drum over which the tape is in contact.

Y   Luma signal.

Y/C   Recording system used in S-VHS where chroma and luma are kept separate to avoid decoding stage.

Ys   Luma signal with sync pulses added.

Zenith angle   Angular error which takes a head out of the plane of the tape resulting in poor contact.

Zero-run-length   Parameter specifying the number of contiguous DCT coefficients of value zero in a scanning sequence.

Zig-zag scan   Method of ordering DCT coefficients so that zero values tend to be at the end of the sequence.

Index